CHARLES A. MICCHELLI
IBM T. J. Watson Research Center
Yorktown Heights, New York

Mathematical Aspects of Geometric Modeling

SOCIETY FOR INDUSTRIAL AND APPLIED MATHEMATICS

PHILADELPHIA, PENNSYLVANIA   1995

Printed by Capital City Press, Montpelier, Vermont

Library of Congress Cataloging-in-Publication Data

Micchelli, Charles A.
      Mathematical aspects of geometric modeling / Charles A. Micchelli.
            p.   cm. — (CBMS-NSF regional conference series in applied
      mathematics ; 65)
      Includes bibliographical references and index.
      ISBN 0-89871-331-5
      1. Curves on surfaces—Mathematical models—Congresses.
      2. Surfaces—Mathematical models—Congresses.   I. Title.
      II. Series.
      QA565.M67   1994
      516.3'52—dc20   94-10478

siam. is a registered trademark

Dedicated to Mary and Salvatore, who worked hard for the benefit of others, with gratitude from one of the ABCD boys.

Preface

Less is more - more or less
Ludwig Mies van der Rohe

During the week of December 17–21, 1990 we presented a series of ten lectures on "Curves and Surfaces: An Algorithmic Viewpoint" at Kent State University under the auspice of the CBMS. When we began to write these lecture notes it became clear to us that anything more than a superficial description of concepts and results would lead to an inappropriately long monograph. Hence we decided that it would be better to choose only a portion of the lectures to elaborate on in detail. For these lectures we would strive for completeness so that the reader could minimize the inconvenience of consulting original sources for details.

Each of the five chapters of this monograph contains introductory material followed by more advanced results. In retrospect, the selection of lectures reflects our fondness for spline functions. This was, in no small measure, the result of our attendance at a CBMS lecture on this subject given by I.J. Schoenberg in 1971. His inspirational lectures have influenced much of our own research and hence also this monograph.

Fading memory always makes it hard to remember the origin of mathematical ideas, but certainly we haven't forgotten our debt of gratitude to our many collaborators. They represent an impressive group of talented mathematicians who have given generously of their time to patiently educate us. Much of the insights and contributions contained in this monograph are due to their efforts.

There is one person whom we wish to single out for special thanks: A.S. Cavaretta, a friend and colleague, who organized the conference. Alfred was a master organizer, as anyone present at the conference would attest. He was always attentive to our needs and no detail, however small, was overlooked, from favorite snacks of the principal lecturer to pizza and card party and tasty lunchtime fare for all the participants. His efforts were much appreciated by everyone.

We also want to thank Gail Bostic of the Mathematics and Computer Science Department of Kent State University. Without her hard work to convert our handwritten notes into LaTeX there would be no monograph. In addition, we were fortunate to have the patient and skillful help of Barbara White of the Mathematical Sciences Department of the T.J. Watson Research Center, who typed the many changes we made in preliminary versions of the monograph.

Most of this monograph was written during two summer visits: one to Cambridge University, Department of Applied Mathematics and Theoretical Physics in 1991 and the other to the Universidad de Zaragoza, Departamento de Matematica Aplicada in 1992, supported respectively by SERC and DGICYT grants. To our hosts M.J.D. Powell of Cambridge and M. Gasca of Zaragoza, we are grateful. Also, several colleagues, including P. Barry of the University of Minnesota, M.D. Buhmann of ETH, A.S. Cavaretta of Kent State, W. Dahmen of RWTH, M. Gasca of the University of Zaragoza, T.N.T. Goodman of the University of Dundee, A. Pinkus of the Technion, E. Quak of Texas A&M, and Z. Shen of the National University of Singapore gave generously of their time to read a preliminary version of the monograph. We benefited from their many helpful comments. G. Rodriguez of the University of Cagliari and E. Quak assisted in preparing several of the figures.

Finally, and most importantly, we are indebted to the IBM Corporation for providing us with the privilege to work in a scientifically stimulating environment where we received both the encouragement and freedom necessary to complete this monograph.

Charles A. Micchelli
Mohegan Lake, New York
June 12, 1993

Contents

A Brief Overview

This monograph examines in detail certain concepts that are useful for the modeling of curves and surfaces. Our emphasis is on the mathematical theory that underlies these ideas.

Two principal themes stand out from the rest. The first one, the most traditional and well-trodden, is the use of piecewise polynomial representation. This theme appears in one form or another in all the chapters. The second theme is that of iterative refinement, also called subdivision. Here, simple iterative geometric algorithms produce, in the limit, curves with complex analytic structure. An introduction to these ideas is given in Chapters 1 and 2.

In Chapter 1, we use de Casteljau subdivision for Bernstein–Bézier curves to introduce matrix subdivision, and in Chapter 2 the Lane–Riesenfeld algorithm for computing cardinal splines is tied to stationary subdivision and ultimately leads us to the construction of pre-wavelets of compact support. In Chapter 3 we study concepts of "visual smoothness" of curves and, as a result, embark on a study of certain spaces of piecewise polynomials determined by connection matrices. Chapter 4 explores the intriguing idea of generating smooth multivariate piecewise polynomials as volumes of "slices" of polyhedra. The final chapter concerns evaluation of polynomials by finite recursive algorithms. Again, Bernstein–Bézier polynomials motivate us to look at polarization of polynomial identities, dual bases, and H.P. Seidel's multivariate B-patches, which we join together with the multivariate B-spline of Chapter 4.

CHAPTER 1

Matrix Subdivision

1.0. Introduction.

In this chapter we begin by reviewing a well-known subdivision algorithm due to P. de Casteljau for the computation of a Bernstein–Bézier curve, everywhere on its parameter interval [0,1]. We develop a matrix theoretic point of view about this algorithm, which leads to the formulation of what we call *matrix subdivision*. Various criteria for the convergence of a given matrix subdivision scheme determined by two matrices A_ϵ, $\epsilon \in \{0, 1\}$ are given and properties of the limit curve are developed. Some examples are given that are based on reparameterization of Bernstein–Bézier curves and embeddings in the Bernstein–Bézier manifold.

The end of the chapter is devoted to matrix subdivision schemes determined by totally positive matrices, a property satisfied, in particular, by the matrices associated with de Casteljau subdivision. Generally, we show that the limit curve associated with totally positive matrices has a strong variation diminishing property. Background material on total positivity is included to assist the reader with the details of the proof.

1.1. de Casteljau subdivision.

Almost everything in this chapter has as its source a rather simple iterative procedure due to de Casteljau for computing the Bernstein–Bézier representation of a polynomial curve. Our contributions to this subject are drawn from two papers, Micchelli and Pinkus [MPi] and Micchelli and Prautzsch [MPr].

Before we get to the heart of the matter we review the *control point, control polygon* paradigm used in computer graphics for the representation of curves. We begin with any set of n scalar valued functions $\{f_1, \ldots, f_n\}$ defined on [0, 1]. These functions determine a curve $\mathbf{f} : [0, 1] \to \mathbb{R}^n$ given by

(1.1)
$$\mathbf{f}(t) := (f_1(t), \ldots, f_n(t))^T, \qquad t \in [0, 1].$$

Thus every real $m \times n$ matrix C determines another curve $C\mathbf{f}$ in $\mathbb{R}^m$ defined by

$$(C\mathbf{f})(t) := C\mathbf{f}(t), \qquad t \in [0, 1].$$

The convention is to express the vector $C\mathbf{f}(t)$ in another form. To this end, we let $\mathbf{c}_1, \ldots, \mathbf{c}_n$ be the column vectors of C and observe that

$$(1.2) \qquad (C\mathbf{f})(t) = \sum_{j=1}^{n} \mathbf{c}_j f_j(t), \qquad t \in [0,1].$$

The prevailing terminology is that $\mathbf{c}_1, \ldots, \mathbf{c}_n$ are the *control points* for the curve $C\mathbf{f}$ and its associated *control polygon* is obtained by joining successive control points with linear segments. In other words, the set of vectors $\mathcal{C} = \{\mathbf{c}_1, \ldots, \mathbf{c}_n\}$ "controls" the curve $C\mathbf{f}$ on $[0,1]$ and can be thought of geometrically as a polygonal line in $\mathbb{R}^m$. (See Fig. 1.1 for the case $m = 2$.) We will denote the control polygon parameterized on $[0,1]$ by $\mathbf{C}$. Heuristically, the "shape" of $C\mathbf{f}$ is roughly approximated by the control polygon $\mathbf{C}$, especially under additional assumptions on $\mathbf{f}$.

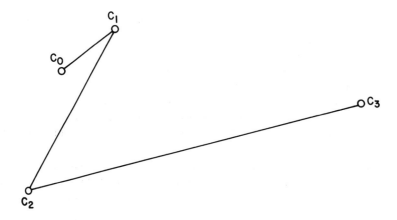

FIG. 1.1.   *Planar control polygon.*

It is sometimes the case that algorithms for evaluating the curve $C\mathbf{f}$ are given as successive alterations of the control polygon $\mathbf{C}$. A particularly important example of this is provided by the Bernstein–Bézier basis

$$(1.3) \qquad b_k(t) := \binom{n}{k} t^k (1-t)^{n-k}, \qquad k = 0, 1, \ldots, n, \qquad t \in [0,1].$$

The Bernstein–Bézier curve $C\mathbf{b}$ can be computed for all $t \in [0,1]$ by a *subdivision procedure* due to de Casteljau. The algorithm begins with the initial control polygon $\mathbf{C}$. New control points are formed by successive averages

$$(1.4) \qquad \begin{cases} \mathbf{d}_r^0 := \mathbf{c}_r, \qquad r = 0, 1, \ldots, n, \\[2mm] \mathbf{d}_r^\ell := \dfrac{1}{2}(\mathbf{d}_r^{\ell-1} + \mathbf{d}_{r+1}^{\ell-1}), \qquad r = 0, 1, \ldots, n - \ell, \qquad \ell = 1, \ldots, n. \end{cases}$$

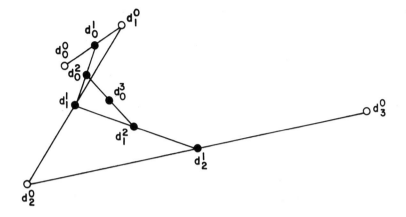

Fig. 1.2.   de Casteljau subdivision for cubic polynomials.

Note that the new control points are obtained by replacing two adjacent control points by their average. We like to refer to such a procedure as *corner cutting* because of its apparent geometric interpretation. We organize the recurrence formula (1.4) in the *de Casteljau tableau.*

$$
\begin{array}{ccccc}
\mathbf{d}_0^0 & \cdot & \cdot & \cdot & \mathbf{d}_n^0 \\
\mathbf{d}_0^1 & \cdot & \cdot & \mathbf{d}_{n-1}^1 & \\
 & \cdot & \cdot & & \\
 & \cdot & \cdot & & \\
 & \cdot & \cdot & & \\
\mathbf{d}_0^n. & & & &
\end{array}
$$

Fig. 1.3.   de Casteljau tableau.

The first step of the algorithm gives us the new control points

$$(1.5) \qquad \mathcal{C}^1 = \left\{\mathbf{d}_0^0, \ldots, \mathbf{d}_0^n, \mathbf{d}_0^n, \ldots, \mathbf{d}_n^0\right\}$$

formed by the first column and the diagonal of the de Casteljau tableau. These control points form a new control polygon $\mathbf{C}^1$, which we divide into two parts, the left and right, parameterized respectively on the interval $[0, 1/2]$ and $[1/2, 1]$. Schematically, we write $\mathbf{C}^1 = (\mathbf{C}_0^1, \mathbf{C}_1^1)$ where $\mathbf{C}_0^1$, $\mathbf{C}_1^1$ are the control polygons determined by the control points $\mathcal{C}_0^1 = \{\mathbf{d}_0^0, \ldots, \mathbf{d}_0^n\}$ and $\mathcal{C}_1^1 = \{\mathbf{d}_0^n, \ldots, \mathbf{d}_n^0\}$, respectively. By choice, $\mathbf{C}_0^1$ and $\mathbf{C}_1^1$ are parametrized on the intervals $[0, 1/2], [1/2, 1]$, respectively and share the common control point $\mathbf{d}_0^n$. This is the first step of the algorithm.

At the next step the same procedure is applied to both $\mathbf{C}_0^1$ and $\mathbf{C}_1^1$. Repeatedly, in this fashion the kth stage of the algorithm produces a control polygon $\mathbf{C}^k$ that naturally splits into 2^k control polygons, one for each of the intervals $[2^{-k}i, 2^{-k}(i+1)]$, $i = 0, 1, \ldots, 2^k - 1$, each consisting of $n+1$ control points in $\mathbb{R}^m$ such that the last control point of any one of these control polygons agrees with the first control point of the next control polygon in the list (provided there is one). In the limit, as $k \to \infty$, $\mathbf{C}^k$ converges to $C\mathbf{b}$ (see Fig. 1.4).

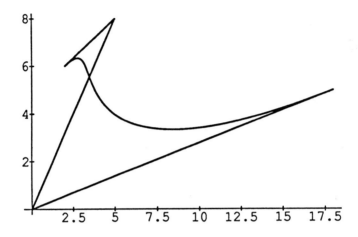

FIG. 1.4. *Convergence of de Casteljau subdivision to a cubic curve.*

To explain this fact let us go back to the first step of the algorithm and observe that $\mathbf{C}_0^1, \mathbf{C}_1^1$ "control" the curve $C\mathbf{b}$ when it is represented in its Bernstein–Bézier form relative to the intervals $[0, 1/2]$, $[1/2, 1]$, respectively. To explain exactly what we mean by this and also set the stage for much of our subsequent discussion in this chapter we introduce two $(n+1) \times (n+1)$ matrices, B_0, B_1 defined by

$$(1.6) \qquad\qquad C_\epsilon^1 := CB_\epsilon^T, \qquad \epsilon \in \{0, 1\}.$$

Here C_ϵ^1 represents the $m \times (n+1)$ matrix whose columns are built from the control points in $\mathcal{C}_\epsilon^1$ and (1.6) expresses the simple fact that the control points in $\mathcal{C}_\epsilon^1$ are obtained as linear combinations of the control points in $\mathcal{C}$.

A comment on notation is appropriate here. If the matrix C has column vectors $\mathbf{c}_0, \ldots, \mathbf{c}_n$ then the column vectors of CB_ϵ^T are $\sum_{j=0}^n (B_\epsilon)_{ij} \mathbf{c}_j$, $i = 0, 1, \ldots, n$. Hence, if we let $\mathbf{c}$ be the vector in $\mathbb{R}^{m(n+1)}$ whose (vector) components are defined as $(\mathbf{c})_{km+i} = (\mathbf{c}_k)_i$, $k = 0, 1, \ldots, n$, $i = 1, \ldots, m$, and likewise interpret $B_\epsilon \mathbf{c}$ as a vector in $\mathbb{R}^{m(n+1)}$ with (vector) components $\sum_{j=0}^n B_{ij} \mathbf{c}_j$, $i = 0, 1, \ldots, n$, then we can identify the control points in $\mathcal{C}_\epsilon^1$ with the components of $B\mathbf{c}$. This will be frequently done throughout our discussion. Also, we find it convenient to denote the vector defined above as $\mathbf{c} = (\mathbf{c}_0, \ldots, \mathbf{c}_n)^T$. We even adhere to this notation when the vectors $\mathbf{c}_i$ have different lengths, say k_i, and define $\mathbf{c}$ as the vector in $\mathbb{R}^k$, $k = k_0 + \cdots + k_n$, with components

$$(\mathbf{c})_{k_0 + \cdots + k_{j-1} + i} = (\mathbf{c}_j)_i, \qquad i = 1, \ldots, k_j, \qquad j = 0, 1, \ldots, n.$$

With a little thought, using the rule to generate Pascal's triangle, it follows (for instance by induction on n) that

$$(1.7) \qquad\qquad (B_0)_{ij} = 2^{-i} \binom{i}{j}, \qquad i, j = 0, 1, \ldots, n,$$

where as usual $\binom{i}{j} = 0$, if $j > i$. Likewise, because the averaging algorithm (1.4) used to generate C_1^1 is the same used for C_0^1, except for the reverse ordering of the control points, we have

$$(1.8) \qquad\qquad\qquad B_1 = P_n B_0 P_n$$

where P_n is the $(n+n) \times (n+1)$ permutation matrix given by

$$(1.9) \qquad\qquad P_n = \begin{pmatrix} 0 & 0 & \cdots & 0 & 1 \\ 0 & \cdot & \cdots & 1 & 0 \\ \cdot & \cdot & \cdots & \cdot & \cdot \\ \cdot & \cdot & \cdots & \cdot & \cdot \\ \cdot & \cdot & \cdots & \cdot & \cdot \\ 0 & 1 & \cdots & \cdot & 0 \\ 1 & 0 & \cdots & 0 & 0 \end{pmatrix}.$$

These formulas lead to the following important observation.

LEMMA 1.1.

$$(1.10) \qquad \mathbf{b}\left(\frac{t+\epsilon}{2}\right) = B_\epsilon^T \mathbf{b}(t), \qquad t \in \mathbb{R}, \qquad \epsilon \in \{0, 1\}.$$

Before we prove this *refinement equation* for the Bernstein–Bézier basis let us show that, as mentioned earlier, it implies that $\mathbf{C}_\epsilon^1$ controls the curve on $[\epsilon/2, (1+\epsilon)/2]$. Specifically, for $t \in [0, 1/2]$ we have from (1.6) and (1.10) (with $\epsilon = 0$ and t replaced by $2t$) that

$$C\mathbf{b}(t) = CB_0^T \mathbf{b}(2t) = C_0^1 \mathbf{b}(2t).$$

That is, there holds the formula

$$(1.11) \qquad\qquad C\mathbf{b}(t) = \sum_{j=0}^n \mathbf{d}_0^j b_j(2t), \qquad t \in [0, 1/2],$$

and similarly we have

$$(1.12) \qquad C\mathbf{b}(t) = \sum_{j=0}^{n} \mathbf{d}_j^{n-j} b_j(2t - 1), \qquad t \in [1/2, 1].$$

Now, for any interval $I = [a, b] \subseteq [0, 1]$ the Bernstein–Bézier polynomials for I are given by

$$b_j(t|I) := b_j((t - a)/(b - a)), \qquad j = 0, 1, \ldots, n,$$

and so indeed (1.11) and (1.12) show that $\mathbf{C}_\epsilon^1$ controls $C\mathbf{b}$ on $[\epsilon/2, (1 + \epsilon)/2]$. Moreover, from (1.3), (1.11), and (1.12) we see that the control point $\mathbf{d}_0^n$ common to $\mathbf{C}_0^1$ and $\mathbf{C}_1^1$ *lies* on the Bernstein–Bézier curve, viz.

$$(1.13) \qquad \mathbf{d}_0^n = C\mathbf{b}(1/2).$$

Let us now prove Lemma 1.1.

    Proof. For $\epsilon = 0$ we have

$$(B_0^T \mathbf{b}(t))_k = \sum_{j=0}^{n} (B_0^T)_{kj} b_j(t)$$

$$= \sum_{j=0}^{n} 2^{-j} \binom{j}{k} \binom{n}{j} t^j (1 - t)^{n-j}$$

$$= \sum_{j=0}^{n-k} 2^{-(j+k)} \frac{n!}{(n - j - k)! j! k!} t^{j+k} (1 - t)^{n-k-j}$$

$$= 2^{-k} t^k \binom{n}{k} \sum_{j=0}^{n-k} \binom{n - k}{j} \left(\frac{t}{2}\right)^j (1 - t)^{n-k-j}$$

$$= 2^{-k} \binom{n}{k} t^k (1 - t/2)^{n-k} = b_k(t/2).$$

This proves (1.10) when $\epsilon = 0$; a similar computation verifies the case $\epsilon = 1$. $\square$

    The *subdivision matrices* $B_\epsilon, \epsilon \in \{0, 1\}$ give us a convenient means to describe all the steps of the de Casteljau algorithm (see Fig. 1.5). As we pointed out, at the first stage of the algorithm $B_0 \mathbf{c}, B_1 \mathbf{c}$ control the curve $C\mathbf{b}$ on the intervals $[0, 1/2], [1/2, 1]$, respectively. For the kth stage, we let $\mathbf{e}^k$ be a typical column vector in $E_k := \{0, 1\} \times \cdots \times \{0, 1\}$, k times. That is, $\mathbf{e}^k = (\epsilon_1, \ldots, \epsilon_k)^T$ and each $\epsilon_r, r = 1, \ldots, k$, is zero or one. We introduce the matrix

$$(1.14) \qquad B_{\mathbf{e}^k} := B_{\epsilon_k} \cdots B_{\epsilon_1}.$$

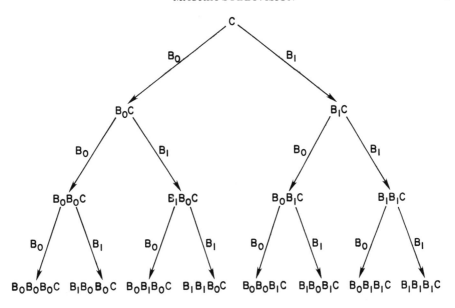

FIG. 1.5. *de Casteljau subdivision.*

Then $B_{\mathbf{e}^k}\mathbf{c}$ controls the curve $C\mathbf{b}$ on the interval $I_{\mathbf{e}^k} := .\epsilon_1 \cdots \epsilon_k + 2^{-k}[0,1]$ where

(1.15)
$$.\epsilon_1 \cdots \epsilon_k := \sum_{j=1}^{k} \epsilon_j 2^{-j}.$$

We call any number of the form (1.15) a *binary* fraction with k digits (each $\epsilon_i, i = 1, \ldots, k$, is a digit). For every k, we order the binary fractions with k digits lexicographically as $d_0^k, \ldots, d_{2^k-1}^k$, so that $d_0 = 0$ and $d_{2^k-1}^k = 1 - 2^{-k}$. The intervals $[d_0^k, d_1^k), \ldots, [d_{2^k-2}^k, d_{2^k-1}^k)$ partition $[0, 1 - 2^{-k})$. The set of all binary fractions is dense in $[0, 1]$ and for every $t \in [0, 1]$ there is a nested sequence of half-open intervals $[d_{j_k}^k, d_{j_k+1}^k)$ that converges to $\{t\}$. Hence every $t \in [0, 1]$ can be written as an infinite binary expansion

(1.16)
$$t = \sum_{j=1}^{\infty} \epsilon_j 2^{-j}, \qquad \epsilon_j \in \{0, 1\}.$$

The convention here is that a point in $[0, 1)$ is a binary fraction if and only if all but a finite number of its digits are zero. Consequently, for $\mathbf{C}^k$ to converge to $C\mathbf{b}$ means that if we form the vectors $\mathbf{e}^k = (\epsilon_1, \ldots, \epsilon_k)^T$ from the binary digits of t given in (1.16) then the *control polygons* $\mathbf{C}_{\mathbf{e}^k}^k$, $k = 1, 2, \ldots$ should converge to the *vector* $C\mathbf{b}(t)$ as $k \to \infty$. Thus *each* of the $n+1$ control points that comprise $\mathbf{C}_{\mathbf{e}^k}^k$ must converge to $C\mathbf{b}(t)$.

For the proof of this result (with an estimate of the rate of convergence) we use two facts about Bernstein polynomials. The first concerns the Bernstein operator $\mathcal{B}_n$, defined for every curve $\mathbf{g} : [0,1] \to \mathbb{R}^m$ as

$$(1.17) \qquad (\mathcal{B}_n\mathbf{g})(t) := \sum_{j=0}^{n} \mathbf{g}\left(\frac{j}{n}\right) b_j(t), \qquad t \in [0,1].$$

Let $\| \cdot \|$ be any norm on $\mathbb{R}^m$ and set

$$(1.18) \qquad \|\mathbf{g}\|_\infty = \max\left\{\|\mathbf{g}(t)\| : t \in [0,1]\right\}.$$

When the maximum above is taken over all t in some subinterval $I \subseteq [0,1]$ we write $\|\mathbf{g}\|_{\infty,I}$. Then there follows the estimate

$$(1.19) \qquad \|\mathcal{B}_n\mathbf{g} - \mathbf{g}\|_\infty \leq \frac{1}{8n}\|\mathbf{g}''\|_\infty,$$

valid for all curves with a continuous second derivative.

A brief comment on the proof of (1.19) is in order. First, from Taylor's theorem with remainder, we have for all $x, t \in [0,1]$

$$\|\mathbf{g}(x) - \mathbf{g}(t) - \mathbf{g}'(t)(t-x)\| \leq \frac{1}{2}(x-t)^2\|\mathbf{g}''\|_\infty.$$

Also, a computation verifies that for any quadratic polynomial Q with leading coefficient q

$$(\mathcal{B}_nQ)(t) = Q(t) + \frac{q}{n}t(1-t).$$

In particular, if Q is linear then $\mathcal{B}_nQ = Q$. Hence

$$\mathcal{B}_n\left((\cdot - t)^2\right)(t) = \frac{1}{n}t(1-t) \leq \frac{1}{4n}$$

for $t \in [0,1]$. Next, observe that for any functions $\mathbf{f}, h$ such that

$$\|\mathbf{f}(t)\| \leq h(t), \qquad t \in [0,1]$$

it follows that $\|(\mathcal{B}_n\mathbf{f})(t)\| \leq (\mathcal{B}_nh)(t)$. Finally, we specialize this inequality to the functions

$$\mathbf{f}(x) := \mathbf{g}(x) - \mathbf{g}(t) - \mathbf{g}'(t)(t-x)$$
$$h(x) = \frac{1}{2}(x-t)^2\|\mathbf{g}''\|_\infty$$

to obtain (1.19).

Thus, from (1.19) it follows that for an arbitrary interval $I = [a, b] \subseteq [0, 1]$

$$(1.20) \qquad \|\mathcal{B}_{n,I}\mathbf{g} - \mathbf{g}\|_{\infty,I} \leq \frac{|I|^2}{8n}\|\mathbf{g}''\|_\infty, \qquad |I| := b - a,$$

where

$$(1.21) \qquad \mathcal{B}_{n,I}\mathbf{g} := \mathcal{B}_n\left(\mathbf{g} \circ L_I^{-1}\right) \circ L_I$$

and L_I is the linear function that maps I onto $[0, 1]$. Also, since $b_0, \ldots, b_n$ are linearly independent on $[0, 1]$, there is a positive constant r (depending only on n) such that for all $\mathbf{c}_0, \ldots, \mathbf{c}_n$ and $k = 0, 1, \ldots, n$

$$(1.22) \qquad r\|\mathbf{c}_k\| \leq \left\|\sum_{j=0}^n \mathbf{c}_j b_j\right\|_\infty.$$

Next, using the refinement equation (1.10) repeatedly, we get for $t \in I_{\mathbf{e}^k}$ that

$$(1.23) \qquad \mathbf{b}(t) = B_{\mathbf{e}^k}^T \mathbf{b}(L_{I_{\mathbf{e}^k}}(t)).$$

Hence, for $t \in I_{\mathbf{e}^k}$ we obtain from (1.17), (1.21) (with $\mathbf{g} = C\mathbf{b}$), and (1.23) that

$$C\mathbf{b}(t) - \mathcal{B}_{n,I_{\mathbf{e}^k}}(C\mathbf{b})(t)$$
$$= \sum_{j=0}^n \left((B_{\mathbf{e}^k}\mathbf{c})_j - (C\mathbf{b})\left(L_{I_{\mathbf{e}^k}}^{-1}(j/n)\right)\right) b_j(L_{I_{\mathbf{e}^k}}(t)).$$

Now, we employ (1.20) (with $\mathbf{g} = C\mathbf{b}$) to the left side of the above equation; to the right-hand side we use (1.22) (with $\mathbf{c}_j = (B_{\mathbf{e}^k}\mathbf{c})_j - (C\mathbf{b})(L_{I_{\mathbf{e}^k}}^{-1}(\frac{j}{n})))$ to obtain

$$(1.24) \qquad \|(B_{\mathbf{e}^k}\mathbf{c})_j - (C\mathbf{b})(t_{j,k})\| \leq (8nr)^{-1}4^{-k}\|C\mathbf{b}''\|_\infty$$

where $t_{j,k} := L_{I_{\mathbf{e}^k}}^{-1}(\frac{j}{n}) = .\epsilon_1 \cdots \epsilon_k + 2^{-k}\frac{j}{n}$. This inequality holds for all $\mathbf{e}^k \in E_k$. When $t = \sum_{j=1}^\infty \epsilon_j 2^{-j}$ both $t_{j,k}$ and t lie in the interval $I_{\mathbf{e}^k}$. Therefore,

$$|t_{j,k} - t| \leq 2^{-k}$$

and we conclude, as asserted above, that *all* the control points in

$$\mathcal{C}_{\mathbf{e}^k}^k = \{(B_{\mathbf{e}^k}\mathbf{c})_0, \ldots, (B_{\mathbf{e}^k}\mathbf{c})_n\}$$

converge to

$$(C\mathbf{b})(t) = \sum_{j=0}^n \mathbf{c}_j b_j(t).$$

We summarize these remarks in

THEOREM 1.1. *For any* $t = \sum_{j=1}^{\infty} \epsilon_j 2^{-j}, \epsilon_j \in \{0,1\}, j = 1, 2, \ldots,$ *any* $\mathbf{c} = (\mathbf{c}_0, \ldots, \mathbf{c}_n)^T \in \mathbb{R}^{m(n+1)}$ *and* $i = 0, 1, \ldots, n$

$$\lim_{k \to \infty} (B_{\epsilon_k} \cdots B_{\epsilon_1} \mathbf{c})_i = \sum_{j=0}^{n} \mathbf{c}_j b_j(t).$$

In the sense of Theorem 1.1 the de Casteljau algorithm converges. Moreover, inequality (1.24) demonstrates that it converges quadratically (relative to the length of the subintervals $I_{\mathbf{e}^k}$).

This way of viewing de Casteljau's algorithm leads us to consider the following concept.

1.2.   Matrix subdivision: Convergence criteria.

DEFINITION 1.1. *Let* A_0, A_1 *be* $n \times n$ *matrices. We say that the matrix sub-division scheme (MSS)* $\{A_{\mathbf{e}^k} : k = 1, 2, \ldots\}$, $A_{\mathbf{e}^k} := A_{\epsilon_k} \cdots A_{\epsilon_1}$, *converges if there exists a nontrivial continuous curve* $\mathbf{f} : [0,1] \to \mathbb{R}^n$ *such that for any* $\mathbf{c} = (\mathbf{c}_1, \ldots, \mathbf{c}_n)^T \in \mathbb{R}^{nm}$ *and* $i = 1, \ldots, n$

$$(1.25) \qquad \lim_{k \to \infty} \left\| (A_{\mathbf{e}^k} \mathbf{c})_i - \sum_{j=1}^{n} \mathbf{c}_j f_j(t_{i,k}) \right\| = 0,$$

where $t_{i,k} = .\epsilon_1 \cdots \epsilon_k + 2^{-k} \frac{i}{n}$, $i = 1, \ldots, n$. *We refer to* $\mathbf{f}$ *as the refinable curve associated with the MSS.*

Note that if an MSS converges for some m it converges for all m. Moreover, as with the de Casteljau algorithm, (1.25) and the continuity of $\mathbf{f}$ imply that for $i = 1, \ldots, n$

$$(1.26) \qquad \lim_{k \to \infty} (A_{\mathbf{e}^k} \mathbf{c})_i = \sum_{j=1}^{n} \mathbf{c}_j f_j(t), \qquad \mathbf{c}_1, \ldots, \mathbf{c}_n \in \mathbb{R}^m,$$

whenever

$$t = \sum_{j=1}^{\infty} \epsilon_j 2^{-j},$$

and conversely, if (1.26) holds, then by the continuity of $\mathbf{f}$ (1.25) follows. Of course, the residuals in (1.25) are a convenient way to measure the rate of convergence of the MSS.

We are now going to provide necessary and sufficient conditions for the MSS generated by two matrices A_0, A_1 to converge. We begin with the following lemma.

LEMMA 1.2. *Let* $\mathbf{e} := (1, 1, \ldots, 1)^T \in \mathbb{R}^n$. *If* A_0, A_1 *generate a convergent MSS scheme then*

$$(1.27) \qquad\qquad A_\epsilon \mathbf{e} = \mathbf{e}, \qquad \epsilon \in \{0, 1\},$$

and

(1.28) $\qquad \mathbf{f}\left(\dfrac{t+\epsilon}{2}\right) = A_\epsilon^T \mathbf{f}(t), \qquad \epsilon \in \{0,1\}, \qquad t \in [0,1].$

Proof. Choose an irrational number $t \in [0,1]$ such that $\mathbf{f}(t) \neq \mathbf{0}$, set $m = 1$ in (1.26), and pick a vector $\mathbf{c} = (c_1,\ldots,c_n)^T \in \mathbb{R}^n$ such that

$$d := \sum_{j=1}^{n} c_j f_j(t) \neq 0.$$

Thus

$$\lim_{k \to \infty} A_{\mathbf{e}^k} \mathbf{c} = d\mathbf{e}$$

where

$$t = \sum_{j=1}^{\infty} \epsilon_j 2^{-j}$$

and $\mathbf{e}^k = (\epsilon_1,\ldots,\epsilon_k)^T$. Now, fix an $\epsilon \in \{0,1\}$. Then there are an infinite number of binary digits of t that equal ϵ, say $\epsilon_{k_j} = \epsilon, j = 1,2,\ldots$. Since

$$\begin{aligned} A_{\mathbf{e}^{k_j}} &= A_\epsilon A_{\epsilon_{k_j}-1} \cdots A_{\epsilon_1} \\ &= A_\epsilon A_{\mathbf{e}^{k_j}-1}, \qquad j = 2,3,\ldots, \end{aligned}$$

we get

$$\begin{aligned} d\mathbf{e} = \lim_{j \to \infty} A_{\mathbf{e}^{k_j}} \mathbf{c} &= A_\epsilon \left(\lim_{j \to \infty} A_{\mathbf{e}^{k_j}-1} \mathbf{c} \right) \\ &= A_\epsilon(d\mathbf{e}) = dA_\epsilon \mathbf{e}, \end{aligned}$$

and so $d(A_\epsilon \mathbf{e} - \mathbf{e}) = 0$. This proves our first claim (1.27), because $d \neq 0$. To prove (1.28), we note that for *any* $t \in [0,1]$ with binary expansion

$$t = \sum_{j=1}^{\infty} \epsilon_j 2^{-j},$$

the binary expansion of $\frac{1}{2}(t+\epsilon)$ is

$$\frac{1}{2}(t+\epsilon) = \frac{\epsilon}{2} + \sum_{j=1}^{\infty} \epsilon_j 2^{-j-1},$$

and so applying (1.26) to the point $2^{-1}(t+\epsilon)$, we have

$$\sum_{j=1}^{n} c_j f_j\left(\frac{t+\epsilon}{2}\right) = \sum_{j=1}^{n} (A_\epsilon \mathbf{c})_j f_j(t)$$

for all $\mathbf{c}_1,\ldots,\mathbf{c}_n \in \mathbb{R}^n$. From this equation (1.28) easily follows. $\square$

Let us record some further facts about convergent MSS. First, from (1.28) we have for any $\mathbf{e}^k \in E_k$ and $t \in [0,1]$

$$(1.29) \qquad\qquad \mathbf{f}(.\epsilon_1 \cdots \epsilon_k + 2^{-k}t) = A_{\mathbf{e}^k}^T \mathbf{f}(t).$$

This implies that for *every* $t \in [0,1]$

$$(1.30) \qquad\qquad \lim_{k \to \infty} A_{\mathbf{e}^k}^T \mathbf{f}(t) = \mathbf{f}(x), \qquad x = \sum_{j=1}^{\infty} \epsilon_j 2^{-j},$$

and, in particular,

$$(1.31) \qquad\qquad \mathbf{f}(t) \neq \mathbf{0}, \qquad t \in [0,1].$$

Next, we set

$$s_k := (1 - 2^{-k})^{-1}.\epsilon_1 \cdots \epsilon_k$$

and observe that $.\epsilon_1 \cdots \epsilon_k + 2^{-k}s_k = s_k$. Thus (1.29) implies that $\mathbf{f}(s_k)$ is an eigenvector of $A_{\mathbf{e}^k}^T$ corresponding to the eigenvalue one. The binary digits of s_k are periodic with period k, and the first k digits are $\epsilon_1, \ldots, \epsilon_k$. Hence, for *any* $\epsilon_1, \ldots, \epsilon_k \in \{0,1\}$ and $k = 1, 2, \ldots$ we get, by specializing (1.26) to $t = s_k$, that

$$\lim_{r \to \infty} A_{\mathbf{e}^k}^r \mathbf{x} = (\mathbf{x}, \mathbf{f}(s_k))\mathbf{e} = (\mathbf{x}^T \mathbf{f}(s_k))\mathbf{e},$$

where $\mathbf{x}$ is an arbitrary vector in $\mathbb{R}^n$ and $(\cdot, \cdot)$ denotes the standard Euclidean inner product on $\mathbb{R}^n$. Consequently, one is the largest eigenvalue of the matrix $A_{\mathbf{e}^k}$, which must be simple, with $\mathbf{e}$ as its right eigenvector and $\mathbf{f}(s_k)$ as its left eigenvector. Also, choosing $\mathbf{x} = \mathbf{e}$ above, and using (1.27) we get that $(\mathbf{e}, \mathbf{f}(s_k)) = 1$. Using the fact that the set $\{s_k : \epsilon_1, \ldots, \epsilon_k \in \{0,1\}, k = 1, 2, \ldots\}$ is dense in $[0,1]$, we obtain the important fact that

$$(1.32) \qquad\qquad (\mathbf{e}, \mathbf{f}(t)) = 1, \qquad t \in [0,1].$$

This equation, with (1.30), leads us to the conclusion that whenever the matrices $A_\epsilon \in \epsilon\{0,1\}$ admit a continuous nontrivial refinable curve $\mathbf{f}$ that is n-dimensional, the corresponding MSS *converges*. To see this, choose any $\mathbf{v} \in \text{span} \{\mathbf{f}(t) : t \in [0,1]\}$. Thus, $\mathbf{v}$ has the form

$$\mathbf{v} = \sum_{i=1}^{N} d_i \mathbf{f}(t_i)$$

for some scalars $d_1, \ldots, d_N$ and points $t_1, \ldots, t_N \in [0,1]$ and, necessarily, by (1.32),

$$(\mathbf{e}, \mathbf{v}) = \sum_{i=1}^{N} d_i.$$

Hence, from (1.30) we get, for any $\mathbf{u} \in \mathbb{R}^n$, that

$$\lim_{k \to \infty} (\mathbf{v}, \, A_{\mathbf{e}^k}\mathbf{u}) = (\mathbf{e}, \mathbf{v})(\mathbf{u}, \mathbf{f}(x)).$$

In other words, when a nontrivial continuous solution of the refinement equation (1.28) has the additional property that

$$\text{span} \, \{\mathbf{f}(t) : t \in [0, 1]\} = \mathbb{R}^n,$$

the corresponding MSS converges.

Finally, there holds a *compatibility* condition between the left eigenvectors $\mathbf{u}_\epsilon := \mathbf{f}(\epsilon)$ of A_ϵ. Namely, from the refinement equation (1.28) (first with $t = 0$ and $\epsilon = 1$ and then $t = 1$ and $\epsilon = 0$), we get

$$(1.33) \qquad\qquad A_1^T \mathbf{u}_0 = A_0^T \mathbf{u}_1.$$

To continue to list necessary conditions for the convergence of an MSS we need some facts pertaining to the $(n - 1) \times n$ (difference) matrix D defined by setting

$$(1.34) \qquad\qquad (D\mathbf{y})_i = y_{i+1} - y_i, \qquad i = 1, \ldots, n - 1,$$

for any $\mathbf{y} = (y_1, \ldots, y_n)^T \in \mathbb{R}^n$. Let $A = (A_{ij})$ be any $n \times n$ matrix such that $A\mathbf{e} = \mathbf{e}$. Then the $(n - 1) \times (n - 1)$ matrix $\mathcal{A} = (\mathcal{A}_{ij})$ defined by

$$(1.35) \qquad \mathcal{A}_{ij} = \sum_{r=j+1}^{n} (A_{i+1,r} - A_{ir}), \qquad i, j = 1, \ldots, n - 1,$$

has the property that

$$(1.36) \qquad\qquad \mathcal{A}D = DA.$$

Using equation (1.27), we let $\mathcal{A}_\epsilon$ be the $(n-1) \times (n-1)$ matrix that corresponds to A_ϵ.

LEMMA 1.3. *Suppose A_0, A_1 generate a convergent MSS scheme and $\| \cdot \|$ is any norm on $\mathbb{R}^{n-1}$. Then there exist constants $M > 0$ and $\rho \in (0, 1)$ such that*

$$(1.37) \qquad\qquad \|\mathcal{A}_{\mathbf{e}^k}\mathbf{y}\| \leq M\rho^k \|\mathbf{y}\|$$

for all $\mathbf{y} \in \mathbb{R}^{n-1}$, $k = 1, 2, \ldots$, and $\mathbf{e}^k \in E_k$.

Proof. From Definition 1.1 there is a constant $M_1 > 0$ such that for all $\mathbf{x} \in \mathbb{R}^n, \epsilon_i \in \{0,1\}, i = 1, \ldots, k$, and $k = 1, 2, \ldots,$

$$\|DA_{\mathbf{e}^k}\mathbf{x}\| \le M_1\|\mathbf{x}\|\omega(\mathbf{f}; 2^{-k}),$$

where

$$\omega(\mathbf{f}; \delta) = \max\{\|\mathbf{f}(t) - \mathbf{f}(s)\| : |t - s| \le \delta\}$$

is the modulus of continuity of $\mathbf{f}$. From (1.36), (applied successively to the matrices $A_{\epsilon_k}, \ldots, A_{\epsilon_1}$) we get

(1.38) $$\|A_{\epsilon_k} \cdots A_{\epsilon_1} D\mathbf{x}\| \le M_1\|\mathbf{x}\|\omega(\mathbf{f}; 2^{-k}).$$

Given any $\mathbf{y} = (y_1, \ldots, y_{n-1})^T \in \mathbb{R}^{n-1}$ choose

$$\mathbf{x}^0 = (0, y_1, y_1 + y_2, \ldots, y_1 + \cdots + y_{n-1})^T.$$

Then $D\mathbf{x}^0 = \mathbf{y}$ and there is a constant M_2 such that

$$\|\mathbf{x}^0\| \le M_2\|\mathbf{y}\|.$$

Hence, for any $\mathbf{y} \in \mathbb{R}^{n-1}$ we conclude from (1.38) (by choosing $\mathbf{x} = \mathbf{x}^0$) that

$$\|A_{\epsilon_k} \cdots A_{\epsilon_1}\mathbf{y}\| \le M_1 M_2\|\mathbf{y}\|\omega(\mathbf{f}; 2^{-k}).$$

Thus, there is an integer q such that for all $\epsilon_1, \ldots, \epsilon_k \in \{0,1\}$, with $k > q$, we get for all $\mathbf{y} \in \mathbb{R}^{n-1}$ that

$$\|A_{\epsilon_k} \cdots A_{\epsilon_1}\mathbf{y}\| \le \frac{1}{2}\|\mathbf{y}\|.$$

Let

$$M_3 := \max\{\|A_{\epsilon_k} \cdots A_{\epsilon_1}\mathbf{y}\| : \|\mathbf{y}\| \le 1, \ \epsilon_1, \ldots, \epsilon_k \in \{0,1\}, \ k < q\}$$

and $\rho = 2^{-1/q}$. Then it follows that (1.37) holds with $M := M_3/\rho^{q-1}$. $\square$

We now show that the necessary conditions described above are also sufficient.

THEOREM 1.2. *Let $A_\epsilon, \epsilon \in \{0,1\}$ be $n \times n$ matrices. Then a necessary and sufficient condition for the corresponding MSS to converge is that*

(1.39)   (i) $A_\epsilon \mathbf{e} = \mathbf{e}, \ \epsilon \in \{0,1\}.$

(1.40)   (ii)   *There exist nonzero vectors $\mathbf{u}_\epsilon$ such that $A_\epsilon^T \mathbf{u}_\epsilon = \mathbf{u}_\epsilon$ and $A_0^T \mathbf{u}_1 = A_1^T \mathbf{u}_0.$*

(1.41)   (iii)   *There are constants $\rho \in (0,1)$ and $M > 0$ such that $\|A_{\epsilon_k} \cdots A_{\epsilon_1}\mathbf{y}\| \le M\rho^k\|\mathbf{y}\|$ for all $k = 1, 2, \ldots, \mathbf{y} \in \mathbb{R}^{n-1}$, and $\epsilon_1, \ldots, \epsilon_k \in \{0,1\}.$*

Proof. We have proved the necessity of (i)–(iii) in Lemma 1.2, Lemma 1.3, and the discussion following Lemma 1.2. For the sufficiency we choose *any* continuously differentiable curve $\mathbf{h} : [0, 1] \to \mathbb{R}^n$ and $n \times n$ matrices $H_\epsilon, \epsilon \in \{0, 1\}$ such that

$$
(1.42) \quad
\begin{array}{llll}
\text{(i)} & \mathbf{h}(\epsilon) & = \mathbf{u}_\epsilon, & \epsilon \in \{0, 1\} \\
& (\mathbf{e}, \mathbf{h}(t)) & = 1, & t \in [0, 1]
\end{array}
$$

$$
(1.43) \quad \text{(ii)} \quad \mathbf{h}\left(\frac{t + \epsilon}{2}\right) = H_\epsilon^T \mathbf{h}(t), \qquad t \in [0, 1].
$$

We will indicate below possible choices of a curve $\mathbf{h}$ and matrices $H_\epsilon, \epsilon \in \{0, 1\}$ that satisfy these conditions.

The curve $\mathbf{h} : [0, 1] \to \mathbb{R}^n$ is the initial curve $\mathbf{h}_0 := \mathbf{h}$ of an iterative procedure defined for $k = 1, 2, \ldots$ by

$$
(1.44) \quad \mathbf{h}_k(x) = \begin{cases} A_0^T \mathbf{h}_{k-1}(2x), & 0 \le x < \dfrac{1}{2}, \\ A_1^T \mathbf{h}_{k-1}(2x - 1), & \dfrac{1}{2} \le x \le 1. \end{cases}
$$

Let us first demonstrate inductively that $\mathbf{h}_k$ is continuous. First, we claim that for all $k = 1, 2, \ldots$ and $\epsilon \in \{0, 1\}$, $\mathbf{h}_k(\epsilon) = \mathbf{u}_\epsilon$. We will also prove this claim inductively. In fact, if it is true that $\mathbf{h}_{k-1}(\epsilon) = \mathbf{u}_\epsilon$, then from (1.44) we get

$$
\mathbf{h}_k(0) = A_0^T \mathbf{h}_{k-1}(0) = A_0^T \mathbf{u}_0 = \mathbf{u}_0
$$

and

$$
\mathbf{h}_k(1) = A_1^T \mathbf{h}_{k-1}(1) = A_1^T \mathbf{u}_1 = \mathbf{u}_1.
$$

To prove the continuity of $\mathbf{h}_k$ we suppose that $\mathbf{h}_{k-1}$ is continuous. Therefore we need only check that the right and left value of $\mathbf{h}_k(t)$ at $t = \frac{1}{2}$ are equal. According to (1.44) and what we just proved above we have

$$
\mathbf{h}_k\left(\frac{1}{2}^-\right) = A_0^T \mathbf{h}_{k-1}(1) = A_0^T \mathbf{u}_1
$$

and

$$
\mathbf{h}_k\left(\frac{1}{2}^+\right) = A_1^T \mathbf{h}_{k-1}(0) = A_1^T \mathbf{u}_0.
$$

Thus the compatibility condition (1.40) implies $\mathbf{h}_k$ is continuous on $[0, 1]$.

Next, we will show that $\{\mathbf{h}_k : k = 1, 2, \ldots\}$ is a Cauchy sequence in $C[0, 1]$, the Banach space of continuous functions on $[0, 1]$ equipped with the maximum norm. We rewrite (1.44) in the equivalent form:

$$
\mathbf{h}_k\left(\frac{t + \epsilon}{2}\right) = A_\epsilon^T \mathbf{h}_{k-1}(t), \qquad t \in [0, 1], \qquad \epsilon \in \{0, 1\}.
$$

Thus for $t \in [0, 1]$ we get that

$$
(1.45) \quad \mathbf{h}_k\left(.\epsilon_1 \cdots \epsilon_k + 2^{-k} t\right) = A_{\mathbf{e}^k}^T \mathbf{h}(t)
$$

where, as before, $.\epsilon_1 \cdots \epsilon_k = \sum_{j=1}^{k} \epsilon_j 2^{-j}$ and $A_{\mathbf{e}^k} = A_{\epsilon_k} \cdots A_{\epsilon_1}$. Set $\sigma := 2^{-k-1}t + .\epsilon_1 \cdots \epsilon_{k+1} = 2^{-k}s + .\epsilon_1 \cdots \epsilon_k$, where $s := \frac{1}{2}(t + \epsilon_{k+1})$. Then by (1.45) (first with k replaced by $k+1$ and then t by s) and (1.43) (with $\epsilon = \epsilon_{k+1}$) we have that

$$\begin{aligned}
\mathbf{h}_{k+1}(\sigma) - \mathbf{h}_k(\sigma) &= A_{\mathbf{e}^{k+1}}^T \mathbf{h}(t) - A_{\mathbf{e}^k}^T \mathbf{h}(s) \\
&= \left(A_{\mathbf{e}^{k+1}}^T - A_{\mathbf{e}^k}^T H_{\epsilon_{k+1}}^T \right) \mathbf{h}(t) \\
&= A_{\mathbf{e}^k}^T \left(A_{\epsilon_{k+1}}^T - H_{\epsilon_{k+1}}^T \right) \mathbf{h}(t).
\end{aligned}$$

Let us estimate the norm of the right-hand side of the last equation. For this purpose, we note that the $n \times n$ matrix

$$S_k := A_{\epsilon_{k+1}} - H_{\epsilon_{k+1}}$$

has the property that $S_k \mathbf{e} = 0$. Thus, it is an easy matter to see that there is a rectangular matrix R_k such that

$$S_k = R_k D,$$

where D is the difference matrix (1.34). Obviously R_k is an $n \times (n-1)$ matrix given explicitly by

$$(R_k)_{ij} = -\sum_{r=1}^{j} (S_k)_{ir}, \qquad i = 1, \ldots, n, \qquad j = 1, \ldots, n-1.$$

Therefore the elements of R_k are bounded independent of k.

Choose any vector $\mathbf{x} \in \mathbb{R}^n$ and consider the inner product of $\mathbf{x}$ with $\mathbf{h}_{k+1}(\sigma) - \mathbf{h}_k(\sigma)$, viz.

$$(\mathbf{x}, \mathbf{h}_{k+1}(\sigma) - \mathbf{h}_k(\sigma)) = (R_k D A_{\epsilon_k} \cdots A_{\epsilon_1} \mathbf{x}, \mathbf{h}(t)).$$

Using our basic relation (1.36), namely $D A_\epsilon = \mathcal{A}_\epsilon D$, we get

$$(\mathbf{x}, \mathbf{h}_{k+1}(\sigma) - \mathbf{h}_k(\sigma)) = (R_k \mathcal{A}_{\epsilon_k} \cdots \mathcal{A}_{\epsilon_1} D \mathbf{x}, \mathbf{h}(t)).$$

Now, using Hölder's inequality, the equivalence of norms on $\mathbb{R}^n$, and our hypothesis (1.41), we conclude that there is a positive constant M_4 such that for all $k = 1, 2, \ldots$

$$|(\mathbf{x}, \mathbf{h}_{k+1}(\sigma) - \mathbf{h}_k(\sigma))| \le M_4 \rho^k \|D\mathbf{x}\| \, \|\mathbf{h}(t)\|.$$

But $\mathbf{h}$ is bounded on $[0, 1]$ and so we have obtained the inequality

$$\|\mathbf{h}_{k+1}(\sigma) - \mathbf{h}_k(\sigma)\| \le M_5 \rho^k, \qquad k = 1, 2, \ldots,$$

where M_5 is some positive constant *independent* of $\epsilon_1, \epsilon_2, \ldots, \epsilon_k \in \{0, 1\}$.

Given any $k = 1, 2, \ldots$ and *any* number σ in $[0, 1]$ we may express it in the form $\sigma = 2^{-k} t + \cdot \epsilon_1 \cdots \epsilon_k$ for some $t \in [0, 1]$ and $\mathbf{e}^k \in E^k$. Hence we have shown that for *all* $\sigma \in [0, 1]$

$$(1.46) \qquad \|\mathbf{h}_{k+1}(\sigma) - \mathbf{h}_k(\sigma)\| \leq M_5 \rho^k, \qquad k = 1, 2, \ldots.$$

From this inequality it follows that $\lim_{k \to \infty} \mathbf{h}_k$ exists (uniformly) and is a continuous function, which we call $\mathbf{f}$. From the equation immediately preceding (1.45), we conclude that the limit curve $\mathbf{f}$ satisfies the refinement equation

$$\mathbf{f}\left(\frac{t + \epsilon}{2}\right) = A_\epsilon^T \mathbf{f}(t), \qquad t \in [0, 1], \qquad \epsilon \in \{0, 1\}.$$

We will now prove that this is indeed the limit of the MSS. But first let us observe that $\mathbf{f}$ is actually *Hölder continuous*. We base the proof on the following inequality. To this end, choose any $t_1, t_2 \in [0, 1]$. Again, for $\mathbf{x} \in \mathbb{R}^n$ and $\sigma_i = \cdot \epsilon_1 \cdots \epsilon_k + 2^{-k} t_i$, $i = 1, 2$, we have

$$(\mathbf{x}, \mathbf{h}_k(\sigma_1) - \mathbf{h}_k(\sigma_2)) = (A_{\mathbf{e}^k} \mathbf{x}, \mathbf{h}(t_1) - \mathbf{h}(t_2)).$$

Next, observe that from (1.42) and (1.39) it follows inductively on k that $(\mathbf{e}, \mathbf{h}_k(t)) = 1$, for $k = 1, \ldots$ and $t \in [0, 1]$. Hence, Hölder's inequality and the equivalence of norms on $\mathbb{R}^n$ yield a constant $M_6 > 0$ such that for any real number d

$$|(\mathbf{x}, \mathbf{h}_k(\sigma_1) - \mathbf{h}_k(\sigma_2))| \leq M_6 \|A_{\mathbf{e}^k} \mathbf{x} - d\mathbf{e}\|_\infty \|\mathbf{h}(t_1) - \mathbf{h}(t_2)\|.$$

Here we use the notation $\|\mathbf{x}\|_\infty := \max\{|x_i| : 1 = i, \ldots, n\}$ where $\mathbf{x} = (x_1, \ldots, x_n)^T \in \mathbb{R}^n$. From the fact that $\mathbf{h} \in C^1[0, 1]$ and by choosing d wisely, there exists another constant $M_7 > 0$ such that

$$|(\mathbf{x}, \mathbf{h}_k(\sigma_1) - \mathbf{h}_k(\sigma_2))| \leq M_7 \|DA_{\mathbf{e}^k} \mathbf{x}\|_\infty |t_1 - t_2|.$$

Finally, we appeal to our basic relation (1.36) and our hypothesis (1.41) to find another constant $M_8 > 0$ such that

$$(1.47) \qquad \|\mathbf{h}_k(\sigma_1) - \mathbf{h}_k(\sigma_2)\| \leq M_8 (2\rho)^k |\sigma_1 - \sigma_2|.$$

As before, by choosing $(\epsilon_1, \ldots, \epsilon_k) \in E_k$ appropriately, this bound holds for any $\sigma_1, \sigma_2 \in [0, 1]$. It now follows, from general principles applied to the estimate (1.46) and (1.47), that h is Hölder continuous. The precise fact we use is the following lemma.

LEMMA 1.4. *Let $\{f_n : n \in \mathbb{Z}_+\}$ ($\mathbb{Z}_+ :=$ nonnegative integers) be a sequence of real-valued continuous functions on $[0, 1]$. Suppose $\rho_0 \in (1, \infty)$, $\rho \in (0, 1)$, and M, N are positive numbers such that for all $x, y \in [0, 1]$, $n \in \mathbb{Z}_+$*
   (i) $|f_n(x) - f_n(y)| \leq M(\rho \rho_0)^n |x - y|$,
   (ii) $|f_{n+1}(x) - f_n(x)| \leq N \rho^n$.

Then there exists a function ϕ continuous on $[0, 1]$ such that

$$\lim_{n \to \infty} f_n(x) = \phi(x)$$

uniformly on $[0, 1]$ and for $\mu \in (0, \infty)$ defined by the equation $\rho = \rho_0^{-\mu}$, we have

$$|\phi(x) - \phi(y)| \leq \delta |x - y|^\mu, \qquad x, y \in [0, 1]$$

for some constant $\delta > 0$.

    Proof. The existence of ϕ depends on noting that $\{f_n(x) : n \in \mathbb{Z}_+\}$ is a Cauchy sequence since by (ii)

$$|f_{n+\ell}(x) - f_n(x)| \leq \frac{N \rho^n}{1 - \rho}$$

for any $n, \ell \in \mathbb{Z}_+$. Thus we get

$$|\phi(x) - f_n(x)| \leq \frac{N \rho^n}{1 - \rho}.$$

Combining this bound with (i) yields, for any $x, y \in [0, 1]$, the inequality

$$|\phi(x) - \phi(y)| \leq \gamma(|x - y|),$$

where

$$\gamma(t) := \inf \left\{ \frac{2N}{1 - \rho} \rho^n + M(\rho_0 \rho)^n t : n \in \mathbb{Z}_+ \right\}, \qquad t \geq 0.$$

    Our goal is to show that $\gamma(t) \leq \delta t^\mu, t \in (0, 1]$ for some positive constant δ. To this end, note that

$$\rho \gamma(\rho_0 t) = \inf \left\{ \frac{2N}{1 - \rho} \rho^{n+1} + M(\rho_0 \rho)^{n+1} t : n \in \mathbb{Z}_+ \right\},$$

and therefore

(1.48) $$\rho \gamma(\rho_0 t) \geq \gamma(t).$$

Moreover, we also have

$$\gamma(t) \leq \frac{2N}{1 - \rho} + Mt.$$

Hence for $\rho_0^{-1} \leq t \leq 1$ we conclude that

(1.49) $$\gamma(t) \leq \delta t^\mu$$

where

$$\delta := \rho_0^\mu \left(\frac{2N}{1 - \rho} + M \right).$$

We now prove that (1.49) persists for all $t \in (0, 1]$ by showing, inductively on $r = 1, 2, \ldots$, that it holds on the interval $[\rho_0^{-r}, 1]$. Suppose we have established it for t in the range $\rho_0^{-r} \le t \le 1$. Consider values of t in the next interval $\rho_0^{-r-1} \le t \le \rho_0^{-r}$. Since in this range $\rho_0^{-r} \le \rho_0 t$, we have by (1.48) and our induction hypothesis that

$$\gamma(t) \le \rho\gamma(\rho_0 t) \le \rho\delta(\rho_0 t)^\mu = \rho\rho_0^\mu \delta t^\mu = \delta t^\mu,$$

because we chose μ so that $\rho\rho_0^\mu = 1$.   □

Although it is not important for us here, we remark that it is easy to see when $\mu \in (0, 1]$ that there is a positive constant δ_1 such that for $t \in [0, 1]$, $\gamma(t) \ge \delta_1 t^\mu$.

Returning to the proof of Theorem 1.2, it remains to be proven that the limit of our sequence $\mathbf{h}_k$, $k = 1, 2, \ldots$ is indeed the limit of the MSS. By the way, since $\mathbf{h}_k(\epsilon) = \mathbf{u}_\epsilon$ for all $k = 1, 2, \ldots$ it follows that $\mathbf{f}(\epsilon) = \mathbf{u}_\epsilon$, as well. Thus there is no doubt that $\mathbf{f}$ is a nonzero curve.

From our inequality (1.46), it follows that for all $t \in [0, 1]$ and $k = 1, 2, \ldots$

$$(1.50) \qquad \|\mathbf{h}_k(t) - \mathbf{f}(t)\| \le \frac{M_5}{1 - \rho}\rho^k.$$

Next, we need to observe that by hypothesis (iii) we have for all $\mathbf{e}^k \in E_k$

$$(1.51) \qquad \lim_{k \to \infty} DA_{\mathbf{e}^k}^k = \mathbf{0}.$$

Now, for every set of control points $\mathcal{C} = \{\mathbf{c}_1, \ldots, \mathbf{c}_n\}$, it follows directly from (1.50) (with $t = t_{i,k}$) and (1.45) (with $t = \frac{i}{n}, i = 1, \ldots, n$) that there is a constant $K > 0$ such that for $k = 1, 2, \ldots$, and $i = 1, \ldots, n$

$$(1.52) \qquad \left\| \sum_{j=1}^n (A_{\mathbf{e}^k}\mathbf{c})_j h_j\left(\frac{i}{n}\right) - \sum_{j=1}^n \mathbf{c}_j f_j(t_{i,k}) \right\| \le K\rho^k.$$

Recall that by our choice of h (see property (i), (1.42)) we have

$$\sum_{j=1}^n h_j\left(\frac{i}{n}\right) = 1$$

for all $i = 1, \ldots, n$. Hence

$$(1.53) \qquad \begin{aligned} &\sum_{j=1}^n (A_{\mathbf{e}^k}\mathbf{c})_j h_j\left(\frac{i}{n}\right) \\ &= (A_{\mathbf{e}^k}\mathbf{c})_i + \sum_{j \ne i}((A_{\mathbf{e}^k}\mathbf{c})_j - (A_{\mathbf{e}^k}\mathbf{c})_i)h_j\left(\frac{i}{n}\right). \end{aligned}$$

From (1.51), we see that the sum on the right-hand side of the equality above goes to zero as $k \to \infty$. Using this observation in (1.52) proves that (1.25) holds and so, indeed, the MSS $\{A_{\mathbf{e}^k} : k = 1, 2, \ldots\}$ converges to $\mathbf{f}$.   □

One final remark is in order for the proof of Theorem 1.2. We need to describe a curve $\mathbf{h} : [0,1] \to \mathbb{R}^n$ and matrices $H_\epsilon, \epsilon \in \{0,1\}$ that satisfy conditions (1.42) and (1.43). Perhaps the simplest choice of $\mathbf{h}$ is obtained by linear interpolation, viz.

$$\mathbf{h}(t) := (1-t)\mathbf{u}_0 + t\mathbf{u}_1, \qquad t \in [0,1].$$

Since $(\mathbf{e}, \mathbf{u}_\epsilon) = 1$ this curve satisfies (1.42). As for (1.43) we merely require matrices $H_\epsilon, \epsilon \in \{0,1\}$, such that $H_\epsilon^T \mathbf{e} = \mathbf{e}$,

$$H_0^T \mathbf{u}_0 = \mathbf{u}_0,$$
$$H_0^T \mathbf{u}_1 = \frac{1}{2}(\mathbf{u}_0 + \mathbf{u}_1),$$

and

$$H_1^T \mathbf{u}_0 = \frac{1}{2}(\mathbf{u}_0 + \mathbf{u}_1),$$
$$H_1^T \mathbf{u}_1 = \mathbf{u}_1.$$

They are easily constructed. In fact, given any two vectors $\mathbf{y}$ and $\mathbf{w}$ such that $(\mathbf{y}, \mathbf{e}) = (\mathbf{w}, \mathbf{e}) = 1$, there is an $n \times n$ matrix R such that

$$R^T \mathbf{e} = \mathbf{e}$$
$$R\mathbf{y} = \mathbf{y}$$
$$R\mathbf{w} = \frac{1}{2}(\mathbf{y} + \mathbf{w}).$$

Specifically, if $\mathbf{y}$ and $\mathbf{w}$ are linearly independent we can find vectors $\mathbf{r}$ and $\mathbf{t}$ such that

$$(\mathbf{r}, \mathbf{y}) = 1, \qquad (\mathbf{r}, \mathbf{w}) = \frac{1}{2}$$
$$(\mathbf{t}, \mathbf{y}) = 0, \qquad (\mathbf{t}, \mathbf{w}) = \frac{1}{2}.$$

Then $R\mathbf{x} := (\mathbf{r}, \mathbf{x})\mathbf{y} + (\mathbf{t}, \mathbf{x})\mathbf{w}$, $\mathbf{x} \in \mathbb{R}^n$, is the desired matrix. When $\mathbf{y}$ and $\mathbf{w}$ are linearly dependent, they must be the same vector and so $R\mathbf{x} := (\mathbf{e}, \mathbf{x})\mathbf{y}, \mathbf{x} \in \mathbb{R}^n$ will do in this case.

Using this construction first for $\mathbf{y} = \mathbf{u}_0$ and $\mathbf{w} = \mathbf{u}_1$ and then for $\mathbf{y} = \mathbf{u}_1$ and $\mathbf{w} = \mathbf{u}_0$ provides matrices $H_\epsilon, \epsilon \in \{0,1\}$, which satisfy (1.42) and (1.43) where $\mathbf{h}$ is the linear function defined above.

1.3.   Matrix subdivision: Reparameterization examples.

Let us now turn our attention to some examples of matrix subdivision schemes. Choose an $x \in (0,1)$ and consider the two 2×2 matrices

(1.54) $$G_0 = \begin{pmatrix} 1 & 0 \\ 1-x & x \end{pmatrix}, \qquad G_1 = \begin{pmatrix} 1-x & x \\ 0 & 1 \end{pmatrix}.$$

As we shall see later, the MSS associated with these matrices has a refinable curve that naturally arises in a simple variation of the de Casteljau subdivision. Let us begin our discussion of this example by using Theorem 1.2 to establish the convergence of the MSS and then develop some properties of the associated refinable function. To this end, we observe that the corresponding 1×1 difference matrices are given by

$$\mathcal{G}_0 = (x), \qquad \mathcal{G}_1 = (1 - x).$$

Hence condition (1.41) of Theorem 1.2 is satisfied with $M = 1$ and $\rho = \max(x, 1 - x)$.

Both G_0 and G_1 are stochastic matrices and so, in particular, (1.39) holds. Also, it is clear that

$$\mathbf{u}_0 = \begin{pmatrix} 1 \\ 0 \end{pmatrix}, \qquad \mathbf{u}_1 = \begin{pmatrix} 0 \\ 1 \end{pmatrix}$$

and so

$$G_1^T \mathbf{u}_0 = (1 - x, x)^T = G_0^T \mathbf{u}_1.$$

Thus (1.40) is satisfied and, consequently, Theorem 1.2 implies that the matrices (1.54) provide a convergent matrix subdivision scheme whose limit curve we shall denote by $\mathbf{g}(\cdot | x) : [0, 1] \to \mathbb{R}^2$.

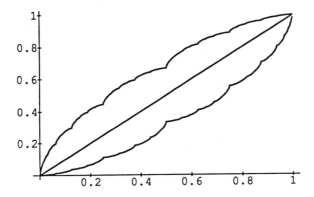

FIG. 1.6. *The function $g_2(\cdot | x)$ for $x = \frac{1}{3}, 1, \frac{2}{3}$.*

According to (1.32), $g_1(t|x) + g_2(t|x) = 1$, $t \in [0, 1]$. Therefore we can use the refinement equation (1.28) (corresponding to the matrices $G_\epsilon, \epsilon \in \{0, 1\}$, in (1.54)) to see that the function $g_2(\cdot | x)$ satisfies the equations

(1.55) $$g_2\left(2^{-1}t | x\right) = x \, g_2\left(t | x\right), \qquad 0 \leq t \leq 1,$$

(1.56) $$g_2\left(2^{-1}(t + 1) | x\right) = x + (1 - x)g_2(t|x), \qquad 0 \leq t \leq 1.$$

To solve these equations for $g_2(\cdot|x)$, we introduce two auxiliary functions

$$(1.57) \qquad U(t|x) = \begin{cases} 0, & 0 \le t < \dfrac{1}{2}, \\ x, & \dfrac{1}{2} \le t < 1, \end{cases}$$

and

$$(1.58) \qquad V(t|x) = \begin{cases} x, & 0 \le t < \dfrac{1}{2}, \\ 1 - x, & \dfrac{1}{2} \le t < 1. \end{cases}$$

We extend each of these functions to $[0, \infty)$ as *one periodic* functions, viz.

$$U(t + 1|x) = U(t|x), \qquad V(t + 1|x) = V(t|x), \qquad t \ge 0.$$

From now on we streamline our notation and drop our explicit indication that these functions depend on x. Note that both U and V (as functions of t) are step functions that are continuous from the right. We also extend the function g_2 to all of $[0, \infty)$ as a one periodic function and call this extension G. As $g_2(0) = 0$ and $g_2(1) = 1$ the function G is discontinuous at each positive integer. Moreover, it is straightforward to show that equations (1.55) and (1.56) are equivalent to the (one) equation

$$(1.59) \qquad G(t) = U(t) + V(t)G(2t), \qquad t \ge 0.$$

In fact, for any nonnegative integer ℓ and $t \in [\ell, \ell + \frac{1}{2})$, (1.59) becomes

$$g_2(t - \ell) = xg_2(2(t - \ell))$$

and for $t \in [\ell + \frac{1}{2}, \ell + 1)$, it becomes

$$g_2(t - \ell) = x + (1 - x)g_2\left(2(t - \ell) - 1\right),$$

both of which are valid because of (1.55) and (1.56).

By definition, we have arranged for the function G to be *bounded* for all $t \ge 0$. Since $V(t)$ is always in $(0, 1)$ for all $t \ge 0$, we may solve (1.59) iteratively and obtain the series expansion

$$(1.60) \qquad G(t) = \sum_{k=0}^{\infty} V_k(t)U\left(2^k t\right), \qquad t \ge 0,$$

where

$$V_k(t) := \begin{cases} 1, & k = 0, \\ V(t) \cdots V\left(2^{k-1}t\right), & k \ge 1. \end{cases}$$

This proves, by the way, that the solution of the refinement equation for the matrices in (1.54) is unique (a fact that is generally true because of (1.30)). As a consequence, we note that for $x = \frac{1}{2}$, $g_2(t|\frac{1}{2}) := t$ is the unique solution to (1.55) and (1.56). Hence we conclude that

$$(1.61) \qquad\qquad \mathbf{g}\left(t \,\Big|\, \frac{1}{2}\right) = (1 - t, t)^T, \qquad t \in [0, 1].$$

In general, we have from (1.60) that

$$g_2(t) = \sum_{k=0}^{\infty} V_k(t) U(2^k t), \qquad t \in [0, 1].$$

Both g_1 and g_2 are nonnegative functions since the elements of matrix A_ϵ, $\epsilon \in \{0, 1\}$ are nonnegative. In addition, we claim that g_2 is a *strictly increasing* function on $[0, 1]$ (and hence g_1 is strictly decreasing). First, we prove that both g_1 and g_2 are positive on $(0, 1)$ (and hence are both strictly less than one on $(0, 1)$).

According to the refinement equation (for $\epsilon = 0$)

$$g_1\left(\frac{t}{2}\right) = g_1(t) + (1 - x) g_2(t),$$

and so if $g_1\left(\frac{t}{2}\right) = 0$ then $\mathbf{g}(t) = \mathbf{0}$. This contradicts (1.31), which says *any* refinable curve never vanishes and so $g_1(t) > 0$ for $t \in [0, 2^{-1}]$. If $t \in [2^{-1}, 1)$ we set $d_\ell := 1 - 2^{-\ell}$, $\ell = 1, 2, \ldots$, and choose $\ell \geq 1$ such that $t \in [d_\ell, d_{\ell+1})$. Then the refinement equation for the curve $\mathbf{g}$ and what we already proved above give us the inequality

$$g_1(t) = (1 - x) g_1(2t - 1) = \cdots = (1 - x)^\ell g_1\left(2^\ell t - 2^\ell + 1\right) > 0.$$

Consequently, we have demonstrated that $g_1(t)$ is positive for $t \in [0, 1)$.

Next, let us show by a similar argument that g_2 is positive on $(0, 1]$. We have from (1.56) that

$$g_2\left(\frac{t + 1}{2}\right) = x g_1(t) + g_2(t)$$

and so, arguing just as above, $g_2(t) > 0$ for $t \in [2^{-1}, 1)$. When $t \in (0, 2^{-1}]$ we choose an $\ell \geq 1$ such that $t \in (2^{-\ell-1}, 2^{-\ell}]$. Then (1.55) and the positivity of $g_2(t)$ on $[2^{-1}, 1]$ give us this time the inequality

$$g_2(t) = x g_2(2t) = \cdots = x^\ell g_2\left(2^\ell t\right) > 0.$$

Now we are ready to prove that g_2 is increasing. To this end, we consider the 2×2 determinant

$$W(t_1, t_2) := \begin{vmatrix} g_1(t_1) & g_1(t_2) \\ g_2(t_1) & g_2(t_2) \end{vmatrix}$$

for $0 \le t_1 < t_2 \le 1$. Our assertion that g_2 is increasing is equivalent to the statement that

$$(1.62) \qquad\qquad W(t_1, t_2) > 0, \qquad 0 < t_1 < t_2 < 1.$$

Since

$$\mathbf{g}(\epsilon) = \mathbf{u}_\epsilon, \qquad \epsilon \in \{0, 1\},$$

and by what we already proved, it is certain that

$$(1.63) \qquad W(t_1, t_2) > 0, \qquad t_1 = 0 < t_2 < 1 \quad \text{or} \quad 0 < t_1 < t_2 = 1.$$

We will show that (1.62) follows from (1.63) and the refinement equation for $\mathbf{g}$.

As in the proof of (1.63), there naturally arises a distinction of cases. Using the refinement equation for $\mathbf{g}$, the following equations for W result:

$$(1.64) \qquad\qquad W(t_1, t_2) = x W(2t_1, 2t_2), \qquad 0 \le t_1 < t_2 \le \frac{1}{2},$$

while for $0 \le t_1 \le \frac{1}{2} \le t_2 \le 1$,

$$(1.65) \qquad \begin{aligned} W(t_1, t_2) \;=\; & x g_1(2t_1) g_1(2t_2 - 1) + (1 - x) g_2(2t_1) g_2(2t_2 - 1) \\ & + g_1(2t_1) g_2(2t_2 - 1), \end{aligned}$$

and

$$(1.66) \qquad W(t_1, t_2) = (1 - x) W(2t_1 - 1, 2t_2 - 1), \qquad \frac{1}{2} \le t_1 < t_2 \le 1.$$

The last equation is a consequence of the formula

$$\begin{vmatrix} g_1(t_1) & g_1(t_2) \\ g_2(t_1) & g_2(t_2) \end{vmatrix} = \begin{vmatrix} g_1(2t_1) + (1 - x)g_2(2t_1) & (1 - x)g_1(2t_2 - 1) \\ x g_2(2t_1) & x g_1(2t_2 - 1) + g_2(2t_2 - 1) \end{vmatrix}.$$

Therefore, from (1.65) and what we already proved, it follows that $W(t_1, t_2) > 0$ for $0 \le t_1 \le 2^{-1} \le t_2 \le 1$ and $t_1 < t_2$. The other cases are covered by an argument similar to that used to show the positivity of g_1 and g_2 on $(0, 1)$. Specifically, for any t_1, t_2 with $0 \le t_1 < t_2 < 1$ but *not* satisfying $0 \le t_1 \le 2^{-1} \le t_2 \le 1$, there is a *largest* positive integer k such that t_1 and t_2 lie in an interval of the form $[d_j^k, d_{j+1}^k)$, $0 \le j \le 2^k - 1$ (see Fig. 1.7). Here d_j^k, d_{j+1}^k are two consecutive binary fractions with k digits. This implies that

$$t_1 - d_j^k < 2^{-k-1} \quad \text{and} \quad t_2 - d_j^k \ge 2^{-k-1},$$

because otherwise t_1 and t_2 would lie between two consecutive binary fractions with $k + 1$ digits. To be precise, we know that for some ℓ, $\ell = 0, 1, \ldots, 2^{k+1} - 1$, $d_j^k = d_\ell^{k+1}$, $d_{j+1}^k = d_{\ell+2}^{k+1}$, and $d_{\ell+1}^{k+1} = 2^{-1}\left(d_j^k + d_{j+1}^k\right)$. If $t_1 - d_j^k \ge 2^{-k-1}$

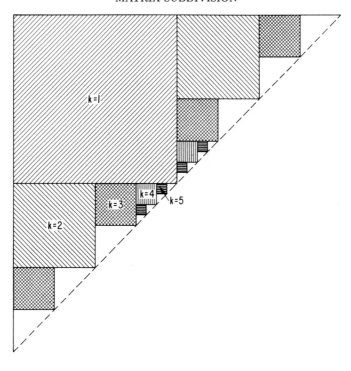

FIG. 1.7. *The proof of* (1.62).

then t_1 and t_2 are both in the interval $[d^{k+1}_{\ell+1}, \ d^{k+1}_{\ell+2})$, while if $t_2 - d^k_j < 2^{-k-1}$ then t_1 and t_2 are in $[d^{k+1}_{\ell}, \ d^{k+1}_{\ell+1})$.

Now, suppose $d^k_j = .\epsilon_1 \cdots \epsilon_k$ for some $\epsilon_1, \ldots, \epsilon_k \in \{0, 1\}$ and $p = \#\{r : \epsilon_r = 0, r = 1, \ldots, k\}$. Then equations (1.64) and (1.66), used repeatedly, imply that

$$(1.67) \qquad W\left(t_1, t_2\right) = x^p(1 - x)^{k-p} W\left(2^k\left(t_1 - d^k_j\right), \ 2^k\left(t_2 - d^k_j\right)\right).$$

Therefore we may use (1.65) to conclude that the right-hand side of equation (1.67) is positive. This conclusion more than adequately covers our assertion (1.62) and establishes that g_1 is decreasing and g_2 is increasing.

Our interest in the curve $\mathbf{g}(\cdot|x) : [0, 1] \to \mathbb{R}^2$, comes from its usefulness in identifying the limit of the following variation of de Casteljau subdivision. We have in mind to replace the midpoints in (1.4) with points that are some fixed distance $x \in (0, 1)$ along each segment of the previous control polygon. That is, we replace (1.4) by the recurrence formula

$$(1.68) \quad \mathbf{c}^\ell_r = (1 - x)\mathbf{c}^{\ell-1}_r + x\mathbf{c}^{\ell-1}_{r+1}, \qquad r = 0, 1, \ldots, n - \ell, \qquad \ell = 1, \ldots, n.$$

As it turns out, this scheme only affects a *reparameterization* of the Bernstein–Bézier curve $C\mathbf{b} : [0,1] \to \mathbb{R}^m$. To show that this is indeed the case we observe that the subdivision matrices $B_\epsilon(x), \epsilon \in \{0,1\}$ for (1.68) are

$$(1.69) \qquad (B_0(x))_{ij} = \binom{i}{j} x^j (1-x)^{i-j}, \qquad i,j = 0,1,\ldots,n$$

and

$$(1.70) \qquad B_1(x) = P_n B_0(1-x) P_n,$$

where P_n is the permutation matrix defined in (1.9). Note that the $B_\epsilon(x)$ reduce to the matrices (1.7)–(1.8) for de Casteljau subdivision when $x = 2^{-1}$.

To identify the subdivision matrices above we use the fact that

$$\mathbf{c}_r^\ell = \sum_{j=0}^{\ell} (B_0(x))_{\ell j}\, \mathbf{c}_{r+j}, \qquad \ell = 0,1,\ldots,n,$$

where we have set $\mathbf{c}_r^0 := \mathbf{c}_r, \ r = 0,1,\ldots,n$. In particular, we have

$$\mathbf{c}_0^\ell = \sum_{j=0}^{\ell} (B_0(x))_{\ell j}\, \mathbf{c}_j, \qquad \ell = 0,1,\ldots,n$$

and because $(B_0(x))_{n-\ell,j-\ell} = (B_0(1-x))_{n-\ell,n-j} = (B_1(x))_{\ell j}$ we get

$$\mathbf{c}_\ell^{n-\ell} = \sum_{j=0}^{n-\ell} (B_0(x))_{n-\ell,j}\, \mathbf{c}_{\ell+j}$$

$$= \sum_{j=\ell}^{n} (B_0(x))_{n-\ell,j-\ell}\, \mathbf{c}_j$$

$$= \sum_{j=0}^{n} (B_1(x))_{\ell j}\, \mathbf{c}_j, \qquad \ell = 0,1,\ldots,n.$$

The formula given above for $\mathbf{c}_r^\ell$ can be proved by induction on ℓ. We postpone the details of the proof to the beginning of Chapter 5 (see (5.4)) because the recurrence formula (1.68) provides the foundation for the material presented there. Note that the rightmost control point of the left control polygon and the leftmost control point of the right control polygon, produced by the first stage of the subdivision algorithm using the matrices $\{B_\epsilon(x) : \epsilon \in \{0,1\}\}$, agree with the value of the Bernstein–Bézier curve $C\mathbf{b}$ at x.

To identify the limit curve

$$\mathbf{b}(\cdot|x) : [0,1] \to \mathbb{R}^{n+1},$$

we pick any real vector $\mathbf{d} = (d_1, d_2)$ in $\mathbb{R}^2$ and use it to define a vector in $\mathbb{R}^{n+1}$ by

$$\mathbf{y}(\mathbf{d}) := \left(d_1^n, d_1^{n-1} d_2, \ldots, d_1 d_2^{n-1}, d_2^n \right)^T.$$

It follows from (1.69) and (1.70) that for all $\mathbf{d} \in \mathbb{R}^2$

(1.71) $B_\epsilon(x)\mathbf{y}(\mathbf{d}) = \mathbf{y}\,(G_\epsilon\mathbf{d})\,, \qquad \epsilon \in \{0,1\}$

where G_ϵ, $\epsilon \in \{0,1\}$ are the 2×2 matrices defined by (1.54). For the derivation of (1.71) it is convenient to use the fact that $P_n\mathbf{y}(\mathbf{d}) = \mathbf{y}(P_2\mathbf{d})$.

Next, we iterate (1.71) to get for every $\mathbf{e}^k \in E_k$ the formula

(1.72) $B_{\mathbf{e}^k}\mathbf{y}(\mathbf{d}) = \mathbf{y}\,(G_{\mathbf{e}^k}\mathbf{d})\,.$

Every vector $\mathbf{y} \in \mathbb{R}^{n+1}$ can be written as a finite linear combination of vectors in the set $\{\mathbf{y}(\mathbf{d}) : \mathbf{d} \in \mathbb{R}^2\}$. Thus, we see from (1.72) that $\lim_{k \to \infty} B_{\mathbf{e}^k}$ exists and the limit curve $\mathbf{b}(\cdot|x)$ satisfies

$$(\mathbf{b}(\cdot|x),\ \mathbf{y}(\mathbf{d}))\,\mathbf{e} = (\mathbf{g}(\cdot|x),\mathbf{d})^n\,\mathbf{e},$$

wher $\mathbf{e} = (1,1,\ldots,1)^T \in \mathbb{R}^{n+1}$. Hence, by the binomial theorem we get

$$\mathbf{b}(t|x) = \mathbf{b}(\mathbf{g}_2(t|x)), \qquad t \in [0,1],\ x \in (0,1).$$

This means that $g_2(\cdot|x) : [0,1] \to [0,1]$ acts as a reparameterization of the Bernstein–Bézier curves $C\mathbf{b}$. That is, the *curves* $C\mathbf{b}(\cdot|x)$ and $C\mathbf{b}(\cdot)$ are the same for any $x \in (0,1)$.

This example has a further implication beyond the simple variation (1.68) of de Casteljau subdivision. It gives us a way of constructing many new convergent MSS whose refinable curves lie on the (multivariate) *Bernstein–Bézier manifold*. To explain what we have in mind, we recall some standard multivariate notation. The set $\mathbb{Z}_+^{d+1}$ represents all vectors $\mathbf{i} = (i_1,\ldots,i_{d+1})^T$ whose components are nonnegative integers. For every $\mathbf{x} = (x_1,\ldots,x_{d+1})^T \in \mathbb{R}^{d+1}$, $\mathbf{x}^{\mathbf{i}}$ is the monomial

$$\mathbf{x}^{\mathbf{i}} := x_1^{i_1} \cdots x_{d+1}^{i_{d+1}}.$$

The standard d-simplex is defined as

$$\Delta^d = \Big\{\mathbf{x} = (x_1,\ldots,x_{d+1})^T : |\mathbf{x}|_1 := x_1 + \cdots + x_{d+1} = 1,$$
$$x_i \geq 0, \qquad i = 1,\ldots,d+1\Big\}$$

and the multivariate Bernstein–Bézier polynomials of degree n (associated with Δ^d) are given by

$$b_{\mathbf{i}}(\mathbf{x}) = \binom{n}{\mathbf{i}}\mathbf{x}^{\mathbf{i}}, \qquad \mathbf{i} \in \Gamma_{d+1,n}.$$

Here $\binom{n}{\mathbf{i}} := \frac{n!}{\mathbf{i}!}$, $\mathbf{i}! = i_1! \cdots i_{d+1}!$ and $\Gamma_{d+1,n} := \{\mathbf{i} : \mathbf{i} \in \mathbb{Z}^{d+1}, |\mathbf{i}|_1 = n\}$. We view the Bernstein–Bézier polynomials as functions on the hyperplane $H_d = \{\mathbf{x} : \mathbf{x} \in \mathbb{R}^{d+1},\ (\mathbf{e},\mathbf{x}) = 1\}$, which contains Δ^d. Also, we choose some ordering of the vectors $\mathbf{i} \in \mathbb{Z}^{d+1}$ with $|\mathbf{i}|_1 = n$ and in this ordering let $\mathbf{b}^{d,n}(\mathbf{x})$ be the vector in $\mathbb{R}^N$, $N := \#\Gamma_{d+1,n} = \binom{n+d}{n}$, whose $\mathbf{i}$th component is $b_{\mathbf{i}}(\mathbf{x})$. Notationally, we write

$$\mathbf{b}^{d,n}(\mathbf{x}) := \left(b_{\mathbf{i}}(\mathbf{x}) : \mathbf{i} \in \Gamma_{d+1,n} \right).$$

Similarly, for any vector $\mathbf{x} \in \mathbb{R}^{d+1}$, we let $\mathbf{y}(\mathbf{x}) := (\mathbf{x}^{\mathbf{i}} : \mathbf{i} \in \Gamma_{d+1,n})$. Geometrically speaking, the map $\Delta^d : \mathbf{x} \to \mathbf{b}^{d,n}(\mathbf{x})$ is a manifold in $\mathbb{R}^N$. The case $d = 1$ corresponds to the Bernstein–Bézier curve $\mathbf{b}$.

Now, let $\{A_\epsilon : \epsilon \in \{0,1\}\}$ be *any* $(d+1) \times (d+1)$ matrices whose MSS converges to the refinable curve $\mathbf{f} : [0,1] \to \mathbb{R}^{d+1}$. These matrices play the role that the 2×2 matrices $\{G_\epsilon : \epsilon \in \{0,1\}\}$ occupied in equation (1.71), which was essential in the analysis of the algorithm (1.68). In the present context we use (1.71) as a device to *define* $N \times N$ matrices. For this purpose, it is helpful to review the notion of the *permanent* of a matrix. Recall that the permanent of a $k \times k$ matrix $A = (a_{i,j})_{i,j=1,\dots,k}$, denoted by per A is defined as

$$\text{per } A = \sum_{\pi \in \mathcal{P}_k} a_{1\pi(1)} \cdots a_{k\pi(k)},$$

where the sum is taken over the set $\mathcal{P}_k$ of all permutations of the set $\{1, \dots, k\}$. We make use of the following interesting formula for permanents. For any $k \times \ell$ matrix C and any $\mathbf{j} \in \Gamma_{\ell,k}$, we let $C(\mathbf{j})$ be the $k \times k$ matrix obtained by repeating the ith column of C j_i times, $i = 1, \dots, \ell$ where $\mathbf{j} = (j_1, \dots, j_\ell)^T$, cf. Goulden and Jackson [GJ, p. 281].

LEMMA 1.5. *For any $k \times \ell$ matrix C*

$$(C\mathbf{x})_1 \cdots (C\mathbf{x})_k = \sum_{\mathbf{i} \in \Gamma_{\ell,k}} \frac{\mathbf{x}^{\mathbf{i}}}{\mathbf{i}!} \text{ per } C(\mathbf{i}), \qquad \mathbf{x} \in \mathbb{R}^\ell.$$

This formula is not hard to prove. It will reappear later in Chapter 5 (as Lemma 5.3), at which time we will provide its proof. An appropriate context for the proof is the notion of *polarization* or *blossoming*, which will be featured prominently in Chapter 5.

We use Lemma 1.5 as follows: let A be any $(d+1) \times (d+1)$ matrix and for $\mathbf{i}, \mathbf{j} \in \Gamma_{d+1,n}$ set $\mathbf{i} = (i_1, \dots, i_{d+1})^T$ and $\mathbf{j} = (j_1, \dots, j_{d+1})^T$. Then for each r, $r = 1, \dots, d+1$, repeat the rth row of the matrix A, i_r times. This procedure creates an $n \times (d+1)$ matrix, which we denote by $A(\mathbf{i})$. Take this matrix and for each s, $s = 1, \dots, d+1$ repeat its sth column j_s times. We call the resulting $n \times n$ matrix $A(\mathbf{i}, \mathbf{j})$. According to Lemma 1.5, with $k = n$, $\ell = d+1$, and $C = A(\mathbf{i})$ we have the formula

$$(A\mathbf{x})^{\mathbf{i}} = \sum_{\mathbf{j} \in \Gamma_{d+1,n}} \frac{\mathbf{x}^{\mathbf{j}}}{\mathbf{j}!} \text{ per } A(\mathbf{i}, \mathbf{j}), \qquad \mathbf{x} \in \mathbb{R}^{d+1}.$$

Thus we see that $\mathbf{y}(A\mathbf{x}) = M\mathbf{y}(\mathbf{x})$ for $\mathbf{x} \in \mathbb{R}^{d+1}$, where

$$M := (\mathbf{j}!^{-1} \text{ per } A(\mathbf{i}, \mathbf{j}))_{\mathbf{i}, \mathbf{j} \in \Gamma_{d+1,n}}.$$

THEOREM 1.3. *Let* $\{A_\epsilon : \epsilon \in \{0,1\}\}$ *be a set of* $(d+1) \times (d+1)$ *matrices which generate a convergent MSS with refinable curve* $\mathbf{f} : [0,1] \to \mathbb{R}^{d+1}$. *Let* M_ϵ *be the* $N \times N$ *matrix defined as*

$$M_\epsilon = \left(\mathbf{j}!^{-1} \text{per } A_\epsilon(\mathbf{i},\mathbf{j}) \right)_{\mathbf{i},\mathbf{j} \in \Gamma_{d+1,n}},$$

where $N = \binom{n+d}{d}$. *Then* $\{M_\epsilon : \epsilon \in \{0,1\}\}$ *determines a convergent MSS with a refinable curve given by*

$$\mathbf{h}(t) = \mathbf{b}^{d,n} \left(\mathbf{f}(t) \right), \qquad t \in [0,1].$$

We remark that (1.32) implies that $\mathbf{f}(t)$ is in H_d for all $t \in [0,1]$.

Proof. By definition, we have for all $\mathbf{x} \in \mathbb{R}^{d+1}$

$$M_\epsilon \mathbf{y}(\mathbf{x}) = \mathbf{y} \left(A_\epsilon \mathbf{x} \right)$$

and hence it follows that for every $\mathbf{e}^k \in E_k$

$$M_{\mathbf{e}^k} \mathbf{y}(\mathbf{x}) = \mathbf{y} \left(A_{\mathbf{e}^k} \mathbf{x} \right).$$

If $t = \sum_{i=1}^\infty \epsilon_i 2^{-i}$, $\mathbf{e}^k = (\epsilon_1, \ldots, \epsilon_k)^T$ we have, by hypothesis, that

$$\lim_{k \to \infty} A_{\mathbf{e}^k} \mathbf{x} = (\mathbf{x}, \mathbf{f}(t)) \, \mathbf{e}.$$

But $\mathbf{y} \left((\mathbf{x}, \mathbf{f}(t)) \mathbf{e} \right) = (\mathbf{x}, \mathbf{f}(t))^n \, \mathbf{e}$ and hence, we obtain by the multinomial theorem the relations

$$\lim_{k \to \infty} \mathbf{y} \left(A_{\mathbf{e}^k} \mathbf{x} \right) = (\mathbf{x}, \mathbf{f}(t))^n \, \mathbf{e} = (\mathbf{h}(t), \mathbf{y}(\mathbf{x})) \mathbf{e}.$$

This leads us to the equation

$$\lim_{k \to \infty} M_{\mathbf{e}^k} \mathbf{y}(\mathbf{x}) = (\mathbf{h}(t), \mathbf{y}(\mathbf{x})) \, \mathbf{e},$$

which is valid for all $t \in [0,1]$ and $\mathbf{x} \in \mathbb{R}^{d+1}$. Since the linear space spanned by the set $\{\mathbf{y}(\mathbf{x}) : \mathbf{x} \in \mathbb{R}^{d+1}\}$ is $\mathbb{R}^N$, we get for every $\mathbf{y} \in \mathbb{R}^N$

$$\lim_{k \to \infty} M_{\mathbf{e}^k} \mathbf{y} = (\mathbf{h}(t), \mathbf{y}) \, \mathbf{e}, \qquad t = \sum_{i=1}^\infty \epsilon_i 2^{-i}. \qquad \square$$

1.4. Stochastic matrices.

We now return to our analysis of general MSS. Our goal is to provide *another* necessary and sufficient condition for the convergence of MSS when the matrices A_0 and A_1 are *stochastic*. The difficult condition to check in Theorem 1.2 is part (iii). When the matrices are stochastic we shall replace it by another condition that can be checked in a *finite* number of steps. Recall that a matrix A is stochastic provided its elements are nonnegative and $A\mathbf{e} = \mathbf{e}$.

This gives us

(1.78) $\delta(A\mathbf{x}) \le \rho(A)\delta(\mathbf{x}),$

where

$$\rho(A) := \frac{1}{2} \max \left\{ \sum_{r=1}^{n} |A_{ir} - A_{jr}| : 1 \le i < j \le n \right\}.$$

When A is a stochastic matrix it follows easily that $\rho(A) \le 1$ and if it has a positive column then $\rho(A) < 1$. In fact, suppose that the kth column of A is positive. We set

$$\rho_0 := \min \{ A_{ik} : 1 \le i \le n \}.$$

Then $0 < \rho_0 \le 1$ and for all $i, j, r = 1, \ldots, n$ we have

$$|A_{ir} - A_{jr}| \le \begin{cases} A_{ir} + A_{jr}, & r \ne k, \\ A_{ik} + A_{jk} - 2\rho_0, & r = k. \end{cases}$$

Thus $\rho(A) \le 1 - \rho_0 < 1$, as claimed.

We apply these inequalities to the matrix $A_{\mathbf{e}^m}$, for any $\mathbf{e}^m \in E_m$. From (1.78) we are assured that there is a positive constant $u < 1$ such that

(1.79) $\delta \left(A_{\mathbf{e}^m} \mathbf{x} \right) \le u \delta(\mathbf{x})$

for every $\mathbf{x} \in \mathbb{R}^n$ and $\mathbf{e}^m \in E_m$. We let M be the maximum of all the numbers $\rho(A_{\mathbf{e}^k})$ where $\mathbf{e}^k \in E_k$, $k = 1, \ldots, m - 1$. Now choose any nonnegative integer ℓ and write it as $\ell = rm + t$ where r and t are integers such that $r \ge 0$ and $0 \le t < m$. Then from (1.79) and the definition of M, we get for all $\mathbf{e}^\ell \in E_\ell$

(1.80) $\delta \left(A_{\mathbf{e}^\ell} \mathbf{x} \right) \le u^r M \delta(\mathbf{x}).$

Next, using (1.77) and (1.80), we get for some positive constant M_1

$$\|D A_{\mathbf{e}^\ell} \mathbf{x}\|_\infty \le M_1 (u^{1/m})^\ell \|D\mathbf{x}\|_\infty$$

and finally, by property (1.36), we obtain

$$\|\mathcal{A}_{\mathbf{e}^\ell} D\mathbf{x}\|_\infty \le M_1 (u^{1/m})^\ell \|D\mathbf{x}\|_\infty.$$

Every vector $\mathbf{y} \in \mathbb{R}^{n-1}$ can be written as $\mathbf{y} = D\mathbf{x}$ where $\mathbf{x} = (x_1, \ldots, x_n)$ with $x_1 := 0$ and for $i = 2, \ldots, n$

$$x_i := \sum_{j=1}^{i-1} y_j.$$

Since $\|\mathbf{x}\|_\infty \leq (n-1)\|\mathbf{y}\|_\infty$, we conclude that

$$\|\mathcal{A}_{\mathbf{e}^\ell}\mathbf{y}\|_\infty \leq (n-1)M_1(u^{1/m})^\ell\|\mathbf{y}\|_\infty.$$

This proves (iii) of Theorem 1.2 with $M = (n-1)M_1$ and $\rho = u^{1/m}$. $\square$

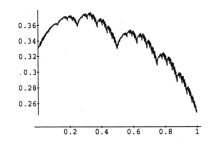

$$(1,0,0)^T$$

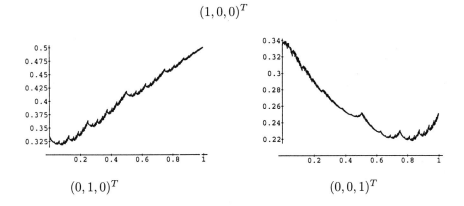

$$(0,1,0)^T \qquad\qquad\qquad (0,0,1)^T$$

$$A_0 = \frac{1}{3}\begin{bmatrix} 2 & 0 & 1 \\ 1 & 2 & 0 \\ 0 & 1 & 2 \end{bmatrix}, \qquad A_1 = \frac{1}{4}\begin{bmatrix} 3 & 1 & 0 \\ 1 & 2 & 1 \\ 1 & 0 & 3 \end{bmatrix}.$$

FIG. 1.8. *Matrix subdivision scheme: an example*

We remark that the first condition of Theorem 1.4 is equivalent to the assertion that $\lim_{k\to\infty} A_{\mathbf{e}^k}\mathbf{x}$ converges for all $\mathbf{x} \in \mathbb{R}^n$ and $\mathbf{e}^k \in E_k$, $k = 1, 2, \ldots$ to a multiple of $\mathbf{e}$. A careful examination of the proof demonstrates this fact. Thus (i) of Theorem 1.4 implies that for any $\epsilon \in \{0, 1\}$ there is a unique $\mathbf{u}_\epsilon \in \mathbb{R}^n$ (with nonnegative components) with $(\mathbf{e}, \mathbf{u}_\epsilon) = 1$ and $A_\epsilon^T\mathbf{u}_\epsilon = \mathbf{u}_\epsilon$. Moreover, the critical integer $m = 2^{n^2}$ can be substantially reduced. This was pointed out to us by Jeff Lagarias of AT&T Bell Laboratories. His comment is useful for applications of Theorem 1.4.

It is desirable to have conditions on the matrices A_0 and A_1 themselves to ensure that (i) of Theorem 1.4 is valid. To this end, we use the following terminology. Given two subsets $\mathcal{R}$ and $\mathcal{S}$ of $\{1, \ldots, n\}$ we say the column of A in $\mathcal{S}$ *connects* the rows of A in $\mathcal{R}$, denoted by $\mathcal{R} \to A \to \mathcal{S}$, provided that for every $i \in \mathcal{R}$ there is a $j \in \mathcal{S}$ such that $A_{ij} > 0$. Therefore the jth column of A is positive means $\{1, \ldots, n\} \to A \to \{j\}$. Moreover, it is easy to see that $\mathcal{R} \to A \to \mathcal{T}$ and $\mathcal{T} \to B \to \mathcal{S}$ implies $\mathcal{R} \to AB \to \mathcal{S}$. Conversely, if $\mathcal{R} \to AB \to \mathcal{S}$ there is a $\mathcal{T} \subseteq \{1, \ldots, n\}$ such that $\mathcal{R} \to A \to \mathcal{T}$ and $\mathcal{T} \to B \to \mathcal{S}$. Using these principles it follows that the jth column of $A_{\epsilon_1} \cdots A_{\epsilon_k}$ is positive if and only if there is a sequence of sets $\mathcal{R}_1, \ldots, \mathcal{R}_{k+1}$ such that $\mathcal{R}_1 = \{1, \ldots, n\}$ and $\mathcal{R}_{k+1} = \{j\}$ and $\mathcal{R}_j \to A_{\epsilon_j} \to \mathcal{R}_{j+1}, j = 1, 2, \ldots, k$. This leads us to two corollaries that provide useful sufficient conditions for the hypothesis (i) of Theorem 1.4 to hold.

COROLLARY 1.1. *Let $A_\epsilon, \epsilon \in \{0, 1\}$, be $n \times n$ nonnegative matrices such that for each $\mathcal{R} \subseteq \{1, \ldots, n\}$, $\#\mathcal{R} > 1$, and $\epsilon \in \{0, 1\}$ there is a set $\mathcal{S} \subseteq \{1, \ldots, n\}$, $0 < \#\mathcal{S} < \#\mathcal{R}$ such that $\mathcal{R} \to A_\epsilon \to \mathcal{S}$. Then for all $\mathbf{e}^{n-1} \in E_{n-1}$, $A_{\mathbf{e}^{n-1}}$ has a positive column.*

Proof. We set $\mathcal{R}_1 = \{1, \ldots, n\}$ and choose any $\mathbf{e}^{n-1} \in E_{n-1}$. Apply the hypothesis to the set $\mathcal{R}_1$ and the matrix A_{ϵ_1}. Thus there is a set $\mathcal{R}_2$ such that $\mathcal{R}_1 \to A_{\epsilon_1} \to \mathcal{R}_2$ and $\#\mathcal{R}_2 \leq n - 1$. Successively applying our hypothesis produces sets $\mathcal{R}_j \to A_{\epsilon_j} \to \mathcal{R}_{j+1}$, $j = 1, \ldots, n - 1$ with $0 < \#\mathcal{R}_{j+1} \leq \#\mathcal{R}_j - 1$. Hence $\#\mathcal{R}_n = 1$ and so there is some i, $1 \leq i \leq n$ such that $\mathcal{R}_n = \{i\}$. Therefore, we conclude that the ith column of $A_{\epsilon_1} \cdots A_{\epsilon_{n-1}} = A_{\mathbf{e}^{n-1}}$ is positive. □

COROLLARY 1.2. *Let $A_\epsilon, \epsilon \in \{0, 1\}$, be $n \times n$ nonnegative matrices such that either*

$$(A_\epsilon)_{11}(A_\epsilon)_{i+1,i} > 0, \qquad i = 1, \ldots, n - 1, \qquad \epsilon \in \{0, 1\}$$

or

$$(A_\epsilon)_{nn}(A_\epsilon)_{i-1,i} > 0, \qquad i = 2, \ldots, n, \qquad \epsilon \in \{0, 1\},$$

then $A_{\mathbf{e}^{n-1}}$ has a positive column.

Proof. In the first case, we let $\mathcal{R}_i = \{1, 2, \ldots, n - i + 1\}$ $i = 1, 2, \ldots, n$. Then $\mathcal{R}_i \to A_{\epsilon_i} \to \mathcal{R}_{i+1}$, $i = 1, 2, \ldots, n - 1$, and so the first column of $A_{\mathbf{e}^{n-1}}$ is positive. In the second case, we let $\mathcal{R}_i = \{i, i + 1, \ldots, n\}$ $i = 1, \ldots, n$ then $\mathcal{R}_i \to A_{\epsilon_i} \to \mathcal{R}_{i+1}$, $i = 1, 2, \ldots, n - 1$ and so the last column of $A_{\mathbf{e}^{n-1}}$ is positive. □

1.5. Corner cutting.

Theorem 1.4 is particularly useful for convergence analysis of *corner-cutting* subdivision algorithms patterned after the de Casteljau algorithm. We describe this next. To explain what we have in mind we interpret the first step of de Casteljau's algorithm in a different way. Instead of two matrices applied to the initial set of control points, we think of it as a sequence of control polygons, one obtained from the previous by a certain type of corner-cutting procedure. Each corner-cutting step is reinterpreted as a matrix operation and the subdivision matrices are viewed as products of simpler (corner-cutting) matrices. Below we have illustrated the case of cubic curves.

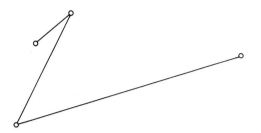

FIG. 1.9.   *Initial control polygon.*

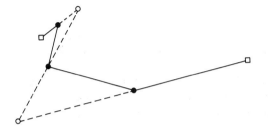

$$\begin{bmatrix} 1 & 0 & 0 & 0 \\ \frac{1}{2} & \frac{1}{2} & 0 & 0 \\ 0 & \frac{1}{2} & \frac{1}{2} & 0 \\ 0 & 0 & \frac{1}{2} & \frac{1}{2} \\ 0 & 0 & 0 & 1 \end{bmatrix}$$

FIG. 1.10.   *Two corners are cut.*

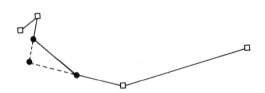

$$\begin{bmatrix} 1 & 0 & 0 & 0 & 0 \\ 0 & 1 & 0 & 0 & 0 \\ 0 & \frac{1}{2} & \frac{1}{2} & 0 & 0 \\ 0 & 0 & \frac{1}{2} & \frac{1}{2} & 0 \\ 0 & 0 & 0 & 1 & 0 \\ 0 & 0 & 0 & 0 & 1 \end{bmatrix}$$

FIG. 1.11.   *One corner is cut.*

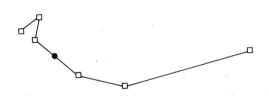

$$\begin{bmatrix} 1 & 0 & 0 & 0 & 0 & 0 \\ 0 & 1 & 0 & 0 & 0 & 0 \\ 0 & 0 & 1 & 0 & 0 & 0 \\ 0 & 0 & \frac{1}{2} & \frac{1}{2} & 0 & 0 \\ 0 & 0 & 0 & 1 & 0 & 0 \\ 0 & 0 & 0 & 0 & 1 & 0 \\ 0 & 0 & 0 & 0 & 0 & 1 \end{bmatrix}$$

FIG. 1.12.   *Break an edge.*

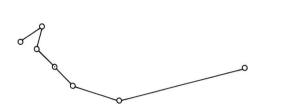

$$\begin{bmatrix} 1 & 0 & 0 & 0 & 0 & 0 & 0 \\ 0 & 1 & 0 & 0 & 0 & 0 & 0 \\ 0 & 0 & 1 & 0 & 0 & 0 & 0 \\ 0 & 0 & 0 & 1 & 0 & 0 & 0 \\ 0 & 0 & 0 & 1 & 0 & 0 & 0 \\ 0 & 0 & 0 & 0 & 1 & 0 & 0 \\ 0 & 0 & 0 & 0 & 0 & 1 & 0 \\ 0 & 0 & 0 & 0 & 0 & 0 & 1 \end{bmatrix}$$

FIG. 1.13.   *Common control point is repeated.*

The last step depicted above affects the doubling of the last midpoint that is shared by the two control polygons consisting of the first four and last four points of the control polygon above.

So we see that the basic step in the de Casteljau algorithm is *corner cutting from both ends*. This has the feature of adding *one* more control point at each step and leaving the ends fixed. Algebraically, it is given by the equations

$$\mathbf{d}_j = \begin{cases} \mathbf{c}_j, & j = 1, \ldots, k, \\ \nu_j \mathbf{c}_{j-1} + (1 - \nu_j)\mathbf{c}_j, & j = k+1, \ldots, n-\ell, \\ \mathbf{c}_{j-1}, & j = n-\ell+1, \ldots, n. \end{cases}$$

The numbers ν_j are assumed to lie in the interval $(0,1)$. In the de Casteljau algorithm ν_j always is chosen to be $\frac{1}{2}$. In the general case above, the first k control points and the last ℓ control points remain unaltered while the remaining are formed by *strict* convex combination of two adjacent control points. We will assume that $k = \ell$ as in de Casteljau subdivision and label the corresponding $(n+1) \times n$ matrix that affects the above operations by $B^{n+1,\ell}$.

When thinking of de Casteljau subdivision for an initial control polygon with n control points we see that there are n successive matrix operations, the last being the doubling of the common control point. Call this last $2n \times (2n-1)$ matrix F. Thus one step of de Casteljau subdivision, modeled by the above matrix operations, is affected by an application of the matrix

(1.81) $A = FB^{2n-1,n-1} \cdots B^{n+2,2} B^{n+1,1}.$

Starting with n control points, $\mathbf{c}_1, \ldots, \mathbf{c}_n$, the new control points $\mathbf{d}_1, \ldots, \mathbf{d}_{2n}$ are given by $\mathbf{d} = A\mathbf{c}$, the nth one being repeated twice. The typical matrix step $B^{n+k,k}$, $1 \le k \le n-1$, takes a control polygon consisting of $n + k - 1$ control points, leaves the first and last k control points the same, and forms $n-k$ convex combinations of adjacent control points ($n-k-1$ corner cuts). The $2n \times n$ matrix A and the $n \times n$ subdivision matrices A_0, A_1 that determine the left and right control polygons, consisting of n control points each, are related by the block decomposition

(1.82)
$$A = \begin{bmatrix} A_0 \\ A_1 \end{bmatrix}.$$

For instance, in the example depicted above for cubic Bernstein–Bézier curves we have

$$A = \begin{bmatrix} 1 & 0 & 0 & 0 \\ \frac{1}{2} & \frac{1}{2} & 0 & 0 \\ \frac{1}{4} & \frac{1}{2} & \frac{1}{4} & 0 \\ \frac{1}{8} & \frac{3}{8} & \frac{3}{8} & \frac{1}{8} \\ \frac{1}{8} & \frac{3}{8} & \frac{3}{8} & \frac{1}{8} \\ 0 & \frac{1}{4} & \frac{1}{2} & \frac{1}{4} \\ 0 & 0 & \frac{1}{2} & \frac{1}{2} \\ 0 & 0 & 0 & 1 \end{bmatrix} = \begin{bmatrix} B_0 \\ B_1 \end{bmatrix}.$$

For the general case (1.81)–(1.82), we list several facts about the matrices A, A_0, and A_1.

(a) *The first row of A_0 is $(1, 0, \ldots, 0)$ and the last row of A_1 is $(0, \ldots, 0, 1)$.*

This is a consequence of the fact that the left control points c_1 and the right control point c_n remain fixed during each alteration of the control polygons until we arrive at the neq control polygon D.

(b) *The matrices A_0 and A_1 are stochastic.*

This is a consequence of the fact that corner cutting consists of forming convex combinations.

(c) *The last row of A_0 is the same as the first row of A_1.*

This was arranged by the final application of the matrix F in (1.81) and forces the repetition of the nth control point of the new control polygon D.

(d) *A_0 and A_1 are nonsingular.*

Let us prove this fact. Suppose $A_0 x = 0$ for $x \in \mathbb{R}^n$. Then the first n components of the vector Ax are zero and, likewise, the first n components of $B^{2n-1,n-1} \cdots B^{n+1,1} x$ are zero. The matrix $B^{2n-1,n-1}$ leaves the first $n - 1$ components of $y := B^{2n-2,n-2} \cdots B^{n+1,1} x$ unchanged and so they must be zero. Since the nth component of $B^{2n-1,n-1} y$ is zero and is also a *strict* convex combination of the $n - 1$th and nth component of y we see that actually the first n components of y are zero. Proceeding in this fashion, inductively on $k = n - 1, \ldots, 1$, we conclude that the vector $B^{n+k,k} \cdots B^{n+1,1} x$ has all its first n components equal to zero. Thus, in particular, the first n components of $B^{n+1,1} x$ are zero and so $x = 0$, as well. An identical argument using the last n components of the appropriate vectors shows that A_1 is nonsingular.

The final and most important property of A is

(e) *A is totally positive.*

Recall that an $n \times k$ matrix A is totally positive (TP) if and only if all its minors are nonnegative. The product of two totally positive matrices is totally positive. This important fact follows from the Cauchy–Binet formula, which we state below for the convenience of the reader. First a bit of notation. The symbol

$$A\begin{pmatrix} i_1, \ldots, i_p \\ j_1, \ldots, j_p \end{pmatrix},$$

where $1 \le i_1 < \cdots < i_p \le n$, $1 \le j_1 < \cdots < j_p \le k$ will denote the determinant of the submatrix of A consisting of rows $i_1, \ldots, i_p$ and columns $j_1, \ldots, j_p$.

LEMMA 1.6 *Let* A, B *be* $n \times m$ *and* $m \times k$ *matrices, respectively, and set* $C = AB$. *Then for any* $1 \le i_1 < \cdots < i_p \le n$, $1 \le j_1 < \cdots < j_p \le k$

$$C\begin{pmatrix} i_1, \ldots, i_p \\ j_1, \ldots, j_p \end{pmatrix} = \sum_{1 \le r_1 < \cdots < r_p \le m} A\begin{pmatrix} i_1, \ldots, i_p \\ r_1, \ldots, r_p \end{pmatrix} B\begin{pmatrix} r_1, \ldots, r_p \\ j_1, \ldots, j_p \end{pmatrix}.$$

For a proof of this formula see Gantmacher [G, p. 9] or Karlin [K, p. 1].

From this lemma we see that the total positivity of A hinges on the total positivity of each of its factors in (1.81). But each factor has nonnegative elements on its diagonal or (lower) secondary diagonal and so it is straightforward to see it is totally positive (see Lemma 2.3 of Chapter 2).

1.6.   Total positivity and variation diminishing curves.

In our next result we show that *any* MSS based on $n \times n$ matrices A_0 and A_1 satisfying conditions **(a)**–**(e)** above leads to a convergent scheme whose refinable curve has an attractive *variation diminishing* property. To state the result we use the following notation. Given a set $\{f_1, \ldots, f_n\}$ of real-valued functions defined on $[0, 1]$ we set, for any integers $1 \le i_1 < \cdots < i_p \le n$, and real numbers $0 \le x_1 < \cdots < x_p \le 1$,

$$F\begin{pmatrix} i_1, \ldots, i_p \\ x_1, \ldots, x_p \end{pmatrix} = \begin{vmatrix} f_{i_1}(x_1) & \cdots & f_{i_1}(x_p) \\ \vdots & & \vdots \\ f_{i_p}(x_1) & \cdots & f_{i_p}(x_p) \end{vmatrix}.$$

We also use the notation

$$F\begin{bmatrix} i_1, \ldots, i_p \\ x_1, \ldots, x_p \end{bmatrix}$$

for the matrix $(f_{i_\ell}(x_m))_{\ell, m=1, \ldots, p}$.

THEOREM 1.5.   *Let* $A_\epsilon, \epsilon \in \{0, 1\}$ *be nonsingular stochastic matrices such that the* $2n \times n$ *matrix*

$$A = \begin{bmatrix} A_0 \\ A_1 \end{bmatrix}$$

is totally positive. Suppose further that the first row of A_0 *is* $(1, 0, \ldots, 0)^T$, *the last row of* A_1 *is* $(0, \ldots, 0, 1)^T$, *and the last row of* A_0 *and the first row of* A_1

are the same. Then there exists a unique continuous solution $\mathbf{f} : [0,1] \to \mathbb{R}^n$ *to the refinement equation*

$$\mathbf{f}\left(\frac{t+\epsilon}{2}\right) = A_\epsilon^T \mathbf{f}(t), \qquad 0 \le t \le 1, \qquad \epsilon \in \{0,1\},$$

satisfying $(\mathbf{e}, \mathbf{f}(t)) = 1, 0 \le t \le 1$. *Furthermore,* $\mathbf{f}$ *is the refinable curve of the convergent MSS,* $\{A_\epsilon : \epsilon \in \{0,1\}\}$, *viz.*

$$\lim_{k \to \infty} A_{\epsilon_k} \cdots A_{\epsilon_1} \mathbf{c} = (\mathbf{c}, \mathbf{f}(t)) \, \mathbf{e}, \qquad t = \sum_{k=1}^{\infty} \epsilon_k \, 2^{-k}, \qquad \epsilon_k \in \{0,1\}.$$

Moreover, for any positive integer $p \le n$

$$(1.83) \qquad\qquad F\begin{pmatrix} i_1, \ldots, i_p \\ x_1, \ldots, x_p \end{pmatrix} \ge 0$$

if $1 \le i_1 < \cdots < i_p \le n$, $0 \le x_1 < \cdots < x_p \le 1$, *where equality holds in* (1.83) *if and only if either* $x_1 = 0, i_1 > 1$ *or* $x_p = 1$, $i_p < n$.

The proof of this theorem is long. Before we get to the details, we will discuss some of its consequences and also auxiliary facts needed in its proof.

There are two essential conclusions in Theorem 1.5. The first says that conditions **(a)**–**(e)** lead to convergent MSS. The second says the associated refinable curve satisfies the strong determinantal property (1.83). In particular, specializing (1.83) to the case $p = n$ and $0 < x_1 < \cdots < x_n < 1$, we conclude that for any vector $\mathbf{y} = (y_1, \ldots, y_n)^T \in \mathbb{R}^n$, there is a vector $\mathbf{d} \in \mathbb{R}^n$ such that

$$(\mathbf{d}, \mathbf{f}(x_i)) = y_i, \qquad i = 1, 2, \ldots, n.$$

Thus we can always interpolate n values at any n distinct points on $(0,1)$ by some linear combination of the components of $\mathbf{f}$.

There are more general corner-cutting schemes than those described by (1.81) that satisfy the conditions of Theorem 1.5. We mention the following. Given the control points $\mathcal{C} = \{\mathbf{c}_1, \ldots, \mathbf{c}_n\} \subset \mathbb{R}^m$ *cutting* $k - 1$ *corners from the right*, $2 \le k \le n - 1$, means forming the new control points $\mathcal{D} = \{\mathbf{d}_1, \ldots, \mathbf{d}_n\} \subset \mathbb{R}^m$ by

$$\mathbf{d}_j = \begin{cases} \mathbf{c}_j, & j = 1, \ldots, n - k, \\ \lambda_j \mathbf{c}_{j-1} + (1 - \lambda_j)\mathbf{c}_j, & j = n - k + 1, \ldots, n, \end{cases}$$

for $0 \le \lambda_j < 1$, $j = n - k + 1, \ldots, n$ and $\lambda_n = 0$. Thus, in matrix terms, $\mathbf{d} = L^k \mathbf{c}$, where $L^k = (L_{ij}^k)_{i,j=1}^n$ is an $n \times n$ *nonsingular, lower triangular, one-banded stochastic matrix* with $L_{i,i-1}^k = 0, i = 2, \ldots, n - k$, for $k = 2, \ldots, n - 1$ (see Fig. 1.14).

Similarly, *cutting* $k - 1$ *corners from the left* has the form $\mathbf{d} = U^k \mathbf{c}$, where the components of $\mathbf{d}$ are given by

$$\mathbf{d}_j = \begin{cases} \mu_j \mathbf{c}_j + (1 - \mu_j)\mathbf{c}_{j+1}, & j = 1, \ldots, k, \\ \mathbf{c}_j, & j = k + 1, \ldots, n, \end{cases}$$

for $0 < \mu_j \le 1, j = 1, \ldots, k$ and $\mu_1 = 1$ (see Fig. 1.15).

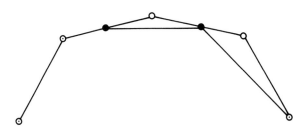

FIG. 1.14.   *Cutting two corners from the right: $n = 5, k = 3$.*

FIG. 1.15.   *Cutting three corners from the left: $n = 5, k = 4$.*

Note that, unlike cutting corners from both ends, cutting corners from the left or right does not increase the number of control points. Moreover, using the same arguments used to verify **(a)**-**(e)**, it follows that the $2n \times n$ matrix

$$(1.84) \quad A = \begin{bmatrix} A_0 \\ A_1 \end{bmatrix} = F\, B^{2n-1,n-1} \cdots B^{n+2,2}\, B^{n+1,1}\, L^2 \cdots L^{n-1}\, U^2 \cdots U^{n-1}$$

also satisfies the hypothesis of Theorem 1.5. The reason we have chosen exactly $n - 1$ corner cuts from the left and the same from the right is that there is a *converse* to our remarks above. Namely, conditions **(a)** through **(e)** above *imply* a factorization of the type given in (1.84). We refer the interested reader to the appropriate section of Micchelli and Pinkus [MPi] (see also Goodman and Micchelli [GM]).

It is important to note that determinantal inequalities of the type (1.83) imply that the curve **f** is *variation diminishing*. For an explanation of this property we need to review some standard notation for counting sign changes of vectors and functions.

Let $\mathbf{x} \in \mathbb{R}^n \backslash \{\mathbf{0}\}$, then $S^-(\mathbf{x})$ is the number of sign changes in the components of the vector $\mathbf{x}$ when zero components are disregarded. On the other hand, $S^+(\mathbf{x})$ is the maximum number of sign changes possible when we assign either $+1$ or -1 to the zero components of $\mathbf{x}$.

For a real-valued continuous function g, defined on $[0, 1]$, $S^-(g)$ is the number of sign changes of g on $[0, 1]$. In other words,

$$S^-(g) := \sup \left\{ S^- \left(g(x_1), \ldots, g(x_n) \right) : 0 < x_1 < \cdots < x_n < 1, \ n = 1, 2, \ldots \right\}.$$

We also use $Z(g)$ for the number of zeros of g on $(0, 1)$.

COROLLARY 1.3. *Let* $\mathbf{f} : [0, 1] \to \mathbb{R}^n$ *be the curve given in Theorem 1.5. Then for every* $\mathbf{c} \in \mathbb{R}^n \backslash \{\mathbf{0}\}$ *we have*

$$S^- \left((\mathbf{c}, \mathbf{f}) \right) \leq S^-(\mathbf{c}).$$

Actually, the determinantal inequalities in Theorem 1.5 imply the stronger inequality

$$(1.85) \qquad\qquad Z \left((\mathbf{c}, \mathbf{f}) \right) \leq S^-(\mathbf{c}).$$

Let us present the proof of (1.85), and hence also of Corollary 1.3. This will give us an opportunity to review some key facts about totally positive matrices that we need in the proof of Theorem 1.5.

A matrix A is called *strictly totally positive* (STP) if and only if all its minors are positive. A very useful STP matrix is $T_n = (T_{ij})_{i,j=1,\ldots,n}$ where

$$(1.86) \qquad\qquad T_{ij} := \exp(x_i y_j), \qquad i, j = 1, \ldots, n,$$

and $x_1 < \cdots < x_n$, $y_1 < \cdots < y_n$. Induction on n is a convenient way to prove that T_n is STP. Suppose any matrix of the form (1.86) with $x_1 < \cdots < x_n$, $y_1 < \cdots < y_n$ is STP. Now, consider any $x_1 < \cdots < x_{n+1}$, $y_1 < \cdots < y_{n+1}$ and the corresponding matrix T_{n+1}. First, we show that the determinant of T_{n+1} is nonzero. If, to the contrary, the determinant of T_{n+1} is zero then there is a nonzero vector $\mathbf{c} = (c_1, \ldots, c_{n+1})^T$ such that the function

$$g(y) := \sum_{j=1}^{n+1} c_j e^{x_j y}$$

vanishes at $y_1, \ldots, y_{n+1}$. Therefore, Rolle's theorem applied to the function $h(y) := e^{-x_{n+1} y} g(y)$ implies that $h'(y)$ vanishes at some n distinct points $y_1' < \cdots < y_n'$. But the matrix T_n in (1.86) corresponding to the values $x_1 - x_{n+1}, \ldots, x_n - x_{n+1}$ and $y_1', \ldots, y_n'$ is nonsingular. This readily implies that $\mathbf{c} = \mathbf{0}$, which is a contradiction. Hence, we conclude that $\det T_{n+1} \neq 0$ for all $x_1 < \cdots < x_{n+1}$, $y_1 < \cdots < y_{n+1}$. As a function of y_{n+1} (with all other values of x's and y's fixed) we have shown that $\det T_{n+1}$ does not change signs in the interval (y_n, ∞). However, by the induction hypothesis, $\lim_{y_{n+1} \to \infty} \det T_{n+1} = \infty$ and so $\det T_{n+1}$ is in fact positive for $y_1 < \cdots < y_{n+1}$ and $x_1 < \cdots < x_{n+1}$.

The matrix T_n is a very useful tool to approximate a TP matrix by STP matrices. Specifically, for any $\delta > 0$ we set

$$G(\delta) = \left(\exp(-(i-j)^2/\delta) \right)_{i,j=1,2,\ldots,n}.$$

This matrix is STP. To see this, we write it in the form

$$G_{ij} = e^{-i^2/\delta} e^{-j^2/\delta} \hat{T}_{ij}, \quad i, \; j = 1, \ldots, n,$$

where

$$\hat{T}_{ij} = e^{2ij/\delta}, \quad i, \; j = 1, \ldots, n,$$

and so, by our remark above, $G(\delta)$ is STP.

Now let A be an $n \times m$ TP matrix of rank m and consider the $n \times m$ matrix

(1.87) $A(\delta) := G(\delta)A.$

By the Cauchy–Binet formula (Lemma 1.6) we obtain the equation

$$A(\delta)\binom{i_1, \ldots, i_s}{j_1, \ldots, j_s} = \sum_{k_1 < \cdots < k_s} G(\delta)\binom{i_1, \ldots, i_s}{k_1, \ldots, k_s} A\binom{k_1, \ldots, k_s}{j_1, \ldots, j_s}.$$

Since A is TP and of rank m and $G(\delta)$ is STP, the sum on the right-hand side of the above equation is positive. Consequently, we conclude that $A(\delta)$ is STP for all $\delta > 0$. Obviously, $\lim_{\delta \to 0+} A(\delta) = A$ and so A can be approximated by STP matrices. (Actually, this remains valid for *any* TP matrix, see Pinkus [Pin, p. 49]). These remarks lead us to the following well-known fact, cf. Karlin [K].

LEMMA 1.7. *Let A be an $n \times m$ TP matrix of rank m with $n > m$. Suppose $\mathbf{x} \in \mathbb{R}^n \backslash \{\mathbf{0}\}$ and $A^T \mathbf{x} = \mathbf{0}$. Then $S^+(\mathbf{x}) \geq m$. Moreover, if A is STP then $S^-(\mathbf{x}) \geq m$.*

Proof. We prove the second claim first. Clearly, $S^-(\mathbf{x}) \geq 1$. Suppose to the contrary that $\ell := S^-(\mathbf{x}) < m$. Then there are integers $0 = i_0 < i_1 < \cdots < i_{\ell-1} < i_\ell = n$ such that for each of the sets $I_r := \{i_{r-1}+1, \ldots, i_r\}$, $r = 1, 2, \ldots, \ell$, there is an $\epsilon_r \in \{-1, 1\}$ such that $x_\ell = \epsilon_r |x_\ell|$ for all $\ell \in I_r$ and $x_\ell \neq 0$ for some $\ell \in I_r$. Next we express the vector $A^T \mathbf{x}$ as $B\mathbf{y}$ where $\mathbf{y} := (\epsilon_1, \ldots, \epsilon_\ell)$, $B^T = CA$ and C is the $\ell \times n$ matrix defined as

$$C = \begin{bmatrix} |x_1| & |x_2| & \cdots & |x_{i_1}| & 0 & \cdots & 0 & \cdots & 0 & \cdots & 0 \\ 0 & 0 & \cdots & 0 & |x_{i_1+1}| & \cdots & |x_{i_2}| & \cdots & 0 & \cdots & 0 \\ \cdot & \cdot & & \cdot & \cdot & & \cdot & & \cdot & & \cdot \\ \cdot & \cdot & & \cdot & \cdot & & \cdot & & \cdot & & \cdot \\ \cdot & \cdot & & \cdot & \cdot & & \cdot & & \cdot & & \cdot \\ 0 & 0 & \cdots & 0 & 0 & \cdots & 0 & \cdots & |x_{i_{\ell-1}+1}| & \cdots & |x_{i_\ell}| \end{bmatrix}.$$

Then $\det B > 0$ because

$$\det B = \Sigma |x_{j_1}| \cdots |x_{j_\ell}| A\binom{j_1, \cdots, j_\ell}{1, \ldots, \ell},$$

where the sum is taken over all integers $j_1, \ldots, j_\ell$ with $j_\ell \in I_\ell, \ell = 1, \ldots, \ell$. This contradicts the fact that $B\mathbf{y} = \mathbf{0}$ and $\mathbf{y} \neq \mathbf{0}$. Hence the second claim is proved.

For the first claim we use the matrix $A(\delta)$, $\delta > 0$, introduced in (1.87). Specifically, we use the fact that

$$A^T(\delta)\mathbf{x}(\delta) = \mathbf{0},$$

where $\mathbf{x}(\delta) := G^{-1}(\delta)\mathbf{x}$. Hence, by what we already proved, it follows that

$$(1.88) \qquad\qquad\qquad S^-(\mathbf{x}(\delta)) \geq m$$

for all $\delta > 0$. Since $\lim_{\delta \to 0^+} \mathbf{x}(\delta) = \mathbf{x}$, we conclude, from (1.88), that $S^+(\mathbf{x}) \geq m$.   □

Lemma 1.7 leads us to a proof of (1.85). Specifically, suppose $\mathbf{c} \in \mathbb{R}^n \backslash \{\mathbf{0}\}$ and the function $g = (\mathbf{c}, \mathbf{f})$ vanishes at m points $x_1 < \cdots < x_m$ in $(0, 1)$. Then, by Theorem 1.5, $m < n$ and the $n \times m$ matrix

$$A = (f_i(x_j)), \qquad i = 1, \ldots, n, \; j = 1, \ldots, m,$$

is strictly totally positive. Moreover, by definition, we have $A^T \mathbf{c} = \mathbf{0}$. Consequently, by Lemma 1.7 we conclude that $S^-(\mathbf{c}) \geq m$.

We now return to the proof of Theorem 1.5. The first part of the proof deals with the existence of the refinable curve $\mathbf{f}$. We begin with several ancillary facts. Their proof requires some additional standard facts about TP matrices, which we briefly review.

Let A be an $n \times n$ matrix. Choose sets $I, J \subseteq \{1, \ldots, n\}$ each with p elements. Order these elements as $i_1 < \cdots < i_p$, $j_1 < \cdots < j_p$, respectively and set

$$A\binom{I}{J} := A\binom{i_1, \ldots, i_p}{j_1, \ldots, j_p}.$$

The next lemma is a basic determinantal identity due to Sylvester.

LEMMA 1.8. *Let A be an $n \times n$ matrix. Given any sets $I, J \subseteq \{1, \ldots, n\}$ with p elements. For all $i \in \{1, \ldots, n\} \backslash I$, $j \in \{1, \ldots, n\} \backslash J$, define the quantities*

$$b_{ij} = A\binom{\{i\} \bigcup I}{\{j\} \bigcup J}$$

and let $B = (b_{ij})$. Then for any sets R in $\{1, \ldots, n\} \backslash I$ and S in $\{1, \ldots, n\} \backslash J$ with q elements we have

$$B\binom{R}{S} = A\binom{I}{J}^{q-1} A\binom{I \bigcup R}{J \bigcup S}.$$

A proof of this useful formula is given in Gantmacher [G, p. 33] and also Karlin [K, p. 4]. We need Sylvester's determinantal identity to prove the following statement used in the sequel.

LEMMA 1.9. *Let A be an $n \times n$ nonsingular TP matrix. Then for every set $I \subseteq \{1, \ldots, n\}$*

$$A\binom{I}{I} > 0.$$

Proof. The proof is by induction on $r := \#I$. For the proof of the case $r = 1$, we rely on the following basic observation concerning TP matrices. Suppose that for some i, k, $i < k \le n$, we have $A_{ii} = 0$ and $A_{ik} > 0$. Then, by considering the 2×2 minor of A $\ell > i$

$$A \begin{pmatrix} i & \ell \\ i & k \end{pmatrix} = -A_{\ell i} A_{ik},$$

we conclude that $A_{\ell i} = 0$ for all $\ell > i$. Similarly, if $A_{ii} = 0$ and $A_{ik} > 0$ for $k < i$ then $A_{\ell i} = 0$ whenever $\ell < i$. Using these facts we sort through several possibilities. If $A_{ii} = 0$ and in row i there is a nonzero element of the matrix A to the left *and* right of the (i,i)th element, then *all* the elements in the ith column are zero. This is not possible since A is nonsingular. Hence, either all the elements in the ith row to the left of i are zero and one positive to the right or vice versa. In the former case, we can use the basic observation above to conclude that all elements below and to the left of the (i,i)th entry are zero. But this would again contradict the nonsingularity of A. Similarly, in the latter case, a contradiction is reached and so we conclude that $A_{ii} > 0$, $i = 1, \ldots, n$.

We now assume the lemma is valid for all subsets I with p integers, $p > 1$. Form the $(n - p) \times (n - p)$ matrix $B = (b_{ij})$ whose elements are defined by

$$b_{ij} = A \begin{pmatrix} \{i\} \cup I \\ \{j\} \cup I \end{pmatrix}, \qquad i, j \notin I.$$

According to Lemma 1.8 and the induction hypothesis, B is totally positive and nonsingular. Hence, by the first part of the proof, we get that the diagonal elements of B are positive. That is,

$$A \begin{pmatrix} \{i\} \cup I \\ \{i\} \cup I \end{pmatrix} > 0.$$

This advances the induction and proves the result.   □

LEMMA 1.10. *Let* $A_\epsilon, \epsilon \in \{0,1\}$, *be nonsingular* $n \times n$ *stochastic matrices such that*

$$A = \begin{bmatrix} A_0 \\ A_1 \end{bmatrix}$$

is totally positive. Then for $1 \le j \le i \le n$, *we have* $(A_0)_{ij}(A_1)_{ji} > 0$ *and the matrix* $A_\epsilon^T, \epsilon \in \{0,1\}$ *has a unique eigenvector* $\mathbf{x}^\epsilon$, *normalized to satisfy* $(\mathbf{e}, \mathbf{x}^\epsilon) = 1$, *corresponding to the eigenvalue one. Moreover,* $\mathbf{x}^\epsilon$ *has nonnegative components.*

Proof. Applying Lemma 1.9 to the matrices A_ϵ, $\epsilon \in \{0,1\}$, we conclude that $(A_\epsilon)_{ii} > 0$, $i = 1, \ldots, n$. Now choose $1 \le j < i \le n$ and observe that the inequality

$$0 \le A \begin{pmatrix} i & j + n \\ j & i \end{pmatrix} = \begin{vmatrix} (A_0)_{ij} & (A_0)_{ii} \\ (A_1)_{jj} & (A_1)_{ji} \end{vmatrix}$$

implies that $(A_0)_{ij}(A_1)_{ji} > 0$. In particular, the first column of A_0 and the last column of A_1 have only positive entries. The remaining conclusions follow from this fact. To this end, fix any $\mathbf{x} \in \mathbb{R}^n$ and $\epsilon \in \{0,1\}$, and consider the sequence of vectors $\mathbf{x}^k := A_\epsilon^k \mathbf{x}$, $k = 0, 1, \ldots$. Set $m^k := \min\{x_i^k : 1 \le i \le n\}$ and $M^k := \max\{x_i^k : 1 \le i \le n\}$ and observe that, since A_ϵ is a stochastic matrix, $\{m^k : k = 0, 1, \ldots\}$ forms a nondecreasing sequence and $\{M^k : k = 0, 1, \ldots\}$ a nonincreasing sequence. According to equation (1.78), we have

$$M^k - m^k \le \rho^k(A_\epsilon)(M^0 - m^0).$$

Since A_ϵ has a positive column, the remark we made after (1.78) implies that $\rho(A_\epsilon) < 1$. Hence, we conclude that $\lim_{k \to \infty} \mathbf{x}^k$ exists and is a vector with all equal components. This means there is a vector $\mathbf{x}^\epsilon \in \mathbb{R}^n$ such that for all $\mathbf{x} \in \mathbb{R}^n$

(1.89) $$\lim_{k \to \infty} A_\epsilon^k \mathbf{x} = (\mathbf{x}^\epsilon, \mathbf{x})\mathbf{e}.$$

Choosing $\mathbf{x} = \mathbf{e}$ above, we see that $(\mathbf{x}^\epsilon, \mathbf{e}) = 1$. Also, replacing $\mathbf{x}$ by $A_\epsilon \mathbf{x}$ in (1.89) we get $(\mathbf{x}^\epsilon, A_\epsilon \mathbf{x}) = (\mathbf{x}^\epsilon, \mathbf{x})$ for all $\mathbf{x} \in \mathbb{R}^n$. Hence, $\mathbf{x}^\epsilon$ is an eigenvector of A_ϵ^T corresponding to the eigenvalue one. Moreover, if $\mathbf{y}$ is *any* other eigenvector of A_ϵ^T corresponding to the eigenvalue one with $(\mathbf{y}, \mathbf{e}) = 1$ then

$$(\mathbf{y}, \mathbf{x}) = \lim_{k \to \infty} ((A_\epsilon^T)^k \mathbf{y}, \mathbf{x})$$

$$= \lim_{k \to \infty} (\mathbf{y}, A_\epsilon^k \mathbf{x}) = (\mathbf{y}, \mathbf{e})(\mathbf{x}^\epsilon, \mathbf{x}) = (\mathbf{x}^\epsilon, \mathbf{x}).$$

That is, $\mathbf{y} = \mathbf{x}^\epsilon$. The fact that $\mathbf{x}^\epsilon$ has nonnegative elements follows from (1.89) and the nonnegativity of the elements of the matrix A_ϵ. □

   LEMMA 1.11. *Let A_ϵ, $\epsilon \in \{0,1\}$, be nonsingular $n \times n$ stochastic matrices such that*

$$A = \begin{bmatrix} A_0 \\ A_1 \end{bmatrix}$$

is totally positive. Suppose further that

(1.90) $$A_1^T \mathbf{x}^0 = A_0^T \mathbf{x}^1,$$

where $\mathbf{x}^\epsilon$, $\epsilon \in \{0,1\}$, is the unique eigenvector of A_ϵ referred to in Lemma 1.10. Then

$$\mathbf{x}^0 = (1, 0, \ldots, 0)^T, \quad \mathbf{x}^1 = (0, \ldots, 0, 1)^T$$

and

$$(A_0)_{1j} = \delta_{1j}, \quad (A_1)_{nj} = \delta_{nj}, \quad (A_0)_{nj} = (A_1)_{1j}, \quad j = 1, \ldots, n.$$

   Proof. Let k be the largest positive integer $\le n$ such that $(\mathbf{x}^0)_k > 0$. For any $j \le k$ we conclude from Lemma 1.10 that $(A_0)_{kj} > 0$ and so

$$(\mathbf{x}^0)_j = \sum_{\ell=1}^n (A_0)_{\ell j}(\mathbf{x}^0)_\ell$$

$$\ge (A_0)_{kj}(\mathbf{x}^0)_k > 0.$$

That is, $(\mathbf{x}^0)_j > 0$ for all $j \leq k$. Similarly, let r be the least positive integer such that $(\mathbf{x}^1)_r > 0$. Then $(\mathbf{x}^1)_j > 0$ for all $j \geq r$.

Let B be the $(n+k) \times n$ submatrix of A consisting of its first $n+k$ rows. According to (1.90) the vector

$$\mathbf{x} = \left(\mathbf{x}^1, -(\mathbf{x}^0)_1, \ldots, -(\mathbf{x}^0)_k\right)^T \in \mathbb{R}^{n+k}$$

satisfies the equation $B^T \mathbf{x} = \mathbf{0}$. Hence, from Lemma 1.7, we conclude that $S^+(\mathbf{x}) \geq n$. By the definition of $\mathbf{x}$, we see that $S^+(\mathbf{x}) = r$ and hence $r = n$. In other words, we conclude that $\mathbf{x}^1 = (0, \ldots, 0, 1)^T$. Similarly, using the vector $\mathbf{y} = (1, -\mathbf{x}^0)^T \in \mathbb{R}^{n+1}$ and the matrix C consisting of the last $n+1$ rows of A we get $C^T \mathbf{y} = \mathbf{0}$. Therefore Lemma 1.7 implies $S^+(\mathbf{y}) \geq n$ while clearly $S^+(\mathbf{y}) = n - k + 1$ and so $k = 1$. □

PROPOSITION 1.1. *Suppose A_ϵ, $\epsilon \in \{0,1\}$ are nonsingular $n \times n$ stochastic matrices such that*

$$A = \begin{bmatrix} A_0 \\ A_1 \end{bmatrix}$$

is totally positive. Then the refinement equation

$$(1.91) \qquad \mathbf{f}\left(\frac{t+\epsilon}{2}\right) = A_\epsilon^T \mathbf{f}(t), \qquad t \in [0,1], \qquad \epsilon \in \{0,1\}$$

has a continuous solution $\mathbf{f}$ with $(\mathbf{e}, \mathbf{f}(0)) \neq 0$ if and only if

$$(A_0)_{1j} = \delta_{1j}, \qquad (A_1)_{nj} = \delta_{nj}, \qquad (A_0)_{nj} = (A_1)_{1j}, \qquad j = 1, \ldots, n.$$

In this case, $(\mathbf{e}, \mathbf{f}(t)) = 1, 0 \leq t \leq 1$, $\mathbf{f}(0) = (1, 0, \ldots, 0)^T$, and $\mathbf{f}(1) = (0, \ldots, 0, 1)^T$.

Proof. First suppose that A_0 and A_1 satisfy all the conditions above. That is, A_0 and A_1 are nonsingular stochastic matrices such that

$$(A_0)_{1j} = \delta_{1j}, (A_1)_{nj} = \delta_{nj}, (A_0)_{nj} = (A_1)_{1j}, \qquad j = 1, \ldots, n$$

and A defined above is totally positive. Then $\mathbf{x}^0 = (1, 0, \ldots, 0)^T$ and $\mathbf{x}^1 = (0, \ldots, 0, 1)^T$ are the unique eigenvectors of A_0^T, A_1^T corresponding to the eigenvalue one (since both matrices have positive columns) and

$$A_1^T \mathbf{x}^0 = A_0^T \mathbf{x}^1.$$

Hence, Theorem 1.4 implies that

$$\lim_{k \to \infty} A_{\epsilon_k} \cdots A_{\epsilon_1} \mathbf{x} = (\mathbf{x}, \mathbf{f}(t))\mathbf{e}, \qquad t = \sum_{k=1}^{\infty} 2^{-k}\epsilon_k, \quad \mathbf{x} \in \mathbb{R}^n,$$

where $\mathbf{f}$ is a continuous curve on $[0,1]$ satisfying $(\mathbf{e}, \mathbf{f}(t)) = 1, 0 \leq t \leq 1$, and the refinement equation (1.91).

Conversely, suppose $A_\epsilon, \epsilon \in \{0,1\}$, are nonsingular $n \times n$ stochastic matrics such that A given above is totally positive and $\mathbf{f}$ satisfies the refinement equation (1.91) with $(\mathbf{e}, \mathbf{f}(0)) \neq 0$. Therefore, it follows that

$$(1.92) \qquad \mathbf{f}(t) = \lim_{k \to \infty} A_{\epsilon_1}^T \cdots A_{\epsilon_k}^T \mathbf{f}(x), \qquad t = \sum_{k=1}^{\infty} \epsilon_k 2^{-k},$$

for any $x, t \in [0, 1]$; see (1.30). Equation (1.92) implies that $(\mathbf{e}, \mathbf{f}(x))$ is a nonzero constant independent of x. Hence $u := (\mathbf{e}, \mathbf{f}(0)) = (\mathbf{e}, \mathbf{f}(1)) \neq 0$. Thus, if we set $\mathbf{x}^0 := u^{-1}\mathbf{f}(0)$ and $\mathbf{x}^1 := u^{-1}\mathbf{f}(1)$, we get $\mathbf{x}^0, \mathbf{x}^1 \neq \mathbf{0}$. Moreover, (1.91) also guarantees that $A_1^T\mathbf{x}^0 = A_0^T\mathbf{x}^1$ as well as the fact that $A_\epsilon^T\mathbf{x}^\epsilon = \mathbf{x}^\epsilon, \epsilon \in \{0,1\}$. Therefore, we may apply Lemma 1.11 to A_0 and A_1 and conclude that

$$(A_0)_{1j} = \delta_{1j}, (A_1)_{nj} = \delta_{nj}, (A_0)_{nj} = (A_1)_{1,j}, \qquad j = 1, \ldots, n,$$

thereby establishing all the desired properties of A_0 and A_1. $\square$

We have now accumulated sufficient information to prove Theorem 1.5.

Proof of Theorem 1.5. We use induction on p to establish (1.83). Let us first consider the case $p = 1$. This case requires us to establish the inequalities:

$$f_1(t) > 0, \quad \text{if and only if } t \in [0, 1),$$
$$f_i(t) > 0, \quad \text{if and only if } t \in (0, 1), \qquad 2 \leq i \leq n - 1,$$

and

$$f_n(t) > 0, \quad \text{if and only if } t \in (0, 1].$$

We have already pointed out that $A_\epsilon^T\mathbf{f}(\epsilon) = \mathbf{f}(\epsilon), \epsilon \in \{0,1\}$ and so by Lemma 1.11 we have $\mathbf{f}(0) = (1, 0, \ldots, 0)^T$ and $\mathbf{f}(1) = (0, \ldots, 0, 1)^T$. Also, choosing $\mathbf{x} = \mathbf{0}$ in (1.92) and using the fact that $A_\epsilon, \epsilon \in \{0,1\}$, are nonnegative matrices, we get $\mathbf{f}(t) \geq \mathbf{0}$ for all $t \in [0, 1]$. To show $f_1(t) > 0$ for $t \in [0, 1)$, we expand t in its binary representation

$$t = \sum_{k=1}^{\infty} \epsilon_k 2^{-k}.$$

Choose the least integer $\ell \geq 1$ such that $\epsilon_r = 1$, $r < \ell$. Then $\epsilon_\ell = 0$ and $y_\ell := 2^{\ell-1}(t - 1/2 - \cdots - 1/2^{\ell-1}) \in [0, \frac{1}{2}]$. For $\ell = 1$ we get $y_1 = t \in [0, \frac{1}{2}]$ and therefore (1.91) implies that

$$f_1(t) = \sum_{k=1}^{n} (A_0)_{k1} f_k(2t).$$

If $f_1(t) = 0$, then, by Lemma 1.10, it follows that $\mathbf{f}(2t) = \mathbf{0}$ and therefore by (1.92) (with $x = 2t$) $\mathbf{f} = \mathbf{0}$, which is a contradiction. Hence, we have established that $f_1(t) > 0$ for $t \in [0, \frac{1}{2}]$. When $\ell \geq 2$ we use the equation

$$f_1(t) = \sum_{k=1}^{n} (A_1^{\ell-1})_{k1} f_k(y_\ell),$$

which implies that

$$f_1(t) \geq ((A_1)_{11})^{\ell-1} \, f_1(y_\ell) > 0.$$

That is, indeed $f_1(t) > 0$ for $t \in [0,1)$. Similarly, to show that $f_n(t) > 0$, $t \in (0,1]$, we let ℓ be the least positive integer $\ell \geq 1$ such that $\epsilon_r = 0$, $r < \ell$. Then $\epsilon_\ell = 1$ and $z_\ell := 2^{\ell-1}t \in [\frac{1}{2}, 1]$. If $\ell = 1$, then $t \in [\frac{1}{2}, 1]$ and we use the equation

$$f_n(t) = \sum_{k=1}^{n} (A_1)_{kn} f_k(2t-1)$$

to conclude that $f_n(t) > 0$ because $(A_1)_{kn} > 0, 1 \leq k \leq n$, and $\mathbf{f}(x) \neq \mathbf{0}$ for all $x \in [0,1]$. Thus, we have shown that $f_n(t) > 0$ for $t \in [\frac{1}{2}, 1]$. When $\ell \geq 2$ we use the inequality

$$f_n(t) = \sum_{k=1}^{n} \left(A_0^{\ell-1} \right)_{kn} f_k(z_\ell) \geq ((A_0)_{nn})^{\ell-1} \, f_n(z_\ell) > 0.$$

Let us now consider the other components of $\mathbf{f}$. For $t \in (0, \frac{1}{2})$ and $2 \leq i \leq n-1$, we use the inequality

$$f_i(t) = \sum_{k=1}^{n} (A_0)_{ki} f_k(2t) \geq (A_0)_{ni} f_n(2t) > 0,$$

while for $t \in [\frac{1}{2}, 1)$ we employ the inequality

$$f_i(t) = \sum_{k=1}^{n} (A_1)_{ki} f_k(2t-1) \geq (A_1)_{1i} f_1(2t-1) > 0.$$

These remarks take care of the case $p = 1$ in (1.83).

We now assume inductively that

$$F \begin{pmatrix} i_1, \ldots, i_\ell \\ x_1, \ldots, x_\ell \end{pmatrix} \geq 0,$$

for $1 \leq i_1 < \cdots < i_\ell \leq n$, $0 \leq x_1 < \cdots < x_\ell \leq 1$, and all $\ell \leq r - 1$, where equality holds above if and only if either $x_1 = 0, i_1 > 0$ or $x_\ell = 1, i_\ell < n$. We consider a typical minor of order r

$$F \begin{pmatrix} i_1, \ldots, i_r \\ t_1, \ldots, t_r \end{pmatrix},$$

where $1 \leq i_1 < \cdots < i_r \leq n$ and $0 \leq t_1 < \cdots < t_r \leq 1$. If $t_1 = 0$, we use the fact that $f_i(0) = \delta_{i1}$, $i = 1, 2, \ldots, n$, to conclude that

$$F \begin{pmatrix} i_1, i_2, \ldots, i_r \\ 0, t_2, \ldots, t_r \end{pmatrix} = \delta_{i_1 1} F \begin{pmatrix} i_2, \ldots, i_r \\ t_2, \ldots, t_r \end{pmatrix},$$

and similarly, if $t_r = 1$, we have the equation

$$F\begin{pmatrix} i_1, \ldots, i_{r-1}, i_r \\ t_1, \ldots, t_{r-1}, 1 \end{pmatrix} = \delta_{i_r n} F\begin{pmatrix} i_1, \ldots, i_{r-1} \\ t_1, \ldots, t_{r-1} \end{pmatrix}.$$

Therefore the induction hypothesis allows us to assume that $0 < t_1 < \cdots < t_r < 1$.

The first possibility that we consider is the case that $\frac{1}{2} = t_1 < \cdots < t_r < 1$. In this case, we use the Cauchy–Binet formula and the refinement equation to obtain

$$F\begin{pmatrix} i_1, \ldots, i_r \\ t_1, \ldots, t_r \end{pmatrix} = \sum_{1 \le j_1 < \cdots < j_r \le n} A_1 \begin{pmatrix} j_1, \ldots, j_r \\ i_1, \ldots, i_r \end{pmatrix} F\begin{pmatrix} j_1, \ldots, j_r \\ 2t_1 - 1, \ldots, 2t_r - 1 \end{pmatrix}$$

$$= \sum_{2 \le j_2 < \cdots < j_r \le n} A_1 \begin{pmatrix} 1, j_2, \ldots, j_r \\ i_1, \ldots, i_r \end{pmatrix} F\begin{pmatrix} j_2, \ldots, j_r \\ 2t_2 - 1, \ldots, 2t_r - 1 \end{pmatrix}.$$

We claim that for some integers $j_2^0, \ldots, j_r^0$ with $2 \le j_2^0 < \cdots < j_r^0 \le n$ we have the inequality

(1.93) $$A_1 \begin{pmatrix} 1, j_2^0, \ldots, j_r^0 \\ i_1, \ldots, i_r \end{pmatrix} > 0.$$

To prove this we consider the two cases $i_1 > 1$ and $i_1 = 1$ separately. If $1 < i_1$ then the last r rows of the $(r+1) \times r$ matrix

$$T := A_1 \begin{bmatrix} 1, i_1, \ldots, i_r \\ i_1, \ldots, i_r \end{bmatrix}$$

are linearly independent (by Lemma 1.9) and the first nonzero (by Lemma 1.10). Thus there exist integers $j_2^0, \ldots, j_r^0$, $2 \le j_2^0 < \cdots < j_r^0 \le n$, $j_\ell^0 \in \{i_1, \ldots, i_r\}$, $2 \le \ell \le r$, such that the first row and rows $j_2^0, \ldots, j_r^0$ of T are linearly independent. When $i_1 = 1$ we set $j_k^0 = i_k, k = 2, \ldots, r$ and conclude that in either case (1.93) holds. Therefore, because A_1 is TP the induction hypothesis gives us the inequality

$$F\begin{pmatrix} i_1, \ldots, i_r \\ t_1, \ldots, t_r \end{pmatrix} \ge A_1 \begin{pmatrix} 1, j_2^0, \ldots, j_r^0 \\ i_1, \ldots, i_r \end{pmatrix} F\begin{pmatrix} j_2^0, \ldots, j_r^0 \\ 2t_2 - 1, \ldots, 2t_r - 1 \end{pmatrix} > 0.$$

Similarly, if $\frac{1}{2}$ is the right endpoint of $[t_1, t_r]$ we use the equation

$$F\begin{pmatrix} i_1, \ldots, i_r \\ t_1, \ldots, t_r \end{pmatrix} = \sum_{1 \le j_1 < \cdots < j_{r-1} \le n-1} A_0 \begin{pmatrix} j_1, \ldots, j_{r-1}, n \\ i_1, \ldots, i_r \end{pmatrix} F\begin{pmatrix} j_1, \ldots, j_{r-1} \\ 2t_1, \ldots, 2t_{r-1} \end{pmatrix}.$$

Just as above, we choose integers $j_1^0, \ldots, j_{r-1}^0 \in \{i_1, \ldots, i_r\}$, $1 \le j_1^0 < \cdots < j_{r-1}^0 \le n-1$, such that

$$A_0 \begin{pmatrix} j_1^0, \ldots, j_{r-1}^0, n \\ i_1, \ldots, i_r \end{pmatrix} > 0,$$

and so by the induction hypothesis it follows that

$$F\begin{pmatrix} i_1, \ldots, i_r \\ t_1, \ldots, t_r \end{pmatrix} \geq A_0 \begin{pmatrix} j_1^0, \ldots, j_{r-1}^0, n \\ i_1, \ldots, i_r \end{pmatrix} F\begin{pmatrix} j_1^0, \ldots, j_{r-1}^0 \\ 2t_1, \ldots, 2t_{r-1} \end{pmatrix} > 0.$$

The case when $\frac{1}{2} \in (t_1, t_r)$ is more involved. In this case, we begin by choosing an integer ℓ, $1 \leq \ell < r$, such that

$$t_1 < \cdots < t_\ell \leq \frac{1}{2} < t_{\ell+1} < \cdots < t_r.$$

(When $\ell = 1$, we need only consider the possibility that $t_1 < \frac{1}{2} < t_2$ because we already considered the case $t_1 = \frac{1}{2}$.) We now use the refinement equation and factor the $n \times r$ matrix

$$F\begin{bmatrix} 1, \ldots, n \\ t_1, \ldots, t_r \end{bmatrix}$$

in the form

$$F\begin{bmatrix} 1, \ldots, n \\ t_1, \ldots, t_r \end{bmatrix} = A^T C, \qquad A = \begin{bmatrix} A_0 \\ A_1 \end{bmatrix},$$

where C is the $2n \times r$ matrix

$$C := \begin{bmatrix} F\begin{bmatrix} 1, \ldots, n \\ 2t_1, \ldots, 2t_\ell \end{bmatrix} & 0 \\ 0 & F\begin{bmatrix} 1, \ldots, n \\ 2t_{\ell+1} - 1, \ldots, 2t_r - 1 \end{bmatrix} \end{bmatrix}.$$

By the Cauchy–Binet formula we have the equation

$$(1.94) \quad F\begin{pmatrix} i_1, \ldots, i_r \\ t_1, \ldots, t_r \end{pmatrix} = \sum_{1 \leq j_1 < \cdots < j_r \leq 2n} A\begin{pmatrix} j_1, \ldots, j_r \\ i_1, \ldots, i_r \end{pmatrix} C\begin{pmatrix} j_1, \ldots, j_r \\ 1, \ldots, r \end{pmatrix}.$$

If $k := |\{j_1, \ldots, j_r\} \cap \{1, \ldots, n\}| > \ell$ then, by taking linear combinations of its first k rows the matrix

$$C\begin{bmatrix} j_1, \ldots, j_r \\ 1, \ldots, r \end{bmatrix}$$

we conclude that it has a zero determinant. Similarly, we observe that

$$C\begin{pmatrix} j_1, \ldots, j_r \\ 1, \ldots, r \end{pmatrix} = 0$$

if $|\{j_1, \ldots, j_r\} \cap \{n+1, \ldots, 2n\}| > r - \ell$. Therefore (1.94) becomes

$$F\begin{pmatrix} i_1, \ldots, i_r \\ t_1, \ldots, t_r \end{pmatrix} = \sum_{\substack{1 \leq j_1 < \cdots < j_l \leq n \\ 1 \leq k_1 < \cdots < k_{r-l} \leq n}} A\begin{pmatrix} j_1, \ldots, j_l, k_1 + n, \ldots, k_{r-l} \\ i_1, \ldots, i_r \end{pmatrix}$$

$$\times F\begin{pmatrix} j_1, \ldots, j_l \\ 2t_1, \ldots, 2t_l \end{pmatrix} F\begin{pmatrix} k_1, \ldots, k_{r-l} \\ 2t_{l+1} - 1, \ldots, 2t_r - 1 \end{pmatrix}.$$

Next, we point out that the $2r \times r$ matrix

$$A \begin{bmatrix} i_1, \ldots, i_r, i_1 + n, \ldots, i_r + n \\ i_1, \ldots, i_r \end{bmatrix}$$

has the property that its first r rows, as well as its last r rows, are linearly independent. Hence, for any choice of ℓ rows among its first r rows there is a choice of $r - \ell$ row vectors from its last rows for which the resulting set of vectors is linearly independent. Thus for any choice of integers $1 \leq j_1^0 < \cdots < j_\ell^0 \leq n$ in $\{i_1, \ldots, i_r\}$, there are integers $1 \leq k_1^0 < \cdots < k_{r-\ell}^0 \leq n$ in $\{i_1, \ldots, i_r\}$ such that

$$A \begin{pmatrix} j_1^0, \ldots, j_\ell^0, k_1^0 + n, \ldots, k_{r-\ell}^0 + n \\ i_1, \ldots, i_r \end{pmatrix} > 0.$$

When $t_\ell = \frac{1}{2}$ we must qualify our choice by setting $j_\ell^0 = n$. Therefore we obtain by our induction hypothesis the inequality

$$F \begin{pmatrix} i_1, \ldots, i_r \\ t_1, \ldots, t_r \end{pmatrix} \geq A \begin{pmatrix} j_1^0, \ldots, j_\ell^0, k_1^0 + n, \ldots, k_{r-\ell}^0 + n \\ i_1, \ldots, i_r \end{pmatrix}$$

$$\times F \begin{pmatrix} j_1^0, \ldots, j_\ell^0 \\ 2t_1, \ldots, 2t_\ell \end{pmatrix} F \begin{pmatrix} k_1^0, \ldots, k_{r-\ell}^0 \\ 2t_{\ell+1} - 1, \ldots, 2t_r - 1 \end{pmatrix} > 0.$$

There remain the two additional cases $t_r < \frac{1}{2}$ or $t_1 > \frac{1}{2}$. In the first instance (the second can be argued similarly) we consider the binary expansion of the vector $\mathbf{t} := (t_1, \ldots, t_r)^T$, viz.

$$\mathbf{t} = \sum_{k=1}^{\infty} \mathbf{d}^k 2^{-k},$$

where $\mathbf{d}^k = (\epsilon_1^k, \ldots, \epsilon_r^k)^T$, $\epsilon_i^k \in \{0, 1\}$. In the case at hand, $\mathbf{d}^1 = \mathbf{0}$. We let m_1 be the largest integer ≥ 2 such that $\mathbf{d}^k = \mathbf{0}$ for $k < m_1$. Thus $\mathbf{d}^{m_1} \neq \mathbf{0}$ and so its last component must be one. Either the first component of $\mathbf{d}^{m_1}$ is zero or we have $\mathbf{d}^{m_1} = (1, \ldots, 1)^T$. In the latter case we let m_2 be the largest integer greater than m_1 such that $\mathbf{d}^k = (1, \ldots, 1)^T$, $m_1 \leq k < m_2$. Continuing in this way we can find a dyadic fraction $\tau \in [0, 1]$ such that $y_i = 2^\mu (t_i - \tau) \in [0, 1]$, $i = 1, \ldots, r$. That is, the first μ binary digits of each $t_i, i = 1, \ldots, r$, agree with τ, and $\frac{1}{2} \in [y_1, y_r]$. We choose μ to be the *least* integer so that this holds. Therefore

$$(1.95) \quad F \begin{pmatrix} i_1, \ldots, i_r \\ t_1, \ldots, t_r \end{pmatrix} = \sum_{1 \leq j_1 < \cdots < j_r \leq n} G \begin{pmatrix} j_1, \ldots, j_r \\ i_1, \ldots, i_r \end{pmatrix} F \begin{pmatrix} j_1, \ldots, j_r \\ y_1, \ldots, y_r \end{pmatrix},$$

where $G := A_{\epsilon_1^\mu} \cdots A_{\epsilon_1^1}$ and $\epsilon_1^1 = 0$. When $0 < y_1 < y_r < 1$, then we use what we have already proved to conclude that

$$F \begin{pmatrix} j_1, \ldots, j_r \\ y_1, \ldots, y_r \end{pmatrix} > 0$$

for all $1 \leq j_1 < \cdots < j_r \leq n$. Since the matrix G is TP and nonsingular, Lemma 1.9 implies that

$$G\begin{pmatrix} i_1, \ldots, i_r \\ i_1, \ldots, i_r \end{pmatrix} > 0,$$

and so we obtain our desired result. If $y_1 = 0$ and $y_r < 1$ then there is a $j, 1 \leq j \leq \mu$ such that $\epsilon_1^j = 1$ (otherwise $t_1 = 0$, which is not allowed for this case). Thus the matrix G is nonsingular and TP. Moreover, G has only positive elements in its first row because, by Lemma 1.10, the first row of A_1 has only positive elements and also $(A_\epsilon)_{11} > 0$, for $\epsilon \in \{0, 1\}$. Hence, as before, there are integers $1 < j_2^0 < \cdots < j_r^0 \leq n$ in the set $\{i_1, \ldots, i_r\}$ such that

$$G\begin{pmatrix} 1, j_2^0, \ldots, j_r^0 \\ i_1, \ldots, i_r \end{pmatrix} > 0.$$

Also, by the induction hypothesis

$$F\begin{pmatrix} 1, j_2^0, \ldots, j_r^0 \\ y_1, \ldots, y_r \end{pmatrix} = F\begin{pmatrix} j_2^0, \ldots, j_r^0 \\ y_2, \ldots, y_r \end{pmatrix} > 0$$

and so

$$F\begin{pmatrix} i_1, \ldots, i_r \\ t_1, \ldots, t_r \end{pmatrix} \geq G\begin{pmatrix} 1, j_2^0, \ldots, j_r^0 \\ i_1, \ldots, i_r \end{pmatrix} F\begin{pmatrix} 1, j_2^0, \ldots, j_r^0 \\ y_1, \ldots, y_r \end{pmatrix} > 0.$$

When $y_1 > 0$ and $y_1 = 1$ there is a $j, 1 \leq j \leq \mu$ such that $\epsilon_1^j = 0$ and so G has only positive elements in its last row. Thus again we may argue, as before, that there are integers $1 \leq j_1^0 < \cdots < j_{r-1}^0 < n$ in $\{i_1, \ldots, i_r\}$ such that

$$G\begin{pmatrix} j_1^0, \ldots, j_{r-1}^0, n \\ i_1, \ldots, i_r \end{pmatrix} > 0,$$

and conclude by the induction hypothesis that

$$F\begin{pmatrix} j_1^0, \ldots, j_{r-1}^0, n \\ y_1, \ldots, y_r \end{pmatrix} > 0.$$

Together these inequalities imply that

$$F\begin{pmatrix} i_1, \ldots, i_r \\ t_1, \ldots, t_r \end{pmatrix} > 0,$$

once again. In the final case, when $y_1 = 0$ and $y_r = 1$, we need to observe that there are integers $1 < j_2^0 < \cdots < j_{r-1}^0 < n$ such that

(1.96)
$$G\begin{pmatrix} 1, j_2^0, \ldots, j_{r-1}^0, n \\ i_1, \ldots, i_r \end{pmatrix} > 0.$$

Therefore by the induction hypothesis we obtain

$$F\begin{pmatrix} 1, j_2^0, \ldots, j_{r-1}^0, n \\ y_1, y_2, \ldots, y_{r-1}, y_r \end{pmatrix} = F\begin{pmatrix} j_2^0, \ldots, j_{r-1}^0 \\ y_2, \ldots, y_{r-1} \end{pmatrix} > 0.$$

For the proof of (1.96), we let $\mathbf{x}^0, \mathbf{x}^1, \ldots, \mathbf{x}^{r+1}$ be the vectors formed from the rows of the $(r+2) \times r$ matrix

$$G \begin{bmatrix} 1, i_1, \ldots, i_r, n \\ i_1, \ldots, i_r \end{bmatrix}.$$

Observe that the vectors $\mathbf{x}^1, \ldots, \mathbf{x}^r$ are linearly independent, since

$$G \begin{pmatrix} i_1, \ldots, i_r \\ i_1, \ldots, i_r \end{pmatrix} > 0.$$

Also, $\mathbf{x}^0, \mathbf{x}^{r+1}$ are linearly independent, since $G \begin{pmatrix} 1 & n \\ i_1 & i_r \end{pmatrix} > 0$ (this inequality is valid, otherwise the i_1 and i_r columns of G would be linearly dependent). Therefore we may remove two vectors from the set $\{\mathbf{x}^1, \ldots, \mathbf{x}^r\}$ and replace them by $\mathbf{x}^0$ and $\mathbf{x}^r$ to form another set of linearly independent vectors. This proves (1.96) and also the theorem. $\quad\square$

References

[CDM] A. S. CAVARETTA, W. DAHMEN, AND C. A. MICCHELLI, *Stationary subdivision*, Mem. Amer. Math. Soc., 93, #453, 1991.

[CS] E. COHEN AND L. L. SCHUMAKER, *Rates of convergence of control polygons*, Comput. Aided Geom. Design, 2(1985), 229–235.

[D] W. DAHMEN, *Subdivision algorithms converge quadratically*, J. Comput. Appl. Math., 16(1986), 145–158.

[G] F. R. GANTMACHER, *The Theory of Matrices*, Vol. 1, Chelsea Publishing Company, New York, 1960.

[GJ] I. P. GOULDEN AND D. M. JACKSON, *Combinatorial Enumeration*, John Wiley, New York, 1983.

[GM] T. N. T. GOODMAN AND C. A. MICCHELLI, *Corner cutting algorithms for the Bézier representation of free form curves*, Linear Algebra Appl., 99(1988), 225–252.

[K] S. KARLIN, *Total Positivity*, Stanford University Press, Stanford, CA, 1968.

[MPi] C. A. MICCHELLI AND A. PINKUS, *Descartes systems from corner cutting*, Constr. Approx., 7(1991), 161–194.

[MPr] C. A. MICCHELLI AND H. PRAUTZSCH, *Uniform refinement of curves*, Linear Algebra Appl., 114/115(1989), 841–870.

[Pin] A. PINKUS, *n-Widths in Approximation Theory*, Springer-Verlag, Berlin, 1985.

CHAPTER 2

Stationary Subdivision

2.0. Introduction.

The study of subdivision algorithms continues to be our theme in this chapter. Our model problem now is computation of cardinal spline functions defined on the real line. We include background material on cardinal splines sufficient to describe the Lane–Riesenfeld algorithm for computing these functions and for proving the convergence of their algorithm. By viewing this algorithm as successive averaging of control points we are led to the general class of stationary subdivision schemes whose properties are the focus of this chapter.

We reduce the study of the convergence of a stationary subdivision scheme to the convergence of a related matrix subdivision scheme. A simple condition for the convergence of stationary subdivision schemes with a nonnegative mask is given. We then show that, when the Laurent polynomial whose coefficients are the elements of the stationary subdivision scheme has all of its zeros in the left half-plane, the refinable function of the scheme has a strong variation diminishing property. At the end of the chapter we describe how one can easily build orthogonal pre-wavelets from such refinable functions.

2.1. de Rahm–Chaikin algorithm.

Consider the following simple iterative procedure. We begin with an ordered bi-infinite sequence of control points $c_j, j \in \mathbb{Z}$ in the plane. On each segment of the

corresponding control polygon two new control points are constructed a quarter
of the distance from the ends of the segment, as indicated below (see Fig. 2.1).

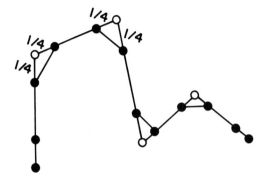

FIG. 2.1. $\frac{1}{4} : \frac{1}{2} : \frac{1}{4}$ *corner cutting.*

The new control points yield a new control polygon and the above procedure is
then repeated. In the limit, a curve is formed (see Fig. 2.2).

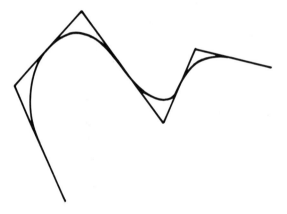

FIG. 2.2. *Limit curve.*

This algorithm was first considered by de Rahm [deR] as a special case of a
family of similar algorithms. (Generally, he chose *any* two points on each segment
whose midpoint is the same as that of the segment itself.) It was later considered
by Chaikin [Ch]. Shortly thereafter, Riesenfeld [Ri] showed that the limit curve

is $C^1(\mathbb{R})$ and is composed of quadratic arcs, a fact observed earlier by de Rahm. The argument of Riesenfeld goes as follows. To each triangle ABC (see Fig. 2.3)

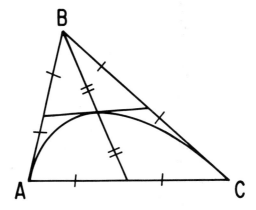

FIG. 2.3. *Triangle and quadratic arc.*

we associate a unique quadratic curve given in Bernstein–Bézier form

(2.1) $Q(t) = A(1-t)^2 + 2Bt(1-t) + Ct^2, \qquad 0 \le t \le 1.$

This quadratic goes through the points A and C and is tangent to the sides AB and BC at these points, respectively. Note also that the line connecting the midpoints of sides AB and BC touches the curve Q at $Q\left(\frac{1}{2}\right) = \frac{1}{4}(A + 2B + C)$ and is also tangent to it there.

Returning to the de Rahm–Chaikin algorithm we observe that at a typical corner (see Fig. 2.4)

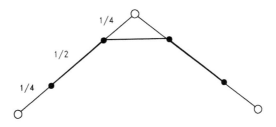

FIG. 2.4. *A corner cut.*

the quadratic Q associated with the midpoints of the sides forming the corner is tangent to the corner cuts made at the next step of the algorithm. Thus, at this stage, three midpoints are on the quadratic. The successive stages of the iteration produce more midpoints on the curve. Since the maximum of the differences between successive control points decreases at each step of the algorithm, the limit curve is Q. The same process occurs, corner to corner. Since midpoints and common tangents are shared, corner to corner, the limit curve is a $C^1(\mathbb{R})$ piecewise quadratic (see Fig. 2.5).

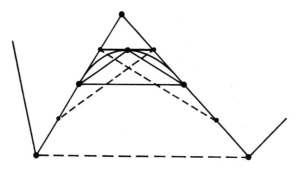

FIG. 2.5.   *Two successive corner cuts.*

The simplicity of the method and the elegant form of the limit curve leads us to wonder about other possibilities. Clearly there are even simpler methods. For instance, between every pair of control points we merely *add* in midpoints. Repeating this process gives more and more points on the control polygon itself and so the limit is a continuous piecewise linear function.

Remarkably, both methods are a special case of a stationary subdivision scheme for computing *cardinal spline functions* due to Lane and Riesenfeld [LR]. Let us now describe their algorithm in detail.

2.2.   Cardinal spline functions.

Let χ be the indicator function of the interval $[0, 1)$, viz.

$$(2.2) \qquad \chi(x) = \begin{cases} 1, & 0 \le x < 1, \\ 0, & \text{otherwise.} \end{cases}$$

From this simple function we build the *nth degree forward B-spline* of Schoenberg [Sc]. (See Fig. 2.6 for the quadratic case.) It is defined recursively on $\mathbb{R}$ as

$$(2.3) \qquad M_n(x) = \begin{cases} \chi(x), & n = 0, \\ \displaystyle\int_0^1 M_{n-1}(x - t)dt, & n \ge 1. \end{cases}$$

This function is central in what follows and so we develop some of its properties.

From (2.3), we obtain for $n \ge 2$ the derivative recurrence formula

$$(2.4) \qquad M_n'(x) = M_{n-1}(x) - M_{n-1}(x - 1), \qquad x \in \mathbb{R}.$$

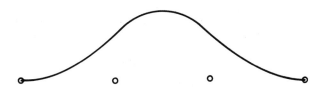

FIG. 2.6. *Second-degree B-spline.*

Moreover, M_n vanishes outside the interval $(0, n+1)$ and is otherwise positive.
Next we recall the identity

(2.5) $$\sum_{j \in \mathbb{Z}} M_n(x - j) = 1, \qquad x \in \mathbb{R}.$$

This can be verified inductively. The case $n = 0$ is clear. Otherwise, from (2.5)
and (2.3) it follows that for $x \in \mathbb{R}$

$$1 = \int_{\mathbb{R}} \chi(x - t)dt \;\; = \;\; \sum_{j \in \mathbb{Z}} \int_{\mathbb{R}} \chi(x - t)M_n(t - j)dt$$
$$= \;\; \sum_{j \in \mathbb{Z}} M_{n+1}(x - j).$$

Also, from (2.4) we conclude that $M_n \in C^{n-1}(\mathbb{R})$, the space of all functions on $\mathbb{R}$ with $n - 1$ continuous derivatives. Moreover, on each of the intervals
$[j, j + 1), M_n$ is a polynomial of degree at most n. This suggests the following definition.

DEFINITION 2.1. *Let n be a positive integer. We let S_n be the linear space of
all functions having $n - 1$ continuous derivatives on $\mathbb{R}$ such that their restriction
to each interval $[j, j + 1), j \in \mathbb{Z}$, is a polynomial of at most degree n. We use
the convention that S_0 denotes the class of all step functions with breakpoints at
the integers that are continuous from the right.*

A function in S_n is called a cardinal spline of degree n. Cardinal spline functions have discontinuities in their nth derivative only at integers. If, in Definition
2.1, the set $\mathbb{Z}$ is replaced by any partition $\{x_j : j \in \mathbb{Z}\}, x_j < x_{j+1}$, of $\mathbb{R}$ the
corresponding functions are called spline functions of degree n with knots at
$x_j, j \in \mathbb{Z}$. Schoenberg [Sc] provides the following representation for cardinal
spline functions.

THEOREM 2.1. *$f \in S_n$ if and only if there exist unique constants $c_j, j \in \mathbb{Z}$
such that*

(2.6) $$f(x) = \sum_{j \in \mathbb{Z}} c_j M_n(x - j), \qquad x \in \mathbb{R}.$$

Note that in (2.6) for each $x \in \mathbb{R}$ there are at most $n + 1$ nonzero summands.

Proof. We prove this representation for f by induction on n. The case $n = 0$ being obvious, we consider $n = 1$. In this case, a direct computation shows that

$$(2.7) \qquad M_1(x) = \begin{cases} x, & 0 \le x \le 1, \\ 2 - x, & 1 \le x \le 2. \end{cases}$$

Thus, for $n = 1$, any sum of the form (2.6) is in S_1. Conversely, given any continuous piecewise linear function f it can be represented on $[j, j+1]$ as

$$\begin{aligned} f(x) &= f(j)(j+1-x) + f(j+1)(x-j) \\ &= f(j)M_1(x-j+1) + f(j+1)M_1(x-j) \\ &= \sum_{\ell \in \mathbb{Z}} f(\ell+1)M_1(x-\ell). \end{aligned}$$

Now suppose the theorem is true for $n \ge 2$ and let $f \in S_{n+1}$. Then $f' \in S_n$ and by induction there are unique constants $c_j, j \in \mathbb{Z}$, such that

$$(2.8) \qquad f'(x) = \sum_{j \in \mathbb{Z}} c_j M_n(x-j), \qquad x \in \mathbb{R}.$$

We choose $d_j, j \in \mathbb{Z}$, such that

$$d_j - d_{j-1} = c_j, \qquad j \in \mathbb{Z}.$$

All such sequences are determined uniquely when d_0 is given. Specifically,

$$(2.9) \qquad d_j = d_0 + v_j, \qquad j \in \mathbb{Z},$$

where

$$(2.10) \qquad v_j = \begin{cases} \displaystyle\sum_{\ell=1}^{j} c_\ell, & j \ge 1, \\[2ex] 0, & j = 0, \\[2ex] \displaystyle-\sum_{\ell=j+1}^{0} c_\ell, & j \le -1. \end{cases}$$

Using (2.4), we see that

$$f(x) = \sum_{j \in \mathbb{Z}} d_j M_{n+1}(x-j) + a,$$

where $a := f(0) - \sum_{j \in \mathbb{Z}} d_j M_{n+1}(-j)$. Therefore by (2.5) we obtain the equation

$$f(x) = \sum_{j \in \mathbb{Z}} (a + d_j) M_{n+1}(x-j).$$

The uniqueness of the representation of f is equally simple. Specifically, if for some constants $y_k, k \in \mathbb{Z}$

$$(2.11) \qquad f(x) = \sum_{k \in \mathbb{Z}} y_k M_{n+1}(x - k),$$

then, by differentiating both sides of (2.11) and using the induction hypothesis, we obtain

$$(2.12) \qquad y_k = v_k + y_0, \qquad k \in \mathbb{Z},$$

while from (2.5) and (2.11) we get the formula

$$(2.13) \qquad f(0) = \sum_{k \in \mathbb{Z}} v_k M_{n+1}(-k) + y_0,$$

which determines y_0.   □

2.3. Lane–Riesenfeld subdivision.

Next we introduce an important idea. We let

$$S_n^1 = \{g : g(x) = f(2x), \ x \in \mathbb{R}, \ f \in S_n\}.$$

The space S_n^1 consists of spline functions of degree n with knots in $\mathbb{Z}/2 = \{j/2 : j \in \mathbb{Z}\}$. This space clearly contains S_n, a fact that leads us to ask, how do we express an $f \in S_n$ written in its cardinal B-spline series (2.6) as an element in S_n^1 (see Fig. 2.7)? That is, given $c_j, j \in \mathbb{Z}$, what are the d_k, $k \in \mathbb{Z}/2$ in the representation

$$(2.14) \qquad \sum_{j \in \mathbb{Z}} c_j M_n(x - j) = \sum_{k \in \mathbb{Z}/2} d_k M_n(2(x - k)), \qquad x \in \mathbb{R}?$$

Note that our preference in (2.14) is to index the B-spline coefficients on the right side of equation (2.14) over the fine knots $\mathbb{Z}/2$.

Let us derive a recurrence formula for d_k in the *degree* n of the B-spline. To this end, we call the left side of (2.14) $f_n(x)$ and we relabel d_k as d_k^n to show its

FIG. 2.7. *Fine knot quadratic B-spline used to build a coarse knot quadratic B-spline.*

dependency on n. Then, according to (2.3), we have

$$f_{n+1}(x) = \int_0^1 f_n(x-t)dt$$

$$= \frac{1}{2}\int_0^1 f_n(x-t/2)dt + \frac{1}{2}\int_0^1 f_n\left(x-t/2-\frac{1}{2}\right)dt$$

$$= \frac{1}{2}\sum_{k\in\mathbb{Z}/2} d_k^n \int_0^1 M_n(2(x-k)-t)dt$$

$$+ \frac{1}{2}\sum_{k\in\mathbb{Z}/2} d_k^n \int_0^1 M_n(2(x-k)-t-1)dt$$

$$= \sum_{k\in\mathbb{Z}/2} \frac{1}{2}\left(d_k^n + d_{k-1/2}^n\right) M_{n+1}(2(x-k)).$$

Therefore, we obtain the recurrence formula

(2.15) $$d_k^{n+1} = \frac{1}{2}\left(d_k^n + d_{k-1/2}^n\right), \qquad k\in\mathbb{Z}/2.$$

For $n = 0$, we see from the representation of f_0 as an element in S_0 that $f_0(x) = c_\ell$ for $\ell \le x < \ell+1$, while as an element in S_0^1 we have $f_0(x) = d_k^0$, for $k \le x < k+\frac{1}{2}$. Thus, we get

(2.16) $$d_\ell^0 = d_{\ell+1/2}^0 = c_\ell, \qquad \ell\in\mathbb{Z}.$$

The formulas (2.15), (2.16) generate d_k^n, $k\in\mathbb{Z}/2$, iteratively. For instance, we have

(2.17) $$d_k^1 = \begin{cases} c_\ell, & k = \ell+\frac{1}{2}, \\ \frac{1}{2}(c_\ell + c_{\ell+1}), & k = \ell+1 \end{cases}$$

and

(2.18) $$d_k^2 = \begin{cases} \frac{3}{4}c_\ell + \frac{1}{4}c_{\ell-1}, & k = \ell+\frac{1}{2}, \\ \frac{1}{4}c_{\ell+1} + \frac{3}{4}c_\ell, & k = \ell+1. \end{cases}$$

If we now think of c_ℓ, $\ell\in\mathbb{Z}$ as control points in the plane then d_k^2, $k\in\mathbb{Z}$, are the control points for the de Rahm–Chaikin algorithm and likewise d_k^1, $k\in\mathbb{Z}$ corresponds to the midpoint method.

This association of the de Rahm–Chaikin algorithm with the B-spline representation of a cardinal spline in S_n as an element in S_n^1 leads us to an alternative proof of the convergence of the de Rahm–Chaikin algorithm that also extends to spline functions of arbitrary degree. To describe the result that we have in mind, it is convenient to re-index our control points produced at the first stage of the algorithm (2.15), (2.16). For this purpose, we set $c_{2k}^1 := d_k^n$ for all $k \in \mathbb{Z}/2$. Thus, (2.15), (2.16) describe a mapping from an initial bi-infinite vector $c := (c_j : j \in \mathbb{Z})$ into the bi-infinite vector $c^1 = (c_j^1 : j \in \mathbb{Z})$. The rth iterate of this mapping, which we denote by c^r, gives the B-spline coefficients for a cardinal spline as an element of the space $S_n^r := \{g : g(x) = f(2^r x), x \in \mathbb{R}, f \in S_n\}$, that is

$$f_n(x) = \sum_{j \in \mathbb{Z}} c_j M_n(x - j) = \sum_{j \in \mathbb{Z}} c_j^r M_n(2^r x - j), \qquad x \in \mathbb{R}.$$

Our next theorem describes the sense in which the iterates of this mapping converge to f_n.

THEOREM 2.2 *For any* $j \in \mathbb{Z}$ *and* $n, r \in \mathbb{Z}_+$ *we have*

(2.19)
$$|f_n(j/2^r) - c_{j-1}^r| \le \frac{n - 1}{2^r} \|\triangle c\|_\infty,$$

where $\| \cdot \|_\infty := \ell^\infty$-*norm on* $\mathbb{Z}$ *and* $(\triangle c)_j := c_{j+1} - c_j, j \in \mathbb{Z}$ *is the forward difference of* c. *Moreover, if* $n \ge 2$, *there is a constant* e_n *such that*

(2.20)
$$\left| f_n\left(\frac{j}{2^r} + \frac{n+1}{2^{r+1}} \right) - c_j^r \right| \le \frac{e_n}{4^r} \|\triangle^2 c\|_\infty,$$

where we set $\triangle^2 c := \triangle(\triangle c)$.

Proof. Thinking geometrically, we first observe that after one step of the algorithm the maximum distance between successive control points of the control polygon is reduced by a half. To see this, we use (2.17) to get the formula

$$d_k^1 - d_{k-1/2}^1 = \begin{cases} \dfrac{1}{2}(c_\ell - c_{\ell-1}), & k = \ell + \dfrac{1}{2}, \\ \dfrac{1}{2}(c_{\ell+1} - c_\ell), & k = \ell + 1, \end{cases}$$

from which we obtain the inequality

$$|d_k^1 - d_{k-1/2}^1| \le \frac{1}{2}\|\triangle c\|_\infty, \qquad k \in \mathbb{Z}/2.$$

Similarly, from (2.18) we have

$$d_k^2 - d_{k-1/2}^2 = \begin{cases} \dfrac{1}{2}(c_\ell - c_{\ell-1}), & k = \ell + \dfrac{1}{2}, \\ \dfrac{1}{4}(c_{\ell+1} - c_{\ell-1}), & k = \ell + 1, \end{cases}$$

and so

$$|d_k^2 - d_{k-1/2}^2| \le \frac{1}{2}\|\triangle c\|_\infty, \qquad k \in \mathbb{Z}/2.$$

In general, the averaging formula (2.15) gives us, for any n, the inequality

$$|d_k^n - d_{k-1/2}^n| \leq \frac{1}{2}\|\triangle c\|_\infty, \qquad k \in \mathbb{Z}/2,$$

so that $\|\triangle c^1\|_\infty \leq \frac{1}{2}\|\triangle c\|_\infty$, and iterating r times provides the bound

$$(2.21) \qquad \qquad \|\triangle c^r\|_\infty \leq \frac{1}{2^r}\|\triangle c\|_\infty, \qquad r \in \mathbb{Z}_+.$$

Since

$$
\begin{aligned}
f_n(j/2^r) &= \sum_{i \in \mathbb{Z}} c_i M_n(j/2^r - i) \\
&= \cdots = \sum_{i \in \mathbb{Z}} c_i^r M_n(j - i) \\
&= \sum_{\ell=1}^{n} c_{j-\ell}^r M_n(\ell),
\end{aligned}
$$

equation (2.5) and the bound (2.21) lead us to the inequality

$$
\begin{aligned}
|f_n(j/2^r) - c_{j-1}^r| &\leq (n-1)\|\triangle c^r\|_\infty \\
&\leq 2^{-r}(n-1)\|\triangle c\|_\infty.
\end{aligned}
$$

This proves (2.19).

The proof of (2.20) is more involved and uses the following additional fact about the B-spline M_n. We require two positive constants A_n and B_n. The first enters into inequality

$$(2.22) \qquad A_n\|c\|_\infty \leq \sup\left\{\left|\sum_{j \in \mathbb{Z}} c_j M_n(x - j)\right| : x \in \mathbb{R}\right\} \leq \|c\|_\infty,$$

valid for all $c = (c_j : j \in \mathbb{Z}) \in \ell^\infty(\mathbb{Z})$. The upper bound in (2.22) follows from (2.5) and the nonnegativity of M_n; the lower bound will be proved momentarily. The other constant we need has to do with the Schoenberg operator

$$(T_r f)(x) := \sum_{j \in \mathbb{Z}} f\left(\frac{j + (n+1)/2}{2^r}\right) M_n(2^r x - j).$$

Namely, if $f \in C^2(\mathbb{R})$, there holds the inequality

$$(2.23) \qquad \qquad \|T_r f - f\|_\infty \leq \frac{B_n}{4^r}\|f''\|_\infty,$$

where $\|\cdot\|_\infty$ is the sup norm on $\mathbb{R}$. We will also prove this well-known bound shortly, but first let's make use of (2.22) and (2.23).

To get (2.19), we first use (2.22) with

$$c_j := f_n\left(\frac{j + (n+1)/2}{2^r}\right) - c_j^r,$$

which gives

$$\left| f_n\left(\frac{j}{2^r} + \frac{n+1}{2^{r+1}}\right) - c_j^r \right|$$

$$\leq A_n^{-1} \sup\left\{ \left|(T_r f_n)(x) - \sum_{j \in \mathbb{Z}} c_j^r M_n(2^r x - j)\right| : x \in \mathbb{R} \right\}.$$

Next, we employ (2.23) to obtain

$$= A_n^{-1}\|T_r f_n - f_n\|_\infty \leq \frac{A_n^{-1}B_n}{4^r}\|f_n''\|_\infty.$$

But then, from (2.4), we have

$$f_n''(x) = \sum_{j \in \mathbb{Z}} \left(\Delta^2 c\right)_{j-2} M_{n-2}(x - j),$$

which proves $\|f_n''\|_\infty \leq \|\Delta^2 c\|_\infty$ and therefore establishes that the constant

$$e_n := \frac{B_n}{A_n}$$

will do in (2.20).

Let us now prove the existence of constants A_n and B_n appearing in (2.22) and (2.23). As for (2.22), we recall the well-known identity

$$(x - t)^n = \sum_{i \in \mathbb{Z}} (i + 1 - t) \cdots (i + n - t) M_n(x - i), \qquad x, t \in \mathbb{R},$$

cf. Schumaker [Sch] or Theorem 3.9 of Chapter 3. This formula can be proved by induction on n. For $n = 1$, it follows from the computation following (2.7). Now suppose $n > 1$ and that the above formula is valid for n replaced by $n - 1$. According to the derivative recursion (2.4) and the induction hypothesis, the derivative with respect to x of both sides of the above equation are equal. Hence, there is a constant $c(t)$ such that

$$(x - t)^n = \sum_{i \in \mathbb{Z}} (i + 1 - t) \cdots (i + n - t) M_n(x - i) + c(t), \qquad x, t \in \mathbb{R}.$$

Since $c(t)$ is *independent* of x, it must be a polynomial of degree at most n in t. We now evaluate the above equation for $x = t = j$, where j is any integer and obtain

$$0 = \sum_{i \in \mathbb{Z}} (i+1-j)\cdots(i+n-j)M_n(j-i) + c(j)$$

$$= \sum_{i \in \mathbb{Z}} (1-i)\cdots(n-i)M_n(i) + c(j)$$

$$= \sum_{i=1}^{n} (1-i)\cdots(n-i)M_n(i) + c(j) = c(j).$$

Therefore, we conclude that $c(t) = 0, t \in \mathbb{R}$, which advances the induction hypothesis.

Thus, for $x \in [0,1]$ and $t \in \mathbb{R}$, we have

$$(x-t)^n = \sum_{i=0}^{n} (1-i-t)\cdots(n-i-t)M_n(x+i).$$

Consequently, on $[0,1]$ the polynomials $M_n(x)$, $M_n(x+1),\ldots,M_n(x+n)$ are linearly independent. Therefore, there is a constant $A_n > 0$ such that for any constants $c_0,\ldots,c_n$

$$A_n \max\{|c_j| : 0 \leq j \leq n\} \leq \max\left\{\left|\sum_{j=0}^{n} c_j M(x+j)\right| : x \in \mathbb{R}\right\}.$$

From this inequality, (2.22) easily follows.

For inequality (2.23) we expand the coefficients of the B-spline series for $(x-t)^n$ in powers of t and equate the coefficients of t^{n-1} and t^{n-2} with like powers of $(x-t)^n$ to conclude that

$$nx = \sum_{i \in \mathbb{Z}} a_i M_n(x-i), \qquad x \in \mathbb{R}$$

and

$$\frac{n(n-1)}{2}x^2 = \sum_{i \in \mathbb{Z}} b_i M_n(x-i), \qquad x \in \mathbb{R},$$

where

$$a_i = n\left\{i + 2^{-1}(n+1)\right\}$$

and

$$b_i = 2^{-1}\left\{\left(\sum_{j=1}^{n}(i+j)\right)^2 - \sum_{j=1}^{n}(i+j)^2\right\}.$$

Using these formulas, it follows that

$$x = \sum_{i \in \mathbb{Z}} \left(i + \frac{n+1}{2} \right) M_n(x - i), \qquad x \in \mathbb{R}$$

and

$$12^{-1}(n+1) = \sum_{i \in \mathbb{Z}} \left(x - i - \frac{n+1}{2} \right)^2 M_n(x - i), \qquad x \in \mathbb{R}.$$

As in the proof of (1.19), Taylor's formula with remainder gives

$$|f(x) - f(t) - f'(t)(x - t)| \leq 2^{-1}(x - t)^2 \|f''\|_\infty, \qquad x, t \in \mathbb{R},$$

and so we conclude that

$$|(T_r f)(x) - f(x)| \leq 4^{-r}(n+1)/24 \|f''\|_\infty, \qquad x \in \mathbb{R},$$

which proves (2.23) with $B_n = (n+1)/24$. $\square$

2.4. Stationary subdivision: Convergence.

We now consider various extensions of the subdivision schemes for computing cardinal splines described above. Our departure point is the averaging formula (2.15). Instead of averaging adjacent control points, we consider the effect of an *arbitrary* convex combination at each step of the computation. Specifically, we begin as in (2.15) and extend the initial control points c_j, $j \in \mathbb{Z}$, to the fine grid, by setting

$$(2.24) \qquad\qquad g_j^0 = g_{j+1/2}^0 = c_j, \qquad j \in \mathbb{Z}.$$

Next, we choose fixed weights $w_1, \ldots, w_n \in (0,1)$ and successively compute

$$(2.25) \qquad g_k^\ell = w_\ell g_k^{\ell-1} + (1 - w_\ell) g_{k-1/2}^{\ell-1}, \qquad k \in \mathbb{Z}/2, \qquad \ell = 1, \ldots, n.$$

After performing n successive convex combinations we obtain the new control points c_j^1, $j \in \mathbb{Z}$, defined as

$$(2.26) \qquad\qquad c_{2k}^1 = g_k^n, \qquad k \in \mathbb{Z}/2.$$

The above process is then repeated r times and we ask: does $\{c_j^r : j \in \mathbb{Z}\}$ converge and what are the properties of the limit function?

   To study this question we first find an alternative expression for the new control points c_j^1, $j \in \mathbb{Z}$. For this purpose, we introduce the polynomials

$$(2.27) \qquad\qquad D(z) = \sum_{j \in \mathbb{Z}} d_j z^j = \prod_{\ell=1}^{n} (w_\ell + (1 - w_\ell)z)$$

and

$$(2.28) \qquad a(z) = (1+z)D(z) = \sum_{j \in \mathbb{Z}} a_j z^j.$$

Note that $d_j = 0$ for $j \notin \{0, 1, \ldots, n\}$ and $a_j = 0$, $j \notin \{0, 1, \ldots, n+1\}$. Using the backward shift operator defined by $(Eg)_k = g_{k-1/2}$, $k \in \mathbb{Z}/2$ for any sequence $g = (g_k : k \in \mathbb{Z}/2)$ we see that (2.25) becomes

$$(2.29) \qquad g^\ell = (w_\ell + (1 - w_\ell)E)g^{\ell-1}, \qquad \ell = 1, \ldots, n,$$

where $g^\ell = (g_k^\ell : k \in \mathbb{Z}/2)$. Therefore we obtain the formula

$$(2.30) \qquad g^n = D(E)g^0,$$

which means that

$$(2.31) \qquad g_k^n = d_0 g_k^0 + d_1 g_{k-1/2}^0 + d_2 g_{k-1}^0 + \cdots, \qquad k \in \mathbb{Z}/2.$$

Consequently, from (2.31) and (2.24), we get

$$g_k^n = \begin{cases} d_0 c_j + (d_1 + d_2)c_{j-1} + \cdots, & k = j, \\ (d_0 + d_1)c_j + (d_2 + d_3)c_{j-1} + \cdots, & k = j + \dfrac{1}{2}. \end{cases}$$

Therefore, since $a_j = d_j + d_{j-1}$, $j \in \mathbb{Z}$, we have

$$g_k^n = \begin{cases} a_0 c_j + a_2 c_{j-1} + \cdots, & k = j, \\ a_1 c_j + a_3 c_{j-1} + \cdots, & k = j + \dfrac{1}{2}, \end{cases}$$

which can be succinctly written as

$$(2.32) \qquad c_i^1 = \sum_{j \in \mathbb{Z}} a_{i-2j} c_j, \quad i \in \mathbb{Z}.$$

Equation (2.32) shows clearly that the iteration (2.24)–(2.26) is stationary, that is, the formula to compute g_{k+1}^n from c_i, $i \in \mathbb{Z}$ is the same as the one to compute g_k^n from c_{i+1}, $i \in \mathbb{Z}$.

Generally, we call $\{a_j : j \in \mathbb{Z}\}$ the *mask* of the stationary subdivision scheme

$$(2.33) \qquad (S_a c)_i = \sum_{j \in \mathbb{Z}} a_{i-2j} c_j$$

and the Laurent polynomial

$$a(z) = \sum_{j \in \mathbb{Z}} a_j z^j, \quad z \in \mathbb{C} \backslash \{0\},$$

$$
\begin{array}{ccccccccc}
 & \vdots & \vdots & \vdots & & & & \\
\cdots & a_6 & a_4 & a_2 & a_0 & a_{-2} & a_{-4} & a_{-6} & \cdots \\
\cdots & a_7 & a_5 & a_3 & a_1 & a_{-1} & a_{-3} & a_{-6} & \cdots \\
\cdots & a_8 & a_6 & a_4 & a_2 & a_0 & a_{-2} & a_{-4} & \cdots \\
\cdots & a_9 & a_7 & a_5 & a_3 & a_1 & a_{-1} & a_{-3} & \cdots \\
 & \vdots & \vdots & \vdots & & & &
\end{array}
$$

FIG. 2.8. *Subdivision operator.*

$$
A_0 \begin{pmatrix} a_0 & a_2 & 0 \\ 0 & a_1 & a_3 \\ 0 & a_0 & a_2 \end{pmatrix}, \qquad
A_1 = \begin{pmatrix} a_1 & a_3 & 0 \\ a_0 & a_2 & 0 \\ 0 & a_1 & a_3 \end{pmatrix}
$$

FIG. 2.9. *Subdivision matrices: $n = 3$.*

its *symbol.* We are interested in when a stationary subdivision scheme (SSS) converges and what can be said about the limit. These questions will first be studied in the generality of finitely supported masks, that is, $\#\{j : a_j \neq 0\} < \infty$, then later specialized to (2.24)–(2.25). Before we begin, we formalize the notion of convergence of the SSS. Theorem 2.2 suggests the following definition.

DEFINITION 2.2. *We say S_a converges uniformly if for every $c \in \ell^\infty(\mathbb{Z})$ (all bounded sequences on $\mathbb{Z}$), there exists a function f continuous on $\mathbb{R}$ (nontrivial for at least one $c \in \ell^\infty(\mathbb{Z})$) such that for every $\epsilon > 0$ there is an $m \in \mathbb{N}$ such that $r \geq m$ implies that*

$$
(2.34) \qquad\qquad |(S_a^r c)_j - f(j/2^r)| < \epsilon, \qquad j \in \mathbb{Z}.
$$

Our first result, taken from Micchelli and Prautzsch [MPr] and also Cavaretta, Dahmen, and Micchelli [CDM], shows how to connect convergence of stationary subdivision to convergence of a related matrix subdivision scheme. To describe this result, we assume n is a positive integer such that $a_j = 0$ for $j < 0$ and $j > n$. We introduce two $n \times n$ matrices A_ϵ, $\epsilon \in \{0,1\}$, defined by

$$
(2.35) \qquad\qquad (A_\epsilon)_{ij} := a_{\epsilon+2j-i}, \qquad 0 \leq i, j \leq n-1.
$$

THEOREM 2.3. *Suppose the matrix subdivision scheme for the matrices in* (2.35) *converges in the sense of Definition 1.1. Then there is a function ϕ*

continuous on $\mathbb{R}$ *such that* $\phi(x) = 0$ *for* $x \leq 0$ *or* $x \geq n$,

$$(2.36) \qquad\qquad \sum_{j \in \mathbb{Z}} \phi(x - j) = 1, \qquad x \in \mathbb{R},$$

and

$$(2.37) \qquad\qquad f(x) = \sum_{j \in \mathbb{Z}} c_j \phi(x - j), \qquad x \in \mathbb{R},$$

is the limit of S_a, *in the sense of* (2.34).

Proof. Assume the MSS converges and has $\mathbf{F} = (F_0, \ldots, F_{n-1})^T : [0, 1] \to \mathbb{R}^n$ as its refinable curve. We will use the components of $\mathbf{F}$ to build ϕ. To this end, we first show that

$$(2.38) \quad F_0(0) = 0, \quad F_{j+1}(0) = F_j(1), \quad j = 0, 1, \ldots, n-2, \quad F_{n-1}(1) = 0.$$

For the proof of these equations we need to recall from Chapter 1 that one is the largest eigenvalue of $A_\epsilon, \epsilon \in \{0, 1\}$, and that it is simple. Moreover,

$$(2.39) \qquad\qquad A_\epsilon \mathbf{e} = \mathbf{e}, \qquad \epsilon \in \{0, 1\}$$

and $\mathbf{F}(\epsilon)$ is the unique left eigenvector of A_ϵ, viz.

$$A_\epsilon^T \mathbf{F}(\epsilon) = \mathbf{F}(\epsilon),$$

normalized so that $(\mathbf{F}(\epsilon), \mathbf{e}) = 1$.

Let us look at the matrix refinement equation (1.28) for $\mathbf{f} = \mathbf{F}$ and the matrices (2.35) when $\epsilon = 0$. This gives us the formula

$$(2.40) \qquad\qquad F_i(0) = \sum_{j=0}^{n-1} a_{2i-j} F_j(0), \qquad i = 0, 1, \ldots, n-1.$$

Keeping in mind the fact that $a_j = 0$, for $j < 0$ or $j > n$, we conclude that a_0 is an eigenvalue of A_0 with eigenvector $(1, 0, \ldots, 0)^T$; consequently, $a_0 \neq 1$. This implies, by specializing (2.40) to $i = 0$, that $F_0(0) = a_0 F_0(0)$. Therefore, it follows that $F_0(0) = 0$.

We now go back to (2.40) and set $F_n(0) := 0$. Thus we have for $i = 0, 1, \ldots, n-2$,

$$\begin{aligned} F_{i+1}(0) &= \sum_{j=0}^{n-1} a_{2i+2-j} F_j(0) \\ &= \sum_{j=0}^{n-1} a_{2i+1-j} F_{j+1}(0). \end{aligned}$$

Notice that, by again using the fact that $a_j = 0$, $j \notin \{0, 1, \ldots, n\}$, we conclude that this equation even holds for $i = n - 1$. This means that the vector $\mathbf{x} :=$

$(F_1(0), \ldots, F_n(0))^T$ is a left eigenvector of A_1 corresponding to the eigenvalue one and also satisfies the equation $(\mathbf{x}, \mathbf{e}) = 1$. Therefore, we conclude that $\mathbf{x} = \mathbf{F}(1)$. This takes care of of the proof of (2.38).

We are now ready to introduce the function ϕ by the recipe,

$$(2.41) \quad \phi(x) = \begin{cases} F_i(x - i), & x \in [i, i+1], \quad i = 0, 1, \ldots, n-1, \\ 0, & x \notin [0, n]. \end{cases}$$

Equation (2.38) ensures that ϕ is continuous and by construction $\phi(x) = 0$ for $x \notin (0, n)$. Next we demonstrate that the matrix refinement equation satisfied by the curve $\mathbf{F}$

$$(2.42) \qquad \mathbf{F}\left(\frac{t + \epsilon}{2}\right) = A_\epsilon^T \mathbf{F}(t), \qquad 0 \le t \le 1,$$

is equivalent to a scalar refinement equation for ϕ

$$(2.43) \qquad \phi(x) = \sum_{j \in \mathbb{Z}} a_j \phi(2x - j), \qquad x \in \mathbb{R}.$$

To see this, we write (2.42) in component form, using our definition (2.41), to obtain for $i = 0, 1, \ldots, n-1$ the relations

$$(2.44) \qquad \phi(t + i) = \sum_{j=0}^{n-1} a_{2i-j} \phi(2t + j), \qquad 0 \le t \le \frac{1}{2}$$

and

$$(2.45) \qquad \phi(t + i) = \sum_{j=0}^{n-1} a_{2i+1-j} \phi(2t - 1 + j), \qquad \frac{1}{2} \le t \le 1.$$

When $i \notin (0, n)$, it is clear the all summands on the right of (2.44) and (2.45) are zero and so both of these equations hold for all $i \in \mathbb{Z}$. Therefore, we get from (2.44), for any i such that $i \le x \le i + \frac{1}{2}$

$$\begin{aligned} \phi(x) &= \sum_{j=0}^{n-1} a_{2i-j} \phi(2x - 2i + j) = \sum_{j \in \mathbb{Z}} a_{2i-j} \phi(2x - 2i + j) \\ &= \sum_{j \in \mathbb{Z}} a_j \phi(2x - j). \end{aligned}$$

Similarly, (2.45) confirms (2.43) on all intervals of the form $[i + \frac{1}{2}, i]$, $i \in \mathbb{Z}$. A function ϕ that satisfies a refinement equation (2.43) is called a refinable function.

Now we can turn our attention to the convergence of S_a. For this purpose, we require two formulas. The first makes direct use of the refinement equation (2.43). Specifically, we let V_0 be the algebraic span of the functions $\phi(\cdot - j)$, $j \in \mathbb{Z}$, that is

$$V_0 = \left\{ \sum_{j \in \mathbb{Z}} c_j \phi(\cdot - j) : c_j \in \mathbb{R}, \qquad j \in \mathbb{Z} \right\},$$

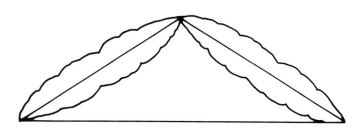

FIG. 2.10.   *Three refinable functions juxtaposed:* $(a_0, a_1, a_2) := (\frac{1}{3}, 1, \frac{2}{3}), (\frac{1}{2}, 1, \frac{1}{2}), (\frac{2}{3}, 1, \frac{1}{3}).$

and let V_1 be the algebraic span of the functions $\phi(2 \cdot -j)$, $j \in \mathbb{Z}$. Then the refinement equation (2.43) implies that

$$V_0 \subseteq V_1,$$

and moreover, every function in V_0 has a representation in the larger space V_1 given by

$$(2.46) \qquad f(x) = \sum_{j \in \mathbb{Z}} c_j \phi(x - j) = \sum_{j \in \mathbb{Z}} (S_a c)_j \phi(2x - j),$$

where $c = (c_j : j \in \mathbb{Z})$.

The second formula we have in mind is the following equation. Given an $i \in \mathbb{Z}_+$, we express it in base two as

$$(2.47) \qquad i = \epsilon_m + 2\epsilon_{m-1} + \cdots + 2^{m-1}\epsilon_1,$$

where $\epsilon_1, \ldots, \epsilon_m$ are either zero or one. Also, for any $c = (c_j : j \in \mathbb{Z})$ we form the vectors

$$(2.48) \qquad \mathbf{x} = (c_{-n+1}, \ldots, c_0)^T$$

and

$$(2.49) \qquad \mathbf{y} = ((S^m c)_{i-n+1}, \ldots, (S^m c)_i)^T.$$

Then we claim that

$$(2.50) \qquad \mathbf{y} = A_{\epsilon_m} \cdots A_{\epsilon_1} \mathbf{x}.$$

For the proof of this formula, we let $k = \frac{1}{2}(i - \epsilon_m) = \epsilon_{m-1} + \cdots + 2^{m-2}\epsilon_1$, so that $i = 2k + \epsilon_m$, and consider the formula

$$(S_a \lambda)_{i-\ell} = \sum_{j \in \mathbb{Z}} a_{i-\ell-2j} \lambda_j, \qquad \ell = 0, 1, \ldots, n - 1.$$

The nonzero summands above correspond to j such that $0 \le i - \ell - 2j \le n$. The largest such j satisfies $2j \le i$ or equivalently $j \le 2^{-1}i = k + 2^{-1}\epsilon_m$, that is, $j \le k$.

Similarly, for the smallest j we have $2j \geq i - (n-1) - n$ or $j \geq k - n + 2^{-1}(\epsilon_m + 1)$, that is, $j \geq k - n + 1$. Thus, for $\ell = 0, 1, \ldots, n-1$

$$(S_a \lambda)_{i-\ell} = \sum_{r=0}^{n-1} a_{2k+\epsilon_m - \ell - 2(k-r)} \lambda_{k-r}$$

$$= \sum_{r=0}^{n-1} (A_{\epsilon_m})_{\ell r} \lambda_{k-r}.$$

Using this equation $m - 1$ times proves (2.50).

We are now ready to prove that S_a converges. Pick any $j \in \mathbb{Z}$, and write it in the form $j = i + 2^m r$ where $r \in \mathbb{Z}$ and i has the form (2.47) for some $\epsilon_1, \ldots, \epsilon_m$. Now, use (2.46) $m - 1$ more times and evaluate the resulting equation at $x = 2^{-m} j$, to obtain the equations

$$f(2^{-m} j) = \sum_{\ell \in \mathbb{Z}} c_\ell \phi \left(2^{-m} j - \ell \right)$$

$$= \sum_{\ell \in \mathbb{Z}} (S_a^m c)_\ell \, \phi(j - \ell)$$

$$= \sum_{\ell=1}^{n-1} (S_a^m c)_{j-\ell} \, \phi(\ell).$$

At this juncture in the proof we need to observe that

$$(2.51) \qquad (S_a^m c)_\ell = \sum_{k \in \mathbb{Z}} a_{\ell - 2^m k}^m c_k, \qquad m = 1, 2, \ldots,$$

where the components of the vector $a^m = \left(a_j^m : j \in \mathbb{Z} \right)$, $m = 1, 2, \ldots$ are defined by the generating function

$$(2.52) \qquad \sum_{j \in \mathbb{Z}} a_j^m z^j = \prod_{j=0}^{m-1} a(z^{2^j}), \qquad z \in \mathbb{C}.$$

Equation (2.51) follows by induction on m. Now from formula (2.51) it is easy to check that for any $\ell \in \mathbb{Z}$, $(S_a^m c)_{j-\ell} = (S_a^m h)_{i-\ell}$, where $h := (c_{r+k} : k \in \mathbb{Z})$. Therefore we conclude that

$$f(2^{-m} j) - (S_a^m c)_j = \sum_{\ell=1}^{n-1} \left((S_a^m h)_{i-\ell} - (S_a^m h)_i \right) \phi(\ell).$$

In the derivation of this equation we used the fact that

$$(2.53) \qquad 1 = (\mathbf{e}, \mathbf{F}(0)) = \sum_{i \in \mathbb{Z}} \phi(i).$$

Moreover, this equation provides us the inequality

(2.54)
$$\left| f\left(2^{-m}j\right) - \left(S_a^m c\right)_j \right|$$
$$\leq R \max\left\{ \left|(S_a^m h)_{i-\ell-1} - (S_a^m h)_{i-\ell}\right| : 0 \leq \ell \leq n-2 \right\},$$

where

$$R := (n-1) \sum_{\ell=1}^{n-1} |\phi(\ell)|.$$

To estimate the upper bound in this inequality, we use the difference matrix D in (1.34) of Chapter 1, and observe that for $i = 0, 1, \ldots, n-2$, (2.49) gives us

$$(S_a^m h)_{i-\ell-1} - (S_a^m h)_{i-\ell} = (DA_{\epsilon_m} \cdots A_{\epsilon_1} \mathbf{v})_\ell = (A_{\epsilon_m} \cdots A_{\epsilon_1} D\mathbf{v})_\ell,$$

where $\mathbf{v} = (c_{-n+r+1}, \ldots, c_r)^T$. Hence, by Theorem 1.2, there are constants $M > 0$ and $\rho \in (0,1)$ such that

$$\left|(S_a^m h)_{i-\ell-1} - (S_a^m h)_{i-\ell}\right| \leq M\rho^m \max\left\{ |c_{r-j-1} - c_{r-j}| : 0 \leq j \leq n-1 \right\}.$$

Combining this inequality with (2.54) gives the inequality

$$\left| f\left(2^{-m}j\right) - \left(S_a^m c\right)_j \right| \leq RM\rho^m \|\Delta c\|_\infty,$$

which proves the convergence of S_a.

The last thing to prove in Theorem 2.3 is (2.36). For this claim we use equation (2.39) to conclude that

(2.55)
$$\sum_{i \in \mathbb{Z}} a_{2i} = \sum_{i \in \mathbb{Z}} a_{2i+1} = 1.$$

Therefore, if we set $e := (1 : j \in \mathbb{Z})$, it then follows, inductively on m, that $S_a^m e = e$. Moreover, for any $c = (c_j : j \in \mathbb{Z})$ and $x \in \mathbb{R}$, iterating (2.46) yields the equation

$$\sum_{i \in \mathbb{Z}} c_i \phi(x - i) = \sum_{i \in \mathbb{Z}} (S_a^m c)_i \, \phi\left(2^m x - i\right).$$

Choosing $c = e$ and $x = 2^{-m}j$ above gives, by (2.53), the formula

$$1 = \sum_{i \in \mathbb{Z}} \phi\left(2^{-m}j - i\right).$$

Since the dyadic fractions $\{2^{-m}j : j \in \mathbb{Z}, \ m \in \mathbb{Z}_+\}$ are dense in $\mathbb{R}$, the continuity and finite support of ϕ prove (2.36) and the theorem. $\square$

When ϕ is the cardinal B-spline of degree $n - 1$, V_0 and V_1 are spline spaces with knots on $\mathbb{Z}$ and $\mathbb{Z}/2$, respectively. Hence, the refinement equation in this case allows us to represent a cardinal spline on the "coarse" knots $\mathbb{Z}$ in terms of one on the "fine" knots $\mathbb{Z}/2$. If we specialize formulas (2.27) and (2.28) to cardinal splines, $(w_\ell = \frac{1}{2}, \ell = 1, \ldots, n)$ then

$$a(z) = 2^{-n}(1 + z)^{n+1}.$$

In other words, $a_j = 2^{-n}\binom{n+1}{j}$, $j = 0, 1, \ldots, n$, is the mask for the refinement equation of the B-spline. That is,

$$M_n(x) = 2^{-n}\sum_{j=0}^{n+1} \binom{n+1}{j} M_n(2x - j), \qquad x \in \mathbb{R}.$$

For instance, Chaikin's algorithm corresponds to quadratic splines and in this case the matrices (2.35) become

$$A_0 = \begin{bmatrix} 1/4 & 3/4 & 0 \\ 0 & 3/4 & 1/4 \\ 0 & 1/4 & 3/4 \end{bmatrix}, \quad A_1 = \begin{bmatrix} 3/4 & 1/4 & 0 \\ 1/4 & 3/4 & 0 \\ 0 & 3/4 & 1/4 \end{bmatrix}.$$

Also $\mathbf{F}(0) = (0, 1/2, 1/2)$, $\mathbf{F}(1) = (1/2, 1/2, 0)^T$, and $A_0^T\mathbf{F}(1) = A_1^T\mathbf{F}(0) = (1/8, 3/4, 1/8)^T$. The cubic spline case gives the matrices

$$A_0 = \begin{bmatrix} 1/8 & 3/4 & 1/8 & 0 \\ 0 & 1/2 & 1/2 & 0 \\ 0 & 1/8 & 3/4 & 1/8 \\ 0 & 0 & 1/2 & 1/2 \end{bmatrix}, \quad A_1 = \begin{bmatrix} 1/2 & 1/2 & 0 & 0 \\ 1/8 & 3/4 & 1/8 & 0 \\ 0 & 1/2 & 1/2 & 0 \\ 0 & 1/8 & 3/4 & 1/8 \end{bmatrix}$$

which yield $\mathbf{F}(0) = (0, 1/6, 2/3, 1/6)^T$, $\mathbf{F}(1) = (1/6, 2/3, 1/6, 0)^T$, and $A_1^T\mathbf{F}(0) = A_0^T\mathbf{F}(1) = 1/48(1, 23, 23, 1)^T$.

The next result provides a converse to Theorem 2.3.

THEOREM 2.4. *Suppose S_a converges. Then the limit f has the form (2.37), where ϕ has the properties described in Theorem 2.3. Moreover, the MSS determined by the matrices (2.35) converges and has*

$$\mathbf{F}(t) = (\phi(t), \phi(t+1), \ldots, \phi(t + n - 1))^T, \qquad 0 \leq t \leq 1,$$

as its refinable curve.

Proof. Let $\delta = (\delta_j : j \in \mathbb{Z})$ where $\delta_j = 0$ for $j \in \mathbb{Z}\backslash\{0\}$ and $\delta_0 = 1$. Hence, there is a function ϕ such that for for any $\epsilon > 0$ there is an m_0 such that $m \geq m_0$ and all $i \in \mathbb{Z}$

$$(2.56) \qquad \left|(S^m\delta)_i - \phi\left(i2^{-m}\right)\right| < \epsilon.$$

From formulas (2.51) and (2.52), it follows that $(S^m\delta)_j = a_j^m = 0$, for $j < 0$ and $j > 2^{m-1}n$. Thus, from (2.56), we easily obtain that $\phi(x) = 0$, for $x \notin (0, n)$, and

$$f(x) = \sum_{j \in \mathbb{Z}} c_j\phi(x - j)$$

for any $c \in \ell^\infty(\mathbb{Z})$.

To prove that the MSS $\{A_\epsilon : \epsilon \in \{0,1\}\}$ converges is straightforward. We pick a $t \in [0,1]$, write it in its binary expansion $t = \cdot \epsilon_1 \epsilon_2 \cdots$, and define $i_m = \epsilon_m + 2\epsilon_{m-1} + \cdots + 2^{m-1}\epsilon_1$. Then, according to (2.34), we have for all $\ell = 0, 1, \ldots, n-1$

$$\lim_{m \to \infty} (S^m c)_{i_m - \ell} = f(t) = \sum_{j=-n+1}^{0} c_j \phi(t+j) = (\mathbf{x}, \mathbf{F}(t)),$$

where $\mathbf{x} = (c_{-n+1}, \ldots, c_0)^T$, while from (2.49) and (2.50) we get

$$\lim_{m \to \infty} A_{\epsilon_m} \cdots A_{\epsilon_1} \mathbf{x} = (\mathbf{x}, \mathbf{F}(t)) \mathbf{e}. \qquad \Box$$

Embodied in Theorem 2.4 are a number of necessary conditions for the convergence of S_a. Namely, the refinable function ϕ satisfies (2.36) and the refinement equation (2.43) and the mask satisfies (2.55). Necessary and sufficient conditions follow by applying Theorem 1.2 to the matrices (2.35).

Our next theorem, from Micchelli and Prautzsch [MPr], is more than sufficient to cover the convergence of the iteration (2.24)–(2.26).

THEOREM 2.5. *Suppose* $(a_j : j \in \mathbb{Z})$ *is a mask such that for some integers* m, ℓ *with* $m - \ell \geq 2$, *we have* $a_j > 0$, *if* $\ell \leq j \leq m$, *and zero otherwise, and*

(2.57)
$$\sum_{j \in \mathbb{Z}} a_{2j} = \sum_{j \in \mathbb{Z}} a_{2j+1} = 1.$$

Then there exists a unique continuous refinable function $\phi \in C(\mathbb{R})$ *of compact support such that*

(2.58)
$$\phi(x) = \sum_{j \in \mathbb{Z}} a_j \phi(2x - j), \qquad x \in \mathbb{R}$$

and

(2.59)
$$\sum_{j \in \mathbb{Z}} \phi(x - j) = 1, \qquad x \in \mathbb{R}.$$

Moreover, the stationary subdivision scheme

$$(S_a c)_j = \sum_{i \in \mathbb{Z}} a_{j-2i} c_i$$

converges to

$$f(x) = \sum_{i \in \mathbb{Z}} c_i \phi(x - i), \qquad x \in \mathbb{R}.$$

Note that the subdivision scheme (2.24)–(2.26) whose mask is given by (2.27)–(2.28) obviously satisfies the hypothesis of Theorem 2.5 with $\ell = 0$ and $m = n+1$ for $n \geq 1$. Therefore, this result establishes the convergence of (2.24)–(2.26).

Proof. We give *two* proofs of this theorem. The first follows Micchelli and Prautzch [MPr] and uses Theorem 2.3 and Theorem 1.4. For this proof we first

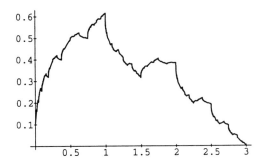

FIG. 2.11. *A refinable function:* $a_0 = .8$, $a_1 = .5$, $a_2 = .2$, $a_3 = .5$, *zero otherwise.*

observe that it suffices to prove Theorem 2.5 when $\ell = 0$. To see this we consider the mask $b_j := a_{j+\ell}$ and observe that $b_j^r = a_{j+(2^r-1)\ell}^r$, $r = 1, 2, \ldots$. Hence, S_a converges if and only if S_b converges and their refinable functions are related by the formula $\phi_b = \phi_a(\cdot + \ell)$. Moreover, $b_j > 0$ if and only if $j = 0, 1, \ldots, m - \ell$.

We now claim that the matrices A_ϵ, $\epsilon \in \{0, 1\}$, defined in (2.35) ($\ell = 0$, $m = n$) have a positive column; in fact the jth column of A_ϵ is positive provided that it is chosen to satisfy $n - 1 \leq \epsilon + 2j \leq n$. This means that (i) of Theorem 1.4 is valid (for $m = 1$). According to our hypothesis A_ϵ, $\epsilon \in \{0, 1\}$, is a stochastic matrix. Hence, by the remark following Theorem 1.4 and the fact that it has a positive column, there is a unique $\mathbf{u}_\epsilon \in \mathbb{R}^n$ (with nonnegative components) such that $(\mathbf{e}, \mathbf{u}_\epsilon) = 1$ and $A_\epsilon^T \mathbf{u}_\epsilon = \mathbf{u}_\epsilon$. To see that $A_0^T \mathbf{u}_1 = A_1^T \mathbf{u}_0$, we argue as we did in the proof of Theorem 2.3. Specifically, since $(A_0)_{0j} = a_0 \delta_{0j}$, $j = 0, 1, \ldots, n - 1$ ($\ell = 0$, $m = n$), we see that $\mathbf{e} = (1, 0, \ldots, 0)^T$ is an eigenvector of A_0 with eigenvalue a_0. As A_0 has a positive column, one must be a simple eigenvalue with corresponding eigenvector $\mathbf{e}$. Moreover, keeping in mind that $n \geq 2$ we conclude that $a_0 \neq 1$. But $(\mathbf{u}_0)_0 = (A_0 \mathbf{u}_0)_0 = a_0 (\mathbf{u}_0)_0$, and so it follows that $(\mathbf{u}_0)_0 = 0$. Now we set $(\mathbf{u}_0)_n = 0$ and introduce the vector $\mathbf{x} = ((\mathbf{u}_0)_1, \ldots, (\mathbf{u}_0)_n)^T$. Then $(\mathbf{e}, \mathbf{x}) = 1$ and a direct computation shows that $A_1^T \mathbf{x} = \mathbf{x}$. This implies that $\mathbf{x} = \mathbf{u}_1$ and also for $i = 0, 1, \ldots, n - 1$ we obtain the relations

$$
\begin{aligned}
\left(A_0^T \mathbf{u}_1\right)_i &= \sum_{j=0}^{n-1} a_{2i-j}(\mathbf{u}_1)_j = \sum_{j=0}^{n-1} a_{2i-j}(\mathbf{u}_0)_{j+1} \\
&= \sum_{j=-1}^{n-2} a_{2i-j}(\mathbf{u}_0)_{j+1} \\
&= \sum_{j=0}^{n-1} a_{2i+1-j}(\mathbf{u}_0)_j = \left(A_1^T \mathbf{u}_0\right)_i.
\end{aligned}
$$

This computation proves (ii) of Theorem 1.4 and hence Theorem 2.3 can be applied to complete our first proof of Theorem 2.5.

Our next proof of Theorem 2.5 bypasses Theorem 1.4 and Theorem 2.3. It is self-contained and uses arguments from Cavaretta, Dahmen, and Micchelli [CDM] (which are especially useful in the analysis of multivariate SSS).

We introduce the linear operator

$$(2.60) \qquad (F_a f)(x) = \sum_{i \in \mathbb{Z}} a_i f(2x - i).$$

As a consequence of (2.57) and the nonnegativity of the mask $(a_j : j \in \mathbb{Z})$, F_a is a bounded linear operator on $C(\mathbb{R})$ with norm two.

Let ψ be any function of compact support. Then $F_a \psi$ is also of compact support. In fact, if $\psi(x) = 0$ for $x \notin (\ell, m)$ then $(F_a \psi)(x) = 0$ outside of (ℓ, m), as well. Thus, for any c_j, $j \in \mathbb{Z}$, one can easily verify the useful formula

$$(2.61) \qquad \sum_{i \in \mathbb{Z}} c_i (F_a \psi)(x - i) = \sum_{i \in \mathbb{Z}} (S_a c)_i \psi(2x - i),$$

which extends for any $r \in \mathbb{Z}_+$ to the equation

$$(2.62) \qquad \sum_{i \in \mathbb{Z}} c_i (F_a^r \psi)(x - i) = \sum_{i \in \mathbb{Z}} (S_a^r c)_i \psi(2^r x - i).$$

Now, pick ψ_0 to be *any* B-spline who support is (ℓ, m); for instance

$$\psi_0(x) = M_{m-\ell-1}(x - \ell)$$

will do nicely. We shall show that the sequence of continuous functions

$$(2.63) \qquad h_r = F_a^r \psi_0, \qquad r \in \mathbb{Z}_+$$

converge uniformly on $\mathbb{R}$.

Central to our analysis of this assertion is the semi-norm

$$(2.64) \qquad \kappa(c) = \max\{|c_j - c_k| : |j - k| < m - \ell\}$$

used in Cavaretta, Dahmen, and Micchelli [CDM]. We pick any other mask b_j, $j \in \mathbb{Z}$ such that

$$(2.65) \qquad \{j : b_j \neq 0\} \subseteq (\ell, m), \qquad b_j \geq 0, \qquad j \in \mathbb{Z},$$

and

$$(2.66) \qquad \sum_{k \in \mathbb{Z}} b_{j-2k} = 1, \qquad j \in \mathbb{Z}.$$

Then for any constant p and any bi-infinite sequence $c = (c_j : j \in \mathbb{Z})$ we have the formula

$$(S_a c - S_b c)_j = \sum_{k \in \mathbb{Z}} (a_{j-2k} - b_{j-2k})(c_k - p).$$

By a judicious choice of p, we get the inequality

(2.67) $$\|S_a c - S_b c\|_\infty \leq \kappa(c).$$

and a similar argument yields

(2.68) $$\kappa(S_a c) \leq \rho(a)\kappa(c),$$

where

(2.69) $$\rho = \rho(a) := \frac{1}{2} \sum_{|j-r|<m-\ell} |a_{j-2k} - a_{r-2k}|.$$

Next we observe that

(2.70) $$\rho < 1.$$

For the proof of this inequality we need to show that for any integers j, r such that $|j - r| < m - \ell$ there is an integer k satisfying

$$\ell \leq j - 2k \leq m \quad \text{and} \quad \ell \leq r - 2k \leq m.$$

This conclusion would imply the *strict* inequality

$$|a_{j-2k} - a_{r-2k}| < a_{j-2k} + a_{r-2k},$$

and therefore establish (2.70). For definiteness, we suppose $j < r$, in other words,

$$0 \leq r - j < m - \ell.$$

Thus, there is an even integer in the interval $[r - m, j - \ell]$ that we call $2k$. This integer k satisfies all the requisite inequalities and consequently we have proved (2.70).

We are now ready to establish the convergence of the sequence (2.63). We let $b = (b_j : j \in \mathbb{Z})$ be the coefficients appearing in the refinement equation for ψ_0, namely

$$\psi_0(x) = \sum_{j \in \mathbb{Z}} b_j \psi_0(2x - j), \qquad x \in \mathbb{R}.$$

The coefficients $b_j, j \in \mathbb{Z}$ satisfy (2.65) and (2.66) and so inequalities (2.67) and (2.68) apply. Moreover, since

(2.71) $$h_r(x) = \sum_{k \in \mathbb{Z}} (S_a^r \delta)_k \psi_0(2^r x - k),$$

we get

$$h_{r+1}(x) - h_r(x) = \sum_{k \in \mathbb{Z}} \left((S_a^{r+1}\delta)_k - S_b(S_a^{r+1}\delta)_k \right) \psi_0(2^r x - k),$$

from which it follows for all $x \in \mathbb{R}, r \in \mathbb{Z}_+$ that

$$|h_{r+1}(x) - h_r(x)| \;\leq\; \|\,(S_a - S_b)\,(S_a^r \delta)\,\|_\infty$$

$$\leq\; \kappa(S_a^r \delta) \leq \rho^r \kappa(\delta).$$

We conclude that $\{h_r : r \in \mathbb{Z}_+\}$ is a Cauchy sequence in $C(\mathbb{R})$. Therefore there is a continuous function ϕ such that

$$\lim_{r \to \infty} h_r(x) = \phi(x), \qquad x \in \mathbb{R}$$

uniformly on $\mathbb{R}$. In particular, $\phi(x) \geq 0$, $x \in \mathbb{R}$ and is zero outside of (ℓ, m). Certainly ϕ satisfies the refinement equation

$$\phi(x) = (F_a\phi)(x) = \sum_{k \in \mathbb{Z}} a_k \phi(2x - k), \qquad x \in \mathbb{R}.$$

Also, (2.59) holds because for any $r \in \mathbb{Z}_+$ and $x \in \mathbb{R}$ we have

$$\sum_{k \in \mathbb{Z}} h_{r+1}(x - k) \;=\; \sum_{k \in \mathbb{Z}} F_a(h_r)(x - k)$$

$$=\; \sum_{k \in \mathbb{Z}} (S_a e)_k h_r(2x - k)$$

$$=\; \sum_{k \in \mathbb{Z}} h_r(2x - k)$$

$$=\; \cdots \;=\; \sum_{k \in \mathbb{Z}} \psi_0(2^r x - k) = 1.$$

There remains only the uniqueness of ϕ and the convergence of the SSS. To this end, we let ψ be any other continuous function of compact support that satisfies (2.58) and (2.59). Then, for $x \in \mathbb{R}$ it follows that

$$|\psi(x) - \phi(x)| \;\leq\; \left| \sum_{k \in \mathbb{Z}} \left((S_a^r \delta)_k - \phi\left(\frac{k}{2^r}\right) \right) \psi(2^r x - k) \right|$$

$$+ \left| \phi(x) - \sum_{k \in \mathbb{Z}} \phi\left(\frac{k}{2^r}\right) \psi(2^r x - k) \right|.$$

The second term above is bounded by a constant times $\omega(\phi; 2^{-r})$, the modulus of continuity of ϕ, while to bound the first term it suffices to show that the SSS converges.

For this purpose, we proceed as before and note that since

$$f(x) = \sum_{i \in \mathbb{Z}} c_i \phi(x - i) = \sum_{i \in \mathbb{Z}} (S^r c)_i \phi(2^r x - i),$$

we have

$$f(j/2^r) = \sum_{i=\ell}^{m} (S^r c)_{j-i} \phi(i)$$

and consequently for $j \in \mathbb{Z}$

$$|f(j/2^r) - (S^r c)_{j-\ell}| \leq \kappa(S^r c)$$

$$\leq \rho^r \kappa(c).$$

Since ϕ is of compact support and $c \in \ell^\infty(\mathbb{Z})$, the function f is uniformly continuous on $\mathbb{R}$ and so (2.34) follows.   □

Remark 2.1. For later use it is helpful to note that

$$\lim_{r \to \infty} h_r(x) = \phi(x)$$

for $x \in \mathbb{R}$, when ψ_0 is chosen to be the indicator function of any interval of the form $[p, p+1)$ contained in $[\ell, m]$. In this case, the mask for ψ_0 satisfies $b_j = 0$, if $j \notin \{2p, 2p+1\}$ and 1 otherwise.

In the remainder of this chapter we develop various properties of the function ϕ constructed above. Generally speaking, these properties center around two issues: smoothness and shape preservation. Principally, we will be concerned with ϕ constructed by the averaging algorithm (2.24)–(2.26). But as we shall see, most of these properties hold with greater generality.

The function ϕ constructed from the *averaging scheme* (2.24)–(2.26) has a variation diminishing property. Specifically, for any $c = (c_j : j \in \mathbb{Z})$ we let $S^-(c)$ be the number of sign changes in the components of c. Likewise, given a continuous function f defined on $\mathbb{R}$ we let

$$S^-(f) = \sup \{S^-(f(x_1), \ldots, f(x_p)) : x_1 < \cdots < x_p, \ p = 1, 2, \ldots\}$$

be the number of sign changes of f on $\mathbb{R}$. With this definition in mind, we claim that

(2.72) $$S^- \left(\sum_{j \in \mathbb{Z}} c_j \phi(\cdot - j) \right) \leq S^-(c).$$

To see this, we argue as follows. Suppose there are points $x_1 < \cdots < x_{p+1}$ such that $f(x_j)(-1)^j > 0$, $j = 1, 2, \ldots, p+1$, where

(2.73) $$f(x) := \sum_{j \in \mathbb{Z}} c_j \phi(x - j).$$

According to (2.34), there are integers $j_{1,r} < \cdots < j_{p+1,r}$ such that

$$(S_a^r c)_{j_\ell, r}(-1)^\ell > 0, \qquad \ell = 1, 2, \ldots, p+1,$$

and in particular, $S^-(S_a^r c) \geq p$. However, because $S_a c$ is obtained from c by successively averaging its components, we have $S^-(S_a c) \leq S^-(c)$ and generally by induction on r it follows that $S^-(S_a^r c) \leq S^-(c)$. Consequently, $S^-(c) \geq p$, as claimed.

Next we point out that ϕ is Hölder continuous. We base our remark on the following general fact already used in Chapter 1, which we restate here.

LEMMA 2.1. *Let* $\{f_n : n \in \mathbb{Z}_+\}$ *be a sequence of real-valued continuous functions on* $\mathbb{R}$. *Suppose* $\rho_0 \in (1, \infty)$, $\rho \in (0, 1)$, *and* M, N *are positive numbers such that for all* $x, y \in \mathbb{R}$, $n \in \mathbb{Z}_+$,

(i) $|f_n(x) - f_n(y)| \leq M(\rho \rho_0)^n |x - y|$, *and*
(ii) $|f_{n+1}(x) - f_n(x)| \leq N \rho^n$.

Then there exists a continuous function ϕ *such that*

$$\lim_{n \to \infty} f_n(x) = \phi(x)$$

uniformly on $\mathbb{R}$, *and for* $\mu \in (0, \infty)$ *defined by* $\rho = \rho_0^{-\mu}$ *we have*

$$|\phi(x) - \phi(y)| \leq \delta |x - y|^\mu, \qquad x, y \in \mathbb{R},$$

for some constant $\delta > 0$.

COROLLARY 2.1. *Let* ϕ *be as described in Theorem 2.5. Then there is a constant* $R > 0$ *such that*

$$|\phi(x) - \phi(y)| \leq R|x - y|^\mu$$

where $\rho = 2^{-\mu}$ *and* ρ *is defined in* (2.69).

Proof. We apply Lemma 2.1 to the sequence $\{h_r(x) : r \in \mathbb{Z}_+\}$ used in the proof of Theorem 2.5. We already noted there that

$$|h_{r+1}(x) - h_r(x)| \leq \rho^r \kappa(\delta), \qquad r \in \mathbb{Z}_+, \qquad x \in \mathbb{R}.$$

Using the derivative recurrence formula for the B-spline, see (2.4) and (2.5), we also get

$$|(h^r)'(x)| \;\leq\; 2^r \sup\{|(S_a^r \delta)_{j+1} - (S_a^r \delta)_j| : j \in \mathbb{Z}\}$$

$$\leq\; 2^r \kappa(S_a^r \delta) \leq (2\rho)^r \kappa(\delta)$$

and so

$$|h_r(x) - h_r(y)| \leq (2\rho)^r \kappa(\delta)|x - y|. \qquad \square$$

Next we recall a fact proved in Micchelli and Pinkus [MPi].

PROPOSITION 2.1. *Let* ϕ *be as described in Theorem 2.5. Then* $\phi(x) \geq 0$ *for all* $x \in \mathbb{R}$ *with strict inequality if and only if* $x \in (\ell, m)$.

Proof. Since ϕ is constructed as a limit of nonnegative functions that vanish outside of (ℓ, m), we need only prove $\phi(x) > 0$ for $x \in (\ell, m)$. For this purpose, choose $x \in [m + \ell - 1, \, m + \ell + 1]$. Then by (2.59), we have

$$1 = \sum_{j \in \mathbb{Z}} \phi(x - j) = \sum_{j=\ell}^{m} \phi(x - j),$$

and so by the refinement equation for ϕ

$$\phi\left(\frac{x}{2}\right) = \sum_{j=\ell}^{m} a_j \phi(x-j) \geq \min\{a_j : \ell \leq j \leq m\}.$$

Hence $\phi(x) > 0$ for $x \in I_0 := 2^{-1}(m+\ell-1, m+\ell+1)$ and the length of I_0 is one. Next we will show how to "propagate" the positivity of ϕ to all of (ℓ, m). To this end, suppose that $\phi(y) > 0$ for some $y \in \mathbb{R}$. Then for any integer $k \in [\ell, m]$, we have

$$\phi\left(\frac{y+k}{2}\right) = \sum_{j=\ell}^{m} a_j \phi(y+k-j)$$
$$\geq a_k \phi(y) > 0.$$

Consequently, if ϕ is positive on an interval (a, b) of length at least one, it is positive on the interval $2^{-1}(a+\ell, b+m)$. Starting with I_0 this process produces intervals $I_r = (a_r, b_r)$, $a_r := 2^{-1}(a_{r-1} + \ell)$, $b_r := 2^{-1}(b_{r-1} + m)$ on which ϕ is positive. Since $\lim_{r \to \infty} a_r = \ell$ and $\lim_{r \to \infty} b_r = m$, the lemma is proved.   □

2.5. Hurwitz polynomials and stationary subdivision.

More can be said about the "shape" of the refinable function ϕ for the averaging scheme (2.24)–(2.26). For instance, we can resolve the following interpolation problem. Given integers $i_1 < \cdots < i_p$, real numbers $x_1 < \cdots < x_p$ and $y_1, \ldots, y_p$ determine a function f of the form

$$f(x) = c_1 \phi(x - i_1) + \cdots + c_p \phi(x - i_p)$$

such that

$$f(x_j) = y_j, \qquad j = 1, \ldots, p.$$

To study this problem we set

$$\phi\begin{pmatrix} x_1, \ldots, x_p \\ i_1, \ldots, i_p \end{pmatrix} := \begin{vmatrix} \phi(x_1 - i_1) & \cdots & \phi(x_1 - i_p) \\ \vdots & & \vdots \\ \phi(x_p - i_1) & \cdots & \phi(x_p - i_p) \end{vmatrix}.$$

Also we say a Laurent polynomial is a *Hurwitz polynomial* provided that it has zeros only in the left half-plane. The following theorem is from Goodman and Micchelli [GM].

THEOREM 2.6. *Let* $a = (a_j : j \in \mathbb{Z})$ *be a mask that satisfies the hypothesis of Theorem 2.5 such that*

$$a(z) = \sum_{j \in \mathbb{Z}} a_j z^j$$

is a Hurwitz polynomial. Then

(2.74) $$S^- \left(\sum_{j \in \mathbb{Z}} c_j \phi(\cdot - j) \right) \leq S^-(c)$$

and

(2.75) $$\phi \begin{pmatrix} x_1, \ldots, x_p \\ i_1, \ldots, i_p \end{pmatrix} \geq 0$$

for all real numbers $x_1 < \cdots < x_p$ and integers $i_1 < \cdots < i_p$. Moreover, strict inequality holds in (2.75) if and only if $x_j - i_j \in (\ell, m)$, $j = 1, \ldots, p$.

Note that the averaging scheme (2.24)–(2.26) surely satisfies the hypothesis of the theorem. In fact, in this case, according to (2.27) and (2.28), $a(z)$ only has negative zeros in $\mathbb{C}$. Using Proposition 2.1, another way of describing Theorem 2.6 is to say that the matrix

$$(\phi(x_j - i_k))_{j,k=1,\ldots,p}$$

is nonsingular if and only if its diagonal elements are nonzero. Also observe that whenever a Laurent polynomial is a Hurwitz polynomial all its coefficients are of the same sign. Therefore the mask $(a_j : j \in \mathbb{Z})$ of Theorem 2.6 has nonnegative entries.

For the proof of Theorem 2.6, we first point out that we already proved (2.74) for the averaging scheme (2.24)–(2.26). The general case covered by Theorem 2.6 can be proved by using the determinantal inequalities (2.75) just as we proved Corollary 1.3. The proof of (2.75) is longer and, as in the proof of Theorem 2.5, we assume without loss of generality that $\ell = 0$ and $m = n + 1$ with $n \geq 1$. We define the bi-infinite matrix

$$A = (A_{ij})_{i,j \in \mathbb{Z}}, \qquad A_{ij} := a_{j-2i}$$

and as usual denote the rth power of D by D^r. The essential step in the proof of (2.75) is the next result of some independent interest.

PROPOSITION 2.2. *Let $a = (a_j : j \in \mathbb{Z})$ be any mask such that $a_j = 0$ if $j \notin \{0, 1, \ldots, n+1\}$, $a_0 a_{n+1} > 0$, and $a(z)$ is a Hurwitz polynomial. Then*

(2.76) $$A^r \begin{pmatrix} i_1, \ldots, i_p \\ j_1, \ldots, j_p \end{pmatrix} \geq 0$$

for all $i_1 < \cdots < i_p, j_1 < \cdots < j_p, n = 1, 2, \ldots$, and strict inequality holds if and only if $0 \leq j_\ell - 2^r i_\ell \leq (2^r - 1)(n + 1)$, $\ell = 1, \ldots, p$.

The proof of this result is somewhat involved. We postpone it until after we demonstrate how it leads to the proof of (2.75).

Proof of Theorem 2.6. First we prove that the determinants in (2.75) are nonnegative. For this purpose, we first observe that (2.75) holds if ϕ were the indicator function of $[0, 1)$. Let us confirm this observation. In this case, there are two possibilities: either $x_1 - i_\ell \notin [0, 1)$ for all $\ell = 1, \ldots, p$, so that the determinant (2.75) is zero, or $x_1 - i_\ell \in [0,1)$ for some ℓ, $1 \le \ell \le p$. Since $i_1 < \cdots < i_p$, then $x_k - i_j \notin [0,1)$ for $k \ge 1$ and $j < \ell$, as well as for $k = 1$ and $j > \ell$. In the first case,

$$x_k - i_j \ge 1 + x_1 - i_\ell \ge 1,$$

while in the second case

$$x_1 - i_j \le x_1 - i_\ell - 1 < 1 - 1 = 0.$$

If $\ell \ge 2$ then the determinant is zero. Otherwise, $\ell = 1$ and the determinant in (2.75) has the same value as the minor corresponding to the last $p - 1$ rows and columns. In this way, we see, inductively on p, that the determinant is either zero or one.

Next we consider the sequence of functions

$$(2.77) \qquad h_{k+1}(x) = (F_a^{k+1}\psi_0)(x) = \sum_{j \in \mathbb{Z}} a_j h_k(2x - j), \qquad r \in \mathbb{Z}_+,$$

where we choose ψ_0 to be the indicator function of $[0,1)$. In view of Remark 2.1 we have $\lim_{r \to \infty} h_r(x) = \phi(x)$, $x \in \mathbb{R}$. Moreover, by (2.46) we have

$$h_{k+1}(x - i) = \sum_{j \in \mathbb{Z}} A_{ij} h_k(2x - j),$$

and so by the Cauchy–Binet formula (see Lemma 1.6 of Chapter 1) we have

$$h_{k+1}\begin{pmatrix} x_1, \ldots, x_p \\ i_1, \ldots, i_p \end{pmatrix} = \sum_{j_1 < \cdots < j_p} A\begin{pmatrix} i_1, \ldots, i_p \\ j_1, \ldots, j_p \end{pmatrix} h_k\begin{pmatrix} 2x_1, \ldots, 2x_p \\ j_1, \ldots, j_p \end{pmatrix}.$$

From this formula, and (2.76) of Proposition 2.2 with $r = 1$, it follows, inductively on k, that

$$h_{k+1}\begin{pmatrix} x_1, \ldots, x_p \\ i_1, \ldots, i_p \end{pmatrix} \ge 0$$

whenever $x_1 < \cdots < x_p$ and $i_1 < \cdots < i_p$. Letting $k \to \infty$ proves the desired inequality (2.75).

To resolve the case of equality in (2.75) we use the full value of Proposition 2.2. We begin with the sufficiency of our conditions. Thus we suppose $0 < x_k - i_k < n + 1$, $k = 1, \ldots, p$ and choose r large enough so that $2^r i_k < 2^r x_k - n - 1$ and $2^r x_k < 2^r i_k + (2^r - 1)(n+1)$, $k = 1, \ldots, p$. Now pick any integers $j_{1,r} < \cdots < j_{p,r}$ such that

$$2^r x_k - n - 1 < j_{k,r} < 2^r x_k, \qquad k = 1, \ldots, p.$$

Consequently, we get the inequalities

$$(2.78) \qquad 0 \le j_{k,r} - 2^r i_k \le (2^r - 1)(n + 1), \qquad k = 1, \ldots, p$$

and

(2.79) $0 < 2^r x_k - j_{k,r} < n + 1, \qquad k = 1, \ldots, p.$

According to the refinement equation (2.58), we have

$$\phi(x - i) \;=\; \sum_{j \in \mathbb{Z}} A_{ij} \phi(2x - j) = \cdots$$

$$=\; \sum_{j \in \mathbb{Z}} A_{ij}^r \phi(2^r x - j).$$

Hence, by (2.76) of Proposition 2.2, inequality (2.75) (which we already proved), and the Cauchy–Binet formula, we get

$$\phi\begin{pmatrix} x_1, \ldots, x_p \\ i_1, \ldots, i_p \end{pmatrix} \geq A^r \begin{pmatrix} i_1, \ldots, i_p \\ j_{1,r}, \ldots, j_{p,r} \end{pmatrix} \phi \begin{pmatrix} 2^r x_1, \ldots, 2^r x_p \\ j_{1,r}, \ldots, j_{p,r} \end{pmatrix}.$$

The minor of A^r above is nonzero by Proposition 2.2 and (2.78). On the other hand, the minor of ϕ has positive diagonal elements by (2.79) and Proposition 2.1, and for $r \to \infty$ it has zero off-diagonal elements, for the same reason. This proves the sufficiency of the theorem.

For the necessity, we first suppose $x_s - i_s \leq 0$ for some $s, 1 \leq s \leq p$. Then $x_r - i_k = 0$ for $k = s, \ldots, p, r = 1, \ldots, s$. Thus the matrix in (2.75) has a zero entry in the (r, k) position, $r = 1, \ldots, s, \ k = s, \ldots, p$. Consequently, the last $p - s + 1$ columns must be linearly dependent and the minor in (2.75) is zero. For a similar reason, it is zero if $x_s - i_s \geq n + 1$ for some $s, 1 \leq s \leq p$. □

We now turn our attention to the proof of Proposition 2.2. The first case to deal with is $r = 1$. This is a result of Kemperman [Ke], which we now prove.

As described earlier, a polynomial

$$d(z) = \sum_{j=0}^{n+1} d_{n+1-j} z^j$$

is said to be a Hurwitz polynomial if all its zeros are in the (open) left half-plane. Recall that all the coefficients of a Hurwitz polynomial are nonzero and necessarily of one sign. We assume for definiteness that

$$d_j > 0, \qquad j = 0, 1, \ldots, n + 1.$$

We adhere to the convention of writing the coefficients of $d(z)$ in reverse order and, in keeping with our previous notation, always make the identification

$$a_j = d_{n+1-j}, \qquad j = 0, 1, \ldots, n + 1.$$

Central to our discussion is the bi-infinite Hurwitz matrix

$$H = (H_{ij})_{i,j \in \mathbb{Z}},$$

where

$$H_{ij} = d_{2j-i}, \qquad i, j \in \mathbb{Z},$$

and d_j is set equal to zero for $j < 0$ or $j > n + 1$. The Roth–Hurwitz criterion states that $d(z)$ is a Hurwitz polynomial if and only if for $p = 1, 2, \ldots, n + 1$,

$$\triangle_p := H \begin{pmatrix} 1, 2, \ldots, p \\ 1, 2, \ldots, p \end{pmatrix} = \begin{vmatrix} d_1 & d_3 & \cdots & d_{2p-1} \\ d_0 & d_2 & \cdots & d_{2p-2} \\ \vdots & \vdots & & \vdots \\ d_{-p+2} & d_{-p+4} & \cdots & d_p \end{vmatrix} > 0,$$

cf. Gantmacher [G, p. 194]. Note that for $p > n + 1$ the last column of the determinant $\triangle_p$ is zero and so $\triangle_p = 0$, for $p = n + 1, \ldots$.

We need the following factorization procedure for Hurwitz matrices.

LEMMA 2.2. *Suppose* $d(z) = \sum_{j=0}^{n} d_{n-j} z^j$, $n \geq 1$, *is a Hurwitz polynomial. Then there is a positive constant c such that the matrix $C = (c_{ij})_{i,j \in \mathbb{Z}}$, defined by*

$$c_{ij} = \begin{cases} 1, & j = i + 1, \\ c, & i = j = \text{even}, \\ 0, & \text{otherwise}, \end{cases}$$

has the property that

(2.80) $$H = CH',$$

where H' is a Hurwitz matrix corresponding to some Hurwitz polynomial of degree $\leq n - 1$.

Proof. Equation (2.80) means that for all $i, j \in \mathbb{Z}$

$$\begin{aligned} d_{2j-i} &= \sum_{k \in \mathbb{Z}} c_{ik} d'_{2j-k} \\ &= c d'_{2j-i} + d'_{2j-i-1}. \end{aligned}$$

Equivalently, we have

$$d_{2m} = c d'_{2m} + d'_{2m-1}, \qquad m \in \mathbb{Z},$$

and also

(2.81) $$d_{2m+1} = d'_{2m}, \qquad m \in \mathbb{Z}.$$

Thus it follows that

(2.82) $$d'_{2m-1} = d_{2m} - c d_{2m+1}.$$

When c is specified, equations (2.81)–(2.82) determine all $d'_j, j \in \mathbb{Z}$. To pin down c, we use the requirement that $d'_j = 0$ for $j < 0$ and so we must have

$$0 = d'_{-1} = d_0 - c d_1,$$

that is, $c = d_0/d_1 > 0$. With this choice of c, we *define* d'_j for all $j \in \mathbb{Z}$ using equations (2.81) and (2.82). It is an easy matter to see that $d'_j = 0$ for $j < 0$ and $j > n - 1$. Therefore, it remains to prove that

$$d'(z) = \sum_{j=0}^{n-1} d'_{n-1-j} z^j$$

is a Roth–Hurwitz polynomial. For this task, we use the Roth–Hurwitz criterion. To this end, we again consider the determinant

$$\triangle_p = \begin{vmatrix} d_1 & d_3 & \cdots & d_{2p-1} \\ d_0 & d_2 & \cdots & d_{2p-2} \\ \vdots & \vdots & & \vdots \\ d_{-p+2} & d_{-p+4} & \cdots & d_p \end{vmatrix}.$$

Multiply the first, third, $\ldots$, rows of $\triangle_p$ by c and subtract from the second, fourth, $\ldots$, rows to get

$$\triangle_p = \begin{vmatrix} d_1 & d_3 & \cdots & d_{2p-1} \\ d_0 - cd_1 & d_2 - cd_3 & \cdots & d_{2p-2} - cd_{2p-1} \\ d_{-1} & d_1 & \cdots & d_{2p-3} \\ d_{-2} - cd_{-1} & d_0 - cd_1 & \cdots & d_{2p-4} - cd_{2p-3} \\ \vdots & \vdots & & \vdots \end{vmatrix}$$

$$= \begin{vmatrix} d'_0 & d'_2 & \cdots & d'_{2p-2} \\ d'_{-1} & d'_1 & \cdots & d'_{2p-3} \\ \vdots & \vdots & & \vdots \\ d'_{-p+1} & & \cdots & d'_{p-1} \end{vmatrix}.$$

The first column of the matrix above has zero elements except for d'_0. Hence we get

$$\triangle_p = d'_0 \triangle'_{p-1} = d_1 \triangle'_{p-1}, \qquad p = 2, 3, \ldots.$$

From this formula and the Roth–Hurwitz criterion, we conclude $d'(z)$ is a Hurwitz polynomial. $\square$

To make use of Lemma 2.2 we use the following lemma.

LEMMA 2.3. *Let* $C = (c_{ij})_{i,j \in \mathbb{Z}}$ *be a one-banded upper triangular matrix, i.e.,* $c_{ij} = 0$, *whenever* $j < i$ *or* $j > i + 1$. *Then*

$$C\begin{pmatrix} i_1, \ldots, i_p \\ j_1, \ldots, j_p \end{pmatrix} = \begin{cases} c_{i_1 j_1} \cdots c_{i_p j_p}, & \text{if } 0 \leq j_r - i_r \leq 1, \ r = 1, \ldots, p, \\ 0, & \text{otherwise.} \end{cases}$$

Actually, the first equation above holds for *all* $i_1, \ldots, i_r, j_1, \ldots, j_r$ because $c_{i_r j_r}$ is zero, if $j_r - i_r < 0$ or $j_r - i_r > 1$.

Proof. For $p = 1$, the formula above is true by hypothesis. Suppose $p \geq 2$ and $j_1 < i_1$. Then $j_1 < i_r, r = 1, \ldots, s$ and so the first column of the determinant is zero. If $i_1 + 1 < j_1$ then the first row is zero. Hence in either case the determinant is zero. When $i_1 = j_1$ then all the elements in the first column are zero except perhaps $c_{i_1 j_1}$. Thus

$$C\begin{pmatrix} i_1, \ldots, i_p \\ j_1, \ldots, j_p \end{pmatrix} = c_{i_1 j_1} C\begin{pmatrix} i_2, \ldots, i_p \\ j_2, \ldots, j_p \end{pmatrix}.$$

Similarly, if $j_1 = i_1 + 1$ we get

$$C\begin{pmatrix} i_1, i_2, \ldots, i_p \\ j_1, j_2, \ldots, j_p \end{pmatrix} = c_{i_1 i_1 + 1} C\begin{pmatrix} i_2, \ldots, i_p \\ j_2, \ldots, j_p \end{pmatrix}.$$

Using these formulas the lemma follows by induction on p. □

LEMMA 2.4 [Ke]. *Let*

$$d(z) = \sum_{j=0}^{n} d_{n-j} z^j$$

be a Hurwitz polynomial. Then the corresponding Hurwitz matrix $H = (d_{2j-i})_{i,j \in \mathbb{Z}}$ *is totally positive, that is,*

$$(2.83) \qquad H\begin{pmatrix} i_1, \ldots, i_p \\ j_1, \ldots, j_p \end{pmatrix} \geq 0$$

for all integers $i_1 < \cdots < i_p, j_1 < \cdots < i_p$ *and strict inequality holds in* (2.83) *if and only if*

$$(2.84) \qquad 0 \leq 2j_\ell - i_\ell \leq n, \qquad \ell = 1, \ldots, p.$$

Proof. The proof is by induction on n. The case $n = 0$ is elementary. Now assume $n \geq 1$ and that the result is true for $n - 1$. Using Lemma 2.2, Lemma 2.3, and the Cauchy–Binet formula we have

$$H\begin{pmatrix} i_1, \ldots, i_p \\ j_1, \ldots, j_p \end{pmatrix} = \sum_{k_1 < \cdots < k_p} c_{i_1 k_1} \cdots c_{i_p k_p} H'\begin{pmatrix} k_1, \ldots, k_p \\ j_1, \ldots, j_p \end{pmatrix},$$

where $c_{i_k} > 0$, if $k = i + 1$ or $k = i =$ an even integer and zero otherwise. By induction, H' is totally positive and therefore so too is H. Also, by induction we have strict equality in (2.83) if and only if there exist integers $k_1, \ldots, k_p$ such that $k_1 < \cdots < k_p$, $0 \leq 2j_\ell - k_\ell \leq n - 1$, $\ell = 1, \ldots, p$ and for each $r = 1, \ldots, p$ either $k_r = i_r + 1$ or $k_r = i_r =$ an even integer. From this fact, (2.84) easily follows. □

Lemma 2.4 proves Proposition 2.2 in the case $r = 1$. To elaborate on the details, we relate the matrices A and H. Since $a_j = d_{n+1-j}$ we get

$$\begin{aligned} A_{ij} = a_{j-2i} &= d_{n+2i-j+1} \\ &= H_{n+1+j, n+1+i}, \end{aligned}$$

and so

$$A\binom{i_1,\ldots,i_p}{j_1,\ldots,j_p} = H\binom{n+1+j_1,\ldots,n+1+j_p}{n+1+i_1,\ldots,n+1+i_p} \geq 0$$

with strict inequality if and only if $0 \leq 2(n+1+i_\ell) - (n+1+j_\ell) \leq n+1$, which simplifies to $0 \leq j_\ell - 2i_\ell \leq n+1$, $\ell = 1,\ldots,p$.

The proof of Proposition 2.2 is finished by induction on r. The details follow next.

Proof of Proposition 2.2. By the Cauchy–Binet formula we have

$$A^{r+1}\binom{i_1,\ldots,i_p}{k_1,\ldots,k_p} = \sum_{j_1<\cdots<j_p} A\binom{i_1,\ldots,i_p}{j_1,\ldots,j_p} A^r\binom{j_1,\ldots,j_p}{k_1,\ldots,k_p}.$$

Thus, by induction on p, we see that all determinants of A^{r+1} are nonnegative and moreover, the determinant of A^{r+1} on the left-hand side of the above equation is positive if and only if there exist integers $j_1 < \cdots < j_p$ such that

$$0 \leq j_\ell - 2i_\ell \leq n+1, \qquad \ell = 1,\ldots,p$$

and

$$0 \leq k_\ell - 2^r j_\ell \leq (2^r - 1)(n+1), \qquad \ell = 1,\ldots,p.$$

Clearly, a necessary condition for the solution of these inequalities is that

$$0 \leq k_\ell - 2^{r+1} i_\ell \leq (2^{r+1} - 1)(n+1).$$

The fact that these inequalities are also sufficient for the existence of $j_1,\ldots,j_p$ satisfying the above inequalities would advance the induction hypothesis and prove Proposition 2.2. This result is proved next. $\square$

LEMMA 2.5. *Let* $r,p \in \mathbb{Z}_+\backslash\{0\}$ *and* $n \in \mathbb{Z}$. *Given integers* $k_1 < \cdots < k_p$ *and* $i_1 < \cdots < i_p$, *there exist integers* $j_1 < \cdots < j_p$ *such that*

(2.85) $0 \leq j_\ell - 2i_\ell \leq n+1, \qquad \ell = 1,\ldots,p$

and

(2.86) $0 \leq k_\ell - 2^r j_\ell \leq (2^r - 1)(n+1), \qquad \ell = 1,\ldots,p$

if and only if

(2.87) $0 \leq k_\ell - 2^{r+1} i_\ell \leq (2^{r+1} - 1)(n+1), \qquad \ell = 1,\ldots,p.$

Proof. It is clear that (2.85) and (2.86) imply (2.87). The converse is proved by induction on p. To this end, we assume that it holds for all positive integers smaller than p. Now choose j_1 to be the smallest integer satisfying the lower bounds on j_1 in both (2.85) and (2.86). Then successively choose j_ℓ, $\ell = 2,\ldots,p$, to be the smallest integer, not only satisfying the lower bound (2.85) and upper bound (2.86), but also greater than $j_{\ell-1}$. We know by induction that there are integers $t_1 < \cdots < t_{p-1}$ that satisfy

$$(2.88) \qquad\qquad 0 \le t_\ell - 2i_\ell \le n+1, \qquad \ell = 1, \ldots, p-1$$

and

$$(2.89) \qquad 0 \le k_\ell - 2^r t_\ell \le (2^r - 1)(n+1), \qquad \ell = 1, \ldots, p-1.$$

Hence by our choice of $j_1, \ldots, j_p$ one obtains $j_\ell \le t_\ell$, $\ell = 1, \ldots, p-1$ and consequently, it suffices to prove that

$$(2.90) \qquad\qquad\qquad j_p \le 2i_p + n + 1$$

and

$$(2.91) \qquad\qquad\qquad j_p \le 2^{-r} k_p$$

to advance the induction hypothesis.

We first observe, without loss of generality, that we can assume that $j_1, \ldots, j_p$ are consecutive integers. Suppose to the contrary that for some $m, 1 \le m \le p-1$, we have $j_{m+1} \ge j_m + 2$. Then j_{m+1} is necessarily the minimum integer that satisfies the inequalities

$$0 \le j_{m+1} - 2i_{m+1} \le n+1$$

and

$$0 \le k_{m+1} - 2^r j_{m+1} \le (2^r - 1)(n+1).$$

Hence we may argue as above and conclude by the induction hypothesis that j_ℓ, $\ell = m+1, \ldots, p$, would satisfy (2.85) and (2.86). In particular, inequalities (2.90) and (2.91) would follow and the proof would end. Thus we need only consider the case that

$$(2.92) \qquad\qquad j_\ell = j_1 + \ell - 1, \qquad \ell = 1, \ldots, p.$$

Let us now check the validity of (2.90). For $p \ge 2$, we have

$$j_p = j_{p-1} + 1 \le 2i_{p-1} + n + 2 \le 2i_p + n + 1,$$

as required. When $p = 1$ we consider, in turn, two possibilities: either $j_1 - 1 < 2i_1$ or $2^r(j_1 - 1) < k_1 - (2^r - 1)(n+1)$. Both inequalities cannot fail, by the definition of j_1. Now, in the first instance we have that $j_1 \le 2i_1 \le 2i_1 + n + 1$, as desired. In the second case, we have by (2.87)

$$2^r(j_1 - 1) + 1 \le k_1 - (2^r - 1)(n+1) \le 2^{r+1}i_1 + 2^r(n+1).$$

Simplifying this inequality gives $j_1 \le 1 - 2^{-r} + 2i_1 + n + 1$, which again proves (2.90).

There remains the proof of (2.91), which we verify by contradiction. Thus we suppose to the contrary that

$$(2.93) \qquad\qquad j_p > 2^{-r} k_p.$$

Using (2.92) and (2.85), inequality (2.93) gives $j_1 + p - 1 > 2i_p \geq 2i_1 + 2p - 2$ or, in other words, $j_1 - 1 \geq 2i_1$. Thus by our choice of j_1, we conclude that

$$(2.94) \qquad\qquad 2^r(j_1 - 1) < k_1 - (2^r - 1)(n + 1).$$

This inequality, with the lower bound on j_p in (2.85) and (2.87), gives the inequalities

$$
\begin{aligned}
2^r(j_1 + p - 1) \;=\;& 2^r j_p \geq 2^{r+1} i_p \\
\geq\;& 2^{r+1}(p - 1) + k_1 - (2^{r+1} - 1)(n + 1) \\
\geq\;& 2^{r+1}(p - 1) + 2^r(j_1 - 1) + 1 + (2^r - 1)(n + 1) \\
& - (2^{r+1} - 1)(n + 1) \\
\geq\;& 2^r(j_1 - 1) + 2^{r+1}(p - 1) - 2^r(n + 1) + 1,
\end{aligned}
$$

and so $p \leq n + 2$. Using (2.94) and the fact that $k_1 \leq k_p - p + 1$, we obtain

$$
\begin{aligned}
2^r j_p = 2^r(j_1 + p - 1) \;\leq\;& 2^r p - 1 + k_p - p + 1 - (2^r - 1)(n + 1) \\
=\;& k_p + (2^r - 1)(p - n - 1),
\end{aligned}
$$

which together with (2.93) implies that $p \geq n + 2$. Hence, we conclude that $p = n + 2$. Next, we choose integers s, t such that

$$k_1 + n = 2^{r+1} s + t$$

and $0 \leq t < 2^{r+1}$. Using (2.87) again, we have

$$2^{r+1} i_1 \geq 2^{r+1} s + t - n - (2^{r+1} - 1)(n + 1),$$

which gives $i_1 + n + 1 \geq s + 2^{-r-1}(t + 1) > s$ and so $i_1 + n \geq s$. Also, by (2.87) we get $2^{-r-1} k_p \geq i_p \geq i_1 + p - 1 = i_1 + n + 1$, which implies that

$$(2.95) \qquad\qquad 2^{-r-1} k_p \geq s + 1.$$

Furthermore, from inequality (2.94) we obtain

$$
\begin{aligned}
2^r(j_1 - 1) \;<\;& 2^{r+1} s + t - n - (2^r - 1)(n + 1) \\
=\;& 2^{r+1} s + t + 1 - 2^r(n + 1),
\end{aligned}
$$

or, equivalently,

$$j_1 - 1 < 2s + 2^{-r}(t + 1) - n - 1 \leq 2s + 1 - n.$$

Therefore (2.95) implies that

$$j_1 \leq 2s + 1 - n \leq 2^{-r} k_p - 1 - n,$$

which means that $j_p = j_1 + p - 1 = j_1 + n + 1 \leq 2^{-r}k_p$. This contradicts (2.93) and proves the result. □

We now return to the issue of smoothness for the function ϕ constructed in Theorem 2.6. There is a simple criterion available to determine how many continuous derivatives ϕ has on $\mathbb{R}$.

THEOREM 2.7. *Let $a = (a_j : j \in \mathbb{Z})$ be a mask that satisfies the hypothesis of Theorem 2.6. Then $\phi \in C^k(\mathbb{R})$ for some $k \geq 1$, if and only if $a(z)$ has a zero of order $k + 1$ at $z = -1$.*

We base the proof of Theorem 2.7 on two facts from Cavaretta, Dahmen, and Micchelli [CDM].

PROPOSITION 2.3. *Let $a = (a_j : j \in \mathbb{Z})$ be a mask such that $|\{j : a_j \neq 0\}| < \infty$. If there exists a nontrivial continuous function f and a vector $c = (c_j : j \in \mathbb{Z})$ such that*

$$\lim_{r \to \infty} \sup_{j \in \mathbb{Z}} \left| (S_a^r c)_j - f(j/2^r) \right| = 0,$$

then $a(-1) = 0$ and $a(1) = 2$.

Proof. We assume without loss of generality that $a_j = 0$ for $j < 0$ and $j > n$. Our hypothesis and Theorem 2.4 imply that the MSS determined by the matrices (2.35) converges. Hence by (1.39) of Theorem 1.2 we have $A_\epsilon \mathbf{e} = \mathbf{e}$, $\epsilon \in \{0, 1\}$, which is easily seen to be equivalent to the desired conclusion. □

We also need the following proposition.

PROPOSITION 2.4. *Let $a = (a_j : j \in \mathbb{Z})$ be a mask such that $|\{j : a_j \neq 0\}| < \infty$. If S_a converges in the sense of (2.34), and has an associated refinable function ϕ then the SSS with the mask $b = (b_j : j \in \mathbb{Z})$ defined by*

$$b(z) = 2^{-1}(1 + z^{-1})a(z)$$

likewise converges and has a refinable function ψ given by

$$\psi(x) = \int_x^{x+1} \phi(t)dt, \qquad x \in \mathbb{R}.$$

Proof. Our first observation is the formula

(2.96) $$\triangle S_b = 2^{-1} S_a \triangle,$$

where $\triangle$ is the forward difference operator defined earlier in Theorem 2.2. This equation means that

$$b_{i+1} - b_i = 2^{-1}(a_{i+2} - a_i), \qquad i \in \mathbb{Z}.$$

Equivalently, multiplying both sides of this equation by z^i, $z \in \mathbb{C}\backslash\{0\}$, and summing over $i \in \mathbb{Z}$ gives

$$(z^{-1} - 1)b(z) = 2^{-1}(z^{-2} - 1)a(z),$$

which simplifies to the definition of $b(z)$ given in Proposition 2.4. Using formula (2.96) repeatedly yields the equation

(2.97) $$\triangle S_b^r = 2^{-r} S_a^r \triangle, \qquad r \in \mathbb{Z}_+\backslash\{0\}.$$

We also need to recall from (2.52) that there is an integer $q > 0$ such that for all $r = 1, 2, \ldots$

$$(S_b^r \delta)_j = (S_a^r \delta)_j = 0, \qquad |j| > q2^r,$$

and so

$$\phi(x) = 0, \quad \text{if } |x| > q.$$

Since S_a converges we conclude that for any $\epsilon > 0$ there is an m such that

$$\left| (S_a^r \triangle \delta)_j - \left(\phi \left(\frac{j}{2^r} + 1 \right) - \phi \left(\frac{j}{2^r} \right) \right) \right| \le \epsilon$$

for $r \ge m$ and $j \in \mathbb{Z}$.

For any $k \in \mathbb{Z}$, replace j by $k + j$ above and sum the left-hand side of the inequality above over all $j \le 0$ for which it is nonzero. There are at most $q'2^r$ such integers, for some constant q', independent of r. Hence we get from (2.97) that

$$\left| 2^r (S_b^r \delta)_{k+1} - \sum_{j=-\infty}^{0} \left(\phi \left(\frac{k}{2^r} + 1 + \frac{j}{2^r} \right) - \phi \left(\frac{k}{2^r} + \frac{j}{2^r} \right) \right) \right| \le q' 2^r \epsilon.$$

We also observe that

$$\left| 2^{-r} \sum_{j=-\infty}^{0} \phi \left(x + \frac{j}{2^r} \right) - \int_{-\infty}^{0} \phi(x + t)dt \right|$$

$$\le \sum_{j=-\infty}^{0} \int_{2^{-r}(j-1)}^{2^{-r}j} \left| \phi \left(x + \frac{j}{2^r} \right) - \phi(x + t) \right| dt.$$

If j gives a nonzero summand above then either $|x + j/2^r| \le q$ or $|x + t| \le q$ for some $t \in [(j-1)/2^r, j/2^r]$. In either case we get

$$|2^r x + j| \le 2^r(q + 1),$$

and so independent of x, there are at most $2q + 3$ nonzero summands. Consequently, we conclude that

$$\left| 2^{-r} \sum_{j=-\infty}^{0} \phi \left(x + \frac{j}{2^r} \right) - \int_{-\infty}^{0} \phi(x + t)dt \right|$$

$$\le \frac{1}{2^r}(2q + 3)\omega \left(\phi; \frac{1}{2^r} \right).$$

Combining these inequalities we obtain for all $k \in \mathbb{Z}$ the inequality

$$\left| (S_b^r \delta)_k - \psi \left(\frac{k}{2^r} \right) \right| \le q'\epsilon + \frac{3}{2^r}(2q + 3)\omega \left(\phi; \frac{1}{2^r} \right).$$

Clearly the upper bound goes to zero as $r \to \infty$. Moreover, for any $c \in \ell^\infty(\mathbb{Z})$ with the function g defined as

$$g(x) = \sum_{j \in \mathbb{Z}} c_j \psi(x - j), \qquad x \in \mathbb{R},$$

we have

$$|(S_b^r c)_j - g(\tfrac{j}{2^r})|$$

$$\leq \|c\|_\infty \sum_{\ell \in \mathbb{Z}} \left| (S_b^r \delta)_{j - 2^r \ell} - \psi\left(\frac{j}{2^r} - \ell\right) \right|.$$

Again, the number of nonzero summands in the sum above is $\leq 2q + 1$ (independent of r) and so the above quantities go to zero as $r \to \infty$. $\quad\square$

We now have at hand the necessary facts to prove Theorem 2.7.

Proof of Theorem 2.7. First, we suppose that $a(z)$ can be factored in the form

$$a(z) = 2^{-k}(1 + z^{-1})^k b(z),$$

where $b(1) = 2$ and $b(-1) = 0$. We define Laurent polynomials $b_0(z)$, $b_1(z), \ldots, b_k(z)$ recursively by setting

$$b_0(z) = b(z)$$

and

$$b_{\ell+1}(z) = 2^{-1}(1 + z^{-1})b_\ell(z), \qquad \ell = 0, 1, \ldots, k - 1.$$

Then each $b_\ell(z)$ satisfies the hypothesis of Theorem 2.6. Therefore, the corresponding subdivision scheme S_{b_ℓ} converges. Let ϕ_ℓ be the associated refinable function. Then from Proposition 2.4 we get

$$\phi_{\ell+1}(x) = \int_x^{x+1} \phi_\ell(t)\,dt, \qquad \ell = 0, 1, \ldots, k - 1.$$

Since $a(z) = b_k(z)$, we have $\phi = \phi_k$ and from the above equation we see that $\phi \in C^k(\mathbb{R})$.

Conversely, when $\phi \in C^k(\mathbb{R})$, we use Proposition 2.3 and factor $a(z)$ in the form

$$a(z) = 2^{-1}(1 + z^{-1})b(z),$$

where b is a Laurent polynomial. Since the zeros of $b(z)$ are certainly in the left half-plane, we conclude from Theorem 2.6 and Proposition 2.4 that S_b converges and has a refinable function in $C^{k-1}(\mathbb{R})$. If $k > 1$ we apply the same procedure to b and inductively arrive at the desired conclusion, by induction on k. $\quad\square$

2.6.   Wavelet decomposition.

Although it is not central to the theme of this chapter we describe an application of Theorem 2.6 to *wavelet decomposition*. Unfortunately, we must refrain from describing the background motivation for our remarks because this would take us too far afield. For an excellent account of this important subject we refer to Chui [C] and Daubechies [D]. We call the function ϕ, constructed in Theorem 2.6, a *ripplet*. Ripplets are therefore created from Hurwitz polynomials. We will follow Micchelli [Mi] and show how easily ripplets lead us to *wavelets*.

To this end, we introduce the sequence $\mu = (\mu_j : j \in \mathbb{Z})$ defined by the equation

$$\mu_j = \int_{\mathbb{R}} \phi(x)\phi(2x + j)dx, \qquad j \in \mathbb{Z},$$

where ϕ and $a = (a_j : j \in \mathbb{Z})$ are as in Theorem 2.6 and assume, without loss of generality, that $\ell = 0$ and $m = n + 1$. Since ϕ is positive on $(0, n + 1)$ and zero otherwise, μ_j is positive for $j \in \{-2n - 1, \ldots, n\}$ and zero elsewhere. The vector μ leads us to the function

$$\psi(x) := \sum_{j \in \mathbb{Z}} (-1)^j \mu_{j-1} \phi(2x - j).$$

It is the study of this function ψ that will occupy us for the remainder of the chapter. Our main goal is to show that the family of functions

$$\psi_{k,j}(z) := \psi(2^k x - j), \qquad k, \, j \in \mathbb{Z}$$

provides an orthogonal decomposition of $L^2(\mathbb{R})$. Moreover, every $f \in L^2(\mathbb{R})$ can be efficiently represented in terms of these functions.

We begin our discussion of the function ψ by observing that $\psi(x)$ is zero for $x \notin (-n, n + 1)$. To describe additional properties of ψ, we need to accumulate some facts about ϕ. For this purpose, we introduce for every $k \in \mathbb{Z}$ the space

$$(2.98) \quad V_k = V_k(\phi) = \left\{ \sum_{j \in \mathbb{Z}} c_j \phi(2^k \cdot -j) : c = (c_j : j \in \mathbb{Z}) \in \ell^2(\mathbb{Z}) \right\}.$$

Each V_k is a closed subspace of $L^2(\mathbb{R})$. For the proof of this fact, we note that since $\phi(x), \ldots,$
$\phi(x + n)$ are linearly independent on $[0, 1]$, it follows that there are constants $K, J > 0$ such that

$$K \sum_{j \in \mathbb{Z}} c_j^2 \leq \int_{-\infty}^{\infty} \left| \sum_{j \in \mathbb{Z}} c_j \phi(x - j) \right|^2 dx \leq J \sum_{j \in \mathbb{Z}} c_j^2.$$

From this inequality it easily follows that V_k is a closed subspace of $L^2(\mathbb{R})$.

In view of the refinement equation (2.58) satisfied by ϕ, the spaces (2.98) are nested, viz.

(2.99) $$V_k \subseteq V_{k+1}, \qquad k \in \mathbb{Z}.$$

Moreover, we claim that

(2.100) $$\bigcap_{k \in \mathbb{Z}} V_k = \{0\}, \qquad \overline{\bigcup_{k \in \mathbb{Z}} V_k} = L^2(\mathbb{R}).$$

To prove the first part of (2.100), we let $g \in V_k$, for all $k \in \mathbb{Z}$. Thus, for every $k \in \mathbb{Z}$, there is a $d = (d_j : j \in \mathbb{Z}) \in \ell^2(\mathbb{Z})$ such that

$$g(x) = \sum_{j \in \mathbb{Z}} d_j \phi(2^k x - j), \qquad x \in \mathbb{R}.$$

Let

$$R := \sup \left\{ \sum_{j \in \mathbb{Z}} |\phi(x - j)|^2 : x \in [0, 1] \right\} < \infty$$

and use the Cauchy–Schwarz inequality to get

$$|g(x)| \leq R \sum_{j \in \mathbb{Z}} d_j^2$$

$$\leq R \, K^{-1} 2^k \int_{\mathbb{R}} |g(t)|^2 dt,$$

so that letting $k \to -\infty$ proves $g = 0$.

The second claim in (2.100) follows by noting that since ϕ is continuous, of compact support, and satisfies (2.59), it follows that whenever f is likewise continuous and of compact support the sequence of functions

$$g_k(x) = \sum_{j \in \mathbb{Z}} f\left(\frac{j}{2^k}\right) \phi(2^k x - j)$$

are also of compact support and converge uniformly to f. Since continuous functions of compact support are dense in $L^2(\mathbb{R})$, the result follows.

Next we let W_k be the orthogonal complement of V_k in V_{k+1}. Two important properties of W_k, $k \in \mathbb{Z}$, follow from equations (2.99) and (2.100), namely,

$$W_k \perp W_{k'}, \qquad k \neq k',$$

and

$$\oplus_{k \in \mathbb{Z}} W_k = L^2(\mathbb{R}).$$

For instance, to prove the first fact we observe that if $k < k'$ then

$$W_k \subseteq V_{k+1} \subseteq V_{k'} \perp W_{k'}.$$

For the second claim we need only show that there is no nontrivial function $f \in L^2(\mathbb{R})$ orthogonal to all $W_k, k \in \mathbb{Z}$. To this end, let $f \in L^2(\mathbb{R})$ be orthogonal to all W_k, $k \in \mathbb{Z}$. Pick $\epsilon > 0$, and choose a $v_{k_0} \in V_{k_0}$ such that

$$\|f - v_{k_0}\| \le \epsilon,$$

where $\| \cdot \|$ is the standard norm on $L^2(\mathbb{R})$,

$$\|f\|^2 = \int_{\mathbb{R}} |f(t)|^2 dt.$$

Hence we conclude that $\|f\|^2 - 2(f, v_{k_0}) \le \epsilon^2$. Next, we write $v_{k_0} = v_{k_0-1} + w_{k_0-1}$, where $v_{k_0-1} \in V_{k_0-1}$ and $w_{k_0-1} \in W_{k_0-1}$. Then

$$\|v_{k_0-1}\| \le \|v_{k_0}\|$$

and

$$\|f\|^2 - 2(f, v_{k_0-1}) \le \epsilon^2.$$

In this way for every integer $k \le k_0$ there is a $v_k \in V_k$ such that

$$\|v_k\| \le \|v_{k_0}\|$$

and

$$\|f\|^2 - 2(f, \ v_k) \le \epsilon^2.$$

Using equation (2.100), and the bound on v_k above, we see that $\lim_{k \to \infty} v_k = 0$, weakly in $L^2(\mathbb{R})$. This implies, by the inequality above, that $\|f\| \le \epsilon$. Since ϵ is arbitrary, it follows that $f = 0$.

For later use, we need to record some facts about the bi-infinite sequences μ defined earlier. For this purpose, we use the following notation. Given any Laurent polynomial

$$y(z) = \sum_{j \in \mathbb{Z}} y_j z^j, \qquad z \in \mathbb{C} \backslash \{0\},$$

we split it into its "even"

$$y_0(z) = \sum_{j \in \mathbb{Z}} y_{2j} z^j$$

and "odd"

$$y_1(z) = \sum_{j \in \mathbb{Z}} y_{2j+1} z^j$$

parts, so that

$$y(z) = y_0(z^2) + z y_1(z^2).$$

We also need the sequence $g = (g_j : j \in \mathbb{Z})$ defined by

$$g_j = \int_{\mathbb{R}} \phi(x)\phi(x-j)dx, \qquad j \in \mathbb{Z}.$$

Using the refinement equation for ϕ we obtain the formula

$$g(z) = a_0(z)\mu_0(z) + za_1(z)\mu_1(z), \qquad z \in \mathbb{C}\backslash\{0\}.$$

Also, referring back to the definition of μ and again using the refinement equation for ϕ likewise provide the companion formula

(2.101) $$\mu(z) = \frac{1}{2}a(z^{-1})g(z), \qquad z \in \mathbb{C}\backslash\{0\}.$$

PROPOSITION 2.5. *Let* $a = (a_j : j \in \mathbb{Z})$ *satisfy the hypothesis of Theorem 2.6. Then*

(2.102) $$g(e^{i\theta}) > 0, \qquad \theta \in \mathbb{R}$$

and $g(z)$ *has only negative zeros on* $\mathbb{C}\backslash\{0\}$. *Furthermore,* $\mu_0(z)$ *and* $\mu_1(z)$ *have no common zeros for* $z \in \mathbb{C}\backslash\{0\}$.

Proof. First we note by the Cauchy–Binet formula and Theorem 2.6 that the bi-infinite matrix $(g_{i-j})_{i,j\in\mathbb{Z}}$ is totally positive. Therefore, by a result of Aissen, Schoenberg, and Whitney [ASW], we conclude that $g(z)$ only has negative zeros. To show that the inequality (2.102) holds, it suffices to demonstrate that $g(e^{i\theta}) \neq 0$ for $\theta \in \mathbb{R}$, since $g(e^{i\theta})$ is real for all $\theta \in \mathbb{R}$ and also positive for $\theta = 0$. Suppose to the contrary that $g(e^{i\theta}) = 0$ for some $\theta \in \mathbb{R}$. Then for all $k \in \mathbb{Z}$

$$0 = \int_{\mathbb{R}} \sum_{j\in\mathbb{Z}} e^{ij\theta}\phi(x-j)\phi(x-k)dx.$$

Multiplying both sides of this equation by $e^{-ik\theta}$ and summing over $k \in \mathbb{Z}$ gives the formula

$$0 = \int_{\mathbb{R}} \left| \sum_{j\in\mathbb{Z}} e^{ij\theta}\phi(x-j) \right|^2 dx.$$

Therefore for $x \in [0,1]$,

$$\sum_{j=-n}^{0} e^{ij\theta}\phi(x-j) = 0,$$

which contradicts the linear independence of the functions $\phi(x), \ldots, \phi(x+n)$ on $[0,1]$.

Finally we show that $\mu_0(z)$, $\mu_1(z)$ have *no common zeros*. Suppose to the contrary that $\mu_0(z) = \mu_1(z) = 0$ for some $z \in \mathbb{C}\backslash\{0\}$. According to (2.101), we have

$$0 = 2\mu_0(z) = a_0(z^{-1})g_0(z) + a_1(z^{-1})g_1(z)$$

and

$$0 = 2\mu_1(z) = a_0(z^{-1})g_1(z) + z^{-1}a_1(z^{-1})g_0(z).$$

We let $\tau^2 := z$ and observe that the determinant of this linear system (in $a_0(z^{-1})$ and $a_1(z^{-1})$) is

$$\begin{aligned}
\triangle &:= z^{-1}g_0^2(z) - g_1^2(z)\\
&= \tau^{-2}g(\tau)g(-\tau).
\end{aligned}$$

Since $g(z)$ has only negative zeros and the vector g has only nonnegative components, it follows that $\triangle \neq 0$. Therefore, we conclude that $a_0(z^{-1}) = a_1(z^{-1}) = 0$ or, equivalently, $a(z^{-1}) = a(-z^{-1}) = 0$. This is a contradiction, since $a(z)$ is a Hurwitz polynomial.   □

We now arrive at the main property of ψ that we wish to describe here.

THEOREM 2.8. *Let $a = (a_j : j \in \mathbb{Z})$ satisfy Theorem 2.6. Then*

$$W_k = V_k(\psi), \qquad k \in \mathbb{Z},$$

and there exist two sequences $b = (b_j : j \in \mathbb{Z})$, $c = (c_j : j \in \mathbb{Z})$ that decay exponentially such that

(2.103) $$\phi(2x - n) = \sum_{j\in\mathbb{Z}}b_{n-2j}\phi(x-j) + \sum_{j\in\mathbb{Z}}c_{n-2j}\psi(x-j)$$

for all $n \in \mathbb{Z}$ and $x \in \mathbb{R}$. Moreover, ψ is the function of minimal support in V_1 that is orthogonal to V_0.

We remark that the orthogonality of the spaces W_k, $k \in \mathbb{Z}$ imply that for any $j, j, k, k' \in \mathbb{Z}$, with $k \neq k'$

$$\int_{\mathbb{R}} \psi\left(2^k x - j\right)\psi\left(2^{k'} x - j'\right)dx = 0.$$

Thus ψ is what is commonly called a *pre-wavelet*.

Proof. First, let us show that $\psi \perp V_0$. To this end, we compute the inner product

$$\int_{\mathbb{R}}\phi(x-j)\psi(x)dx = \sum_{n\in\mathbb{Z}}(-1)^n\mu_{n-1}\int_{\mathbb{R}}\phi(x-j)\phi(2x-n)dx$$

$$= \sum_{n\in\mathbb{Z}}(-1)^n\mu_{n-1}\mu_{2j-n}$$

$$= \sum_{n\in\mathbb{Z}}\mu_{2n-1}\mu_{2j-2n} - \sum_{n\in\mathbb{Z}}\mu_{2n}\mu_{2j-2n-1}$$

$$= 0.$$

Next, using the refinement equation for ϕ and the definition of ψ, we see that to prove (2.103), it suffices to show

$$(2.104) \qquad \sum_{j\in\mathbb{Z}} b_{n-2j} a_{\ell-2j} + \sum_{j\in\mathbb{Z}} c_{n-2j} d_{\ell-2j} = \delta_{n\ell}, \qquad n,\ell \in \mathbb{Z},$$

where we have set

$$d_j := (-1)^j \mu_{j-1}, \qquad j \in \mathbb{Z}.$$

The solution of equation (2.104) is facilitated by the following lemma.

LEMMA 2.6. *Let $a, b, c, d \in \ell^1(\mathbb{Z})$ (absolutely summable sequences) satisfy* (2.104). *Then*

$$(2.105) \quad \Delta(z) := a_0(z) d_1(z) - d_0(z) a_1(z) \neq 0, \qquad z \in D := unit\ disc,$$

$$(2.106) \qquad\qquad b(z) = -z \frac{d(-z^{-1})}{\Delta(z^{-1})}, \qquad z \in D,$$

and

$$(2.107) \qquad\qquad c(z) = z \frac{a(-z^{-1})}{\Delta(z^{-1})}, \qquad z \in D.$$

Conversely, if (2.105) *holds for $a, d \in \ell^1(\mathbb{Z})$ then b, c given by* (2.106), (2.107) *are likewise in $\ell^1(\mathbb{Z})$ and satisfy* (2.104).

Proof. Let $z \in D$, multiply both sides of (2.104) by z^n, and sum over $n \in \mathbb{Z}$. This gives us the equivalent system of equations

$$(2.108) \quad z^\ell = b(z) \left(\sum_{j\in\mathbb{Z}} a_{\ell-2j} z^{2j} \right) + c(z) \left(\sum_{j\in\mathbb{Z}} d_{\ell-2j} z^{2j} \right), \qquad \ell \in \mathbb{Z}.$$

Let $e \in \{0, 1\}$, and write ℓ in the form $\ell = 2r + e$. Then (2.108) simplifies to the matrix equation

$$\begin{pmatrix} b_0(z) & c_0(z) \\ b_1(z) & c_1(z) \end{pmatrix} \begin{pmatrix} a_0(z^{-1}) & a_1(z^{-1}) \\ d_0(z^{-1}) & d_1(z^{-1}) \end{pmatrix} = \begin{pmatrix} 1 & 0 \\ 0 & 1 \end{pmatrix}.$$

Thus we conclude that $\Delta(z) \neq 0$ and also by Cramer's rule we obtain

$$\begin{pmatrix} b_0(z) & c_0(z) \\ b_1(z) & c_1(z) \end{pmatrix} = \frac{1}{\Delta(z^{-1})} \begin{pmatrix} d_1(z^{-1}) & -a_1(z^{-1}) \\ -d_0(z^{-1}) & a_0(z^{-1}) \end{pmatrix}.$$

This formula proves (2.106) and (2.107).

The converse statement of the lemma is obtained by reversing the above steps, using (2.105) and Wiener's lemma, cf. Rudin [R, p. 266], to ensure that b, c given by (2.106), (2.107) are in $\ell^1(\mathbb{Z})$. $\square$

Returning to the proof of Theorem 2.8, we verify that for the choice of d given above that $\triangle \neq 0$. In fact, $d_0(z) = z\mu_1(z), d_1(z) = -\mu_0(z)$ and thus

$$
\begin{aligned}
\triangle(z) &= a_0(z)d_1(z) - d_0(z)a_1(z) \\
&= -a_0(z)\mu_0(z) - za_1(z)\mu_1(z) \\
&= -g(z).
\end{aligned}
$$

Therefore, in view of Proposition 2.5, we conclude that indeed $\triangle(z) \neq 0$ for $z \in D$.

To finish the proof, we shall establish that ψ is the function of minimal support in V_1 which is orthogonal to V_0. By this we mean the following. If $h \in V_1$ is orthogonal to V_0 and vanishes outside of $(-n, m)$ where m is some number less than $n + 1$, then $h = 0$.

Using the linear independence of the integer translates of ϕ, we may as well assume m is an integer and that for some sequence $d = (d_j : j \in \mathbb{Z})$ with $d_j = 0$, whenever $j \notin \{-2n - 1, \ldots, m - 1\}$,

$$
h(x) = \sum_{j \in \mathbb{Z}} d_j \phi(2x - j), \qquad x \in \mathbb{R}.
$$

Now, since $h \perp V_0$, we conclude that

$$
\mu_0(z)d_0(z) + z\mu_1(z)d_1(z) = 0, \qquad z \in \mathbb{C}\backslash\{0\}.
$$

Proposition 2.5 states that $\mu_0(z), \mu_1(z)$ have no common zeros in $\mathbb{C}\backslash\{0\}$. Therefore there exist Laurent polynomials f and r such that

$$
\mu_0(z)r(z) + z\mu_1(z)f(z) = 1, \qquad z \in \mathbb{C}\backslash\{0\},
$$

cf. Walker, [W, Thm. 9.6, p. 25]. Define $q := d_0 f - d_1 r$ and observe that

$$
\begin{pmatrix} \mu_0(z) & z\mu_1(z) \\ f(z) & -r(z) \end{pmatrix} \begin{pmatrix} d_0(z) \\ d_1(z) \end{pmatrix} = \begin{pmatrix} 0 \\ q(z) \end{pmatrix},
$$

or, equivalently, $d(z) = -z\mu(-z)q(z^2)$.

We wish to show that $q(z) = 0$, for all $z \in \mathbb{C}$. Suppose to the contrary that $q \neq 0$. Then, for some integers, $\ell \leq s$, and constants $q_\ell, \ldots, q_s$ we have

$$
q(z) = \sum_{j=\ell}^{s} q_j z^j,
$$

with $q_\ell q_s \neq 0$. This implies that

$$
d(z) = \sum_{-2n-1+2\ell}^{n+2s+1} d_j z^j
$$

with $d_{-2n-1+2\ell} d_{n+2s+1} \neq 0$. However, $d_j = 0$ whenever $j \notin \{-2n - 1, \ldots, m - 1\}$, which leads us to the contradiction that $\ell \geq 0$ and $s \leq -1$. □

A disadvantage of the representation (2.103) is that both sequences b, c are infinitely supported, albeit decaying exponentially, even though both ϕ and ψ are of compact support. The next theorem provides a *nonorthogonal* decomposition of V_1 through V_0 that remedies this difficulty.

THEOREM 2.9. *Let* $a = (a_j : j \in \mathbb{Z})$ *satisfy Theorem 2.6. There exist a function* $\theta \in V_1$, supp $\theta = (0, 2n + 1)$ *and sequences of finite support* $b = (b_j : j \in \mathbb{Z})$, $c = (c_j : j \in \mathbb{Z})$ *such that*

$$\phi(2x - n) = \sum_{j \in \mathbb{Z}} b_{n-2j} \phi(x - j) + \sum_{j \in \mathbb{Z}} c_{n-2j} \theta(x - j)$$

for all $n \in \mathbb{Z}$ *and* $x \in \mathbb{R}$.

Proof. As we already pointed out in the proof of Proposition 2.5, the polynomials $a_0(z), a_1(z)$ have no common zeros for $z \in \mathbb{C}$. Therefore there exist polynomials $d_0(z), d_1(z)$ such that

$$\triangle(z) = a_0(z)d_1(z) - d_0(z)a_1(z) = 1, \qquad z \in \mathbb{C}.$$

Consequently, the function

$$\theta(x) = \sum_{j \in \mathbb{Z}} d_j \phi(2x - j), \qquad x \in \mathbb{R},$$

satisfies the requirements of the theorem. $\quad\square$

An advantage of the decomposition in Theorem 2.6 is that every function of finite support in V_1 is a unique linear combination of functions of finite support from $V(\phi)$ and $V(\theta)$.

References

[ASW]   A. AISSEN, I. SCHOENBERG, AND A. WHITNEY, *On the generating functions of totally positive sequences*, I, J. d'Anal. Math., 2 (1952), 93–103.

[C]   C. K. CHUI, *An Introduction to Wavelets*, Academic Press, Boston, 1992.

[CDM]   A. S. CAVARETTA, W. DAHMEN AND C. A. MICCHELLI, *Stationary subdivision*, Mem. Amer. Math. Soc., 93, #453, 1991.

[Ch]   G. M. CHAIKIN, *An algorithm for high speed curve generation*, Computer Graphics and Image Processing, 3 (1974), 346–349.

[D]   I. DAUBECHIES, *Ten Lectures on Wavelets*, SIAM, Philadelphia, 1992.

[deR]   G. DE RHAM, *Sur une courbe plane*, J. Math. Pures Appl., 39 (1956), 25–42.

[G]   F. R. GANTMACHER, *The Theory of Matrices*, Vol. 2, Chelsea Publishing Company, New York, 1960.

[GM]   T. N. T. GOODMAN AND C. A. MICCHELLI, *On refinement equations determined by Pólya frequency sequences*, SIAM J. Math. Anal., 23 (1992), 766–784.

[Ke]   J. H. B. KEMPERMAN, *A Hurwitz matrix is totally positive*, SIAM J. Math. Anal., 13 (1982), 331–341.

[LR]   J. M. LANE AND R. F. RIESENFELD, *A theoretical development for the computer generation of piecewise polynomial surfaces*, IEEE Transactions on Pattern Analysis and Machine Intelligence, 2 (1980), 34–46.

[Mi]   C. A. MICCHELLI, *Using the refinement equation for the construction of prewavelets*, Numerical Algorithms, 1 (1991), 75–116.

[MPi]   C. A. MICCHELLI AND A. PINKUS, *Descartes systems from corner cutting*, Constr. Approx., 7 (1991), 161–194.

[MPr]   C. A. MICCHELLI AND H. PRAUTZSCH, *Refinement and subdivision for spaces of integer translates of a compactly supported function*, in Numerical Analysis, eds., D. F. Griffiths and G. A. Watson, eds., Pitman Research Notes in Mathematical Series, Longman Scientific and Technical, Harlow, UK, 1987, 192–222.

[R]   W. RUDIN, *Functional Analysis*, McGraw-Hill Book Company, New York, 1993.

[Ri]   R. F. RIESENFELD, *On Chaikin's algorithm*, Computer Graphics and Image Processing, 4 (1975), 304–310.

[Sc]   I. J. SCHOENBERG, *Cardinal Spline Interpolation*, SIAM, Philadelphia, 1973.

[Sch]   L. L. SCHUMAKER, *Spline Functions: Basic Theory*, John Wiley and Sons, New York, 1981.

[W]   R. J. WALKER, *Algebraic Curves*, Princeton University Press, Princeton, N.J., 1950.

CHAPTER 3

Piecewise Polynomial Curves

3.0. Introduction.

Basic concepts from the differential geometry of curves are reviewed. These notions give us a means to develop various methods to join together the segments of piecewise polynomial curves that have geometric significance for the composite curve. We demonstrate that these methods can be conveniently described in terms of (connection) matrices with easily identifiable properties.

To create curves with these desirable properties we study spaces of piecewise polynomials specified by connection matrices and knots. We show that totally positive connection matrices lead to piecewise polynomial spaces with B-spline bases that are variation diminishing. Several algebraic properties of these B-spline bases are obtained, including polynomial identities, dual functionals, and knot insertion formulas.

3.1. Geometric continuity: Reparameterization matrices.

The two previous chapters contain methods useful for the construction of curves by iterative methods. In contrast, in this chapter we focus on modeling curves by a finite number of distinct polynomial curve segments. The distinction between smoothness of a curve in a given parameterization and its *visual* or geometric smoothness motivates the point of view taken here.

It was pointed out early by those interested in geometric modeling, first by Sabin [S], and then by Manning [M], that when modeling a continuous planar curve $\mathbf{x}(t) = (x(t), y(t))$ by cubic splines, each component of $\mathbf{x}(t)$ need not be twice continuously differentiable in its parameterization to obtain a visually pleasing curve. For instance, at a knot of $\mathbf{x}(t)$, say, $t = a$, for $\mathbf{x}$ to have a continuous unit tangent requires only that there be a positive constant m_1 such that

$$\mathbf{x}'(a^+) = m_1 \mathbf{x}'(a^-).$$

Here, as elsewhere in this chapter, we assume that $\mathbf{x}(t)$ is always a regular curve so that the derivatives (right and left) at any point in its parameter domain are nonzero. With this proviso, the unit tangent is given by

$$\mathbf{t}(t) = \mathbf{x}'(t)/(\mathbf{x}'(t), \mathbf{x}'(t))^{1/2},$$

and so by the above equation it follows that $\mathbf{t}(a^+) = \mathbf{t}(a^-)$. Conversely, if $\mathbf{t}$ is continuous at a then indeed $\mathbf{x}'(a^+) = m_1\mathbf{x}'(a^-)$ for some positive constant $m_1 > 0$.

Geometric continuity of order two requires, by definition, not only a continuous unit tangent but also a continuous curvature vector (the second derivative of the curve relative to arc length parameterization). For a planar curve, recall that the curvature vector is given by the expression

$$\mathbf{k}(t) = \frac{\mathbf{x}''(t) - \mathbf{t}(t)(\mathbf{x}''(t), \mathbf{t}(t))}{(\mathbf{x}'(t), \mathbf{x}'(t))}.$$

Therefore, if there is another constant m_2 such that both equations

$$\mathbf{x}'(a^+) = m_1\mathbf{x}'(a^-)$$

and

$$\mathbf{x}''(a^+) = m_1^2\mathbf{x}''(a^-) + m_2\mathbf{x}'(a^-)$$

hold, then indeed both $\mathbf{t}$ and $\mathbf{k}$ are continuous at a, and conversely. These equations were derived in [M].

It was Barsky [B_1] who was the first to introduce a *local* basis for cubic piecewise polynomials that satisfy at each knot a the conditions

$$\begin{pmatrix} f(a^+) \\ f'(a^+) \\ f''(a^+) \end{pmatrix} = \begin{pmatrix} 1 & 0 & 0 \\ 0 & m_1 & 0 \\ 0 & m_2 & m_1^2 \end{pmatrix} \begin{pmatrix} f(a^-) \\ f'(a^-) \\ f''(a^-) \end{pmatrix},$$

where the *same* value of m_1 and m_2 is used, *independent* of the knot a. Barsky called m_1 bias and m_2 tension, and his basis β-*splines*. In [B_2], Barsky examined practical implications of modeling both curves and surfaces by β-splines. He argues that the introduction of bias and tension into the design paradigm gives added flexibility that has geometric significance and usefulness. Thus a curve given in the form (1.2) of Chapter 1,

$$\mathbf{x}(t) = \sum_{j=1}^{n} \mathbf{c}_j f_j(t), \qquad t \in [0, 1],$$

where $f_1, \ldots, f_n$ are β-splines, has the advantage that the control points of $\mathbf{x}$ can be kept *fixed* while the bias and tension can be adjusted to alter the shape of the curve. This was done in Fig. 3.1. First an ordinary cubic spline was used on the control polygon. Then an adjustment of the m_2 value was made (either increased or decreased and indicated by a $+$ or $-$ value, respectively) to finally yield the desired shape.

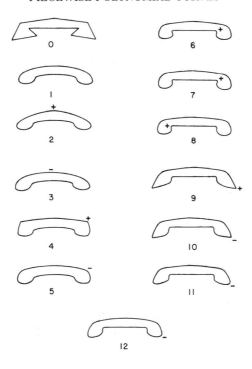

FIG. 3.1. *Modeling with β-splines.*

The first mathematical treatment of β-splines for arbitrary degree was given by Goodman [Go]. At each knot, his connection relations were

$$
\begin{pmatrix}
f(a^+) \\
f'(a^+) \\
f''(a^+) \\
\vdots \\
f^{(n)}(a^+)
\end{pmatrix}
=
\begin{pmatrix}
1 & 0 & \cdot & \cdot & \cdot & \cdot & \cdot & 0 \\
0 & m_1 & \cdot & \cdot & \cdot & \cdot & & \cdot \\
0 & m_2 & m_1^2 & \cdot & \cdot & \cdot & & \cdot \\
\cdot & \cdot & \cdot & \cdot & \cdot & \cdot & \cdot & \cdot \\
\cdot & \cdot & \cdot & \cdot & \cdot & \cdot & \cdot & \cdot \\
\cdot & \cdot & \cdot & \cdot & \cdot & \cdot & \cdot & \cdot \\
0 & 0 & 0 & \cdot & \cdot & \cdot & m_n & m_1^n
\end{pmatrix}
\begin{pmatrix}
f(a^-) \\
f'(a^-) \\
f''(a^-) \\
\vdots \\
f^{(n)}(a^-)
\end{pmatrix},
$$

where he allows $m_1, \ldots, m_n$ to change from knot to knot. He obtained a local basis for his β-spline and studied interpolation by linear combinations of such piecewise polynomials.

In general terms, the principal issue we are concerned with in this chapter is the sense in which different polynomial segments of a curve are joined together and how that is reflected in geometric properties of the curve. Initially we are interested in the behavior of a curve $\mathbf{x}$ defined on $[0, 1]$ at one point a within

its domain of definition. All our curves are required to be continuous and have
continuous derivatives of all orders, up to and including n, to the left and right
of $a \in (0, 1)$. It is convenient to think of the curve

(3.1) $$\mathbf{x} : [0, 1] \to \mathbb{R}^d$$

as composed of two "smooth" segments $\mathbf{x}_-$ and $\mathbf{x}_+$, which denote $\mathbf{x}$ restricted
to $[0, a)$ and $(a, 1]$, respectively. We form the vector of derivatives

(3.2) $$D_n[\mathbf{x}](t) = (\mathbf{x}(t), \ldots, \mathbf{x}^{(n)}(t))^T, \qquad t \neq a.$$

For the value of $D_n[\mathbf{x}](t)$ at $t = a$, we assume that as $t \to a^-$ or $t \to a^+$ the
first n derivatives of $\mathbf{x}$ converge to a limit and denote them by $D_n[\mathbf{x}_-](a)$ and
$D_n[\mathbf{x}_+](a)$, respectively. Now, fix an $(n + 1) \times (n + 1)$ nonsingular matrix M.
We tie together the two segments $\mathbf{x}_-$ and $\mathbf{x}_+$ by requiring that the constraint

(3.3) $$D_n[\mathbf{x}_+](a) = M D_n[\mathbf{x}_-](a)$$

is satisfied. To explain the geometric motivation for such a requirement we
review some basic terminology about the differential geometry of curves. First,
it should be understood that, as mentioned above, we always suppose that both
$\mathbf{x}_-, \mathbf{x}_+$ are parameterized as *regular* curves on $[0, a], [a, 1]$, respectively. This
means that

$$\mathbf{x}'_-(t) \neq 0, \qquad t \in [0, a]$$

and

$$\mathbf{x}'_+(t) \neq 0, \qquad t \in [a, 1].$$

If the matrix M is the identity matrix the curve $\mathbf{x}$ is given by an n-times continuously differentiable regular parameterization throughout $[0, 1]$. Even if this
were not the case $\mathbf{x}$ may be *visually* smooth, that is, lack of smoothness is only
a consequence of a *specific parameterization*, not an *intrinsic* property of the
curve itself. Let us see what happens if we *reparameterize* the left segment of $\mathbf{x}$.

An admissible parameterization of $\mathbf{x}_-$ is an n-times continuously differentiable function r defined on some neighborhood of a. That is, $r : [b, c] \to [b, c]$,
for some $b < c$, $r(a) = a$, and $r'(t) > 0$ for $t \in [b, c]$ where $a \in (b, c)$. The curve
segment

(3.4) $$\mathbf{y}(t) := \mathbf{x}_-(r(t)), \qquad t \in [b, a],$$

formed by composing $\mathbf{x}_-$ with r, joins together with $\mathbf{x}_+$ to form an n-times
continuously regular parameterization of $\mathbf{x}$ if and only if

(3.5) $$\mathbf{y}^{(j)}(a) = \mathbf{x}^{(j)}_+(a), \qquad j = 0, 1, \ldots, n.$$

To interpret these equations in the form (3.3) we need to compute the derivatives
of $\mathbf{y}$ by the chain rule. For this purpose, we introduce the $(n+1) \times (n+1)$ matrix

$R = R(r)$ defined by the property that for all f having n continuous derivatives in a neighborhood of a

(3.6) $$D_n[f \circ r](a) = R D_n[f](a).$$

Here $f \circ r$ denotes the composite function

(3.7) $$(f \circ r)(t) = f(r(t)), \qquad t \in [b, c].$$

We say R is a *reparameterization matrix* whenever there exists a function r which is n-times continuously differentiable in some neighborhood of a such that $r(a) = a$, $r'(a) \neq 0$, satisfying (3.6). Alternatively, R is determined by any real numbers $r_1, \ldots, r_n$ with $r_1 > 0$ where we choose

$$r(t) := a + \sum_{j=1}^{n} r_j \frac{(t-a)^j}{j!}, \qquad t \in [b, c]$$

in (3.6) and b, c sufficiently close to a. A formula for the elements of R is known, the Faa di Bruno formula, cf. Goodman [Go] or Knuth [Kn, p. 50]. Alternatively, they can be computed inductively in the following way. The matrix R in (3.6) is a lower triangular matrix. When we increment n by one to $n + 1$, the new matrix is obtained by appending a new row and column to the previous matrix. Thus it is convenient to think of R in (3.6) as the first $n + 1$ rows and column of an *infinite* lower triangular matrix. For notational simplicity we also use R to denote this matrix. Therefore, by differentiating both sides of (3.6) with respect to a, it follows that the rows of R can be computed successively by setting $R_{ij} = R_{ij}(a)|_{r(a)=a}$ where $R_{0j}(a) = \frac{1}{j!}(r(a) - a)^j$, $j = 0, 1, \ldots$, and

$$R_{i+1,j}(a) = R_{i,j-1}(a) + \frac{\partial R_{i,j}(a)}{\partial a}, \qquad R_{i0}(a) = \delta_{i0}, \qquad i, j = 0, 1, \ldots .$$

In particular, we have for $n = 2$

(3.8) $$R = \begin{pmatrix} 1 & 0 & 0 \\ 0 & r_1 & 0 \\ 0 & r_2 & r_1^2 \end{pmatrix},$$

while for $n = 3$, R is given by

(3.9) $$R = \begin{pmatrix} 1 & 0 & 0 & 0 \\ 0 & r_1 & 0 & 0 \\ 0 & r_2 & r_1^2 & 0 \\ 0 & r_3 & 3r_1 r_2 & r_1^3 \end{pmatrix}.$$

In general, the first column of R is given by $R_{i0} = \delta_{i0}$, $i = 0, 1, \ldots, n$, its second column by $R_{i1} = r_i$, $i = 1, \ldots, n$, and its diagonal elements by

$$R_{ii} = (r_1)^i, \qquad i = 0, 1, \ldots, n.$$

In particular, R is nonsingular. We will say more about this matrix later. At the moment the following geometric characterization of the condition (3.3) for $M = R$ is of interest.

PROPOSITION 3.1. *Let* $\mathbf{x} : [0, 1] \to \mathbb{R}^d$ *be composed of two curve segments*

$$(3.10) \qquad \mathbf{x}(t) = \begin{cases} \mathbf{x}_-(t), & 0 \le t \le a, \\ \mathbf{x}_+(t), & a \le t \le 1, \end{cases}$$

where each segment is given by a regular n-times continuously differentiable parameterization. Then there exists an admissible reparameterization of $\mathbf{x}_-$ *so that* $\mathbf{x}$ *becomes a regular n-times continuously differentiable curve in this parameterization if and only if there exist constants* $r_1, \dots, r_n$, $r_1 > 0$ *such that*

$$(3.11) \qquad D_n[\mathbf{x}_+](a) = R(r)D_n[\mathbf{x}_-](a).$$

Proof. We have essentially proved this simple result. The discussion above establishes the necessity. For the sufficiency we choose

$$r(t) = a + \sum_{j=1}^{n} r_j \frac{(t-a)^j}{j!}$$

and note that r is an admissible parameterization in any sufficiently small neighborhood of a, when $r_1 > 0$.   □

The prevailing terminology in the literature is that a curve satisfying (3.11) is called geometrically continuous of order n, denoted by GC^n. We formally record this definition below.

DEFINITION 3.1. *Let* $\mathbf{x} : [0, 1] \to \mathbb{R}^d$ *be a curve composed of two segments*

$$\mathbf{x}(t) = \begin{cases} \mathbf{x}_-(t), & 0 \le t \le a, \\ \mathbf{x}_+(t), & a \le t \le 1, \end{cases}$$

for some $a \in (0, 1)$ *where each segment is given by a regular n-times continuously differentiable parameterization. Then* $\mathbf{x}$ *is said to be geometrically continuous of order* n *if there is an admissible reparameterization* $r : [b, a] \to [0, a]$ *for some* $b \in (0, a)$ *such that the curve* $\mathbf{y}(t) = \mathbf{x}_-(r(t))$, $t \in [b, a]$, *joins with* $\mathbf{x}_+$ *with* n *continuous derivatives at* $t = a$, *that is,*

$$\mathbf{y}^{(i)}(a) = \mathbf{x}_+^{(i)}(a), \qquad i = 0, 1, \dots, n.$$

3.2.   Curvatures and Frenet equation: Frenet matrices.

Let us go back to the reparameterization matrix R. We summarize some of the most important properties of R:

$$\begin{aligned} R_{ij} &= 0, & 0 \le i < j \le n, \\ R_{i0} &= \delta_{i0}, & 0 \le i \le n, \\ R_{ii} &= (R_{11})^i, & 0 \le i < n, \end{aligned}$$

and

$$R_{11} > 0.$$

These conditions define a larger class of nonsingular matrices which was given a geometric interpretation by Dyn and Micchelli [DM]. To describe this observation we recall some results about the geometry of curves discussed in Spivak [Sp] and also in Gregory [Gr].

We begin with a continuous regular curve $\mathbf{x} : [0,1] \to \mathbb{R}^d$ having d continuous derivatives. Note that the number of continuous derivatives of $\mathbf{x}$ and the dimension of its range are the *same*. Also, we assume that for each $t \in [0,1]$ the set of vectors

$$(3.12) \qquad \left\{ \mathbf{x}^{(1)}(t), \ldots, \mathbf{x}^{(d-1)}(t) \right\}$$

are linearly independent. The *Frenet frame* of the curve $\mathbf{x}$ at the point t is a right-handed orthonormal system of vectors $\{\mathbf{f}_1, \ldots, \mathbf{f}_d\}$ formed from (3.12) by a Gram–Schmidt process. These vectors can be defined recursively. Specifically, for $i = 1, 2, \ldots, d-1$ set

$$(3.13) \qquad \mathbf{f}_i(t) := \mathbf{y}_i(t)/((\mathbf{y}_i(t), \mathbf{y}_i(t)))^{1/2},$$

where $(\ ,\ )$ denotes the standard inner product on $\mathbb{R}^d$ and define

$$(3.14) \qquad \mathbf{y}_i(t) = \mathbf{x}^{(i)}(t) - \sum_{j=1}^{i-1} (\mathbf{x}^{(i)}(t), \mathbf{f}_j(t))\mathbf{f}_j(t), \qquad i = 1, \ldots, d-1.$$

(When $i = 1$, the sum above is assumed to be zero.) The vector $\mathbf{f}_d(t)$ is chosen so that

$$(3.15) \qquad \det(\mathbf{f}_1(t), \ldots, \mathbf{f}_d(t)) := \begin{vmatrix} (\mathbf{f}_1(t))_1 & (\mathbf{f}_2(t))_1 & \cdots & (\mathbf{f}_d(t))_1 \\ \vdots & \vdots & & \vdots \\ (\mathbf{f}_1(t))_d & (\mathbf{f}_2(t))_d & \cdots & (\mathbf{f}_d(t))_d \end{vmatrix} = 1.$$

For each $i = 1, \ldots, d-1$ the Gram–Schmidt process guarantees that the vectors $\mathbf{f}_1(t), \ldots, \mathbf{f}_i(t)$ span the same linear subspace spanned by the vectors $\mathbf{x}^{(1)}(t), \ldots, \mathbf{x}^{(i)}(t)$. We can express this fact as a matrix equation

$$(3.16) \qquad \left(\mathbf{x}^{(1)}(t), \ldots, \mathbf{x}^{(d)}(t) \right)^T = L(t) \left(\mathbf{f}_1(t), \ldots, \mathbf{f}_d(t) \right)^T,$$

where $L(t) := (\ell_{ij}(t))_{i,j=1,\ldots,d}$ is some $d \times d$ lower triangular matrix. According to (3.13) and (3.14), we have

$$(3.17) \qquad \ell_{ii}(t) = ((\mathbf{y}_i(t), \mathbf{y}_i(t)))^{1/2} > 0, \qquad i = 1, \ldots, d-1,$$

while in view of (3.15), the sign of $\ell_{dd}(t)$ is given by the sign of $\det(\mathbf{x}^{(1)}(t), \ldots, \mathbf{x}^{(d)}(t))$. In fact, by (3.16) we get, quite explicitly,

$$(3.18) \qquad \ell_{dd}(t) = \frac{\det (\mathbf{x}^{(1)}(t), \ldots, \mathbf{x}^{(d)}(t))}{\displaystyle\prod_{j=1}^{d-1} \ell_{jj}(t)}.$$

The Frenet frame $\{\mathbf{f}_1(t), \ldots, \mathbf{f}_d(t)\}$ flows according to a system of first-order differential equations. These *Frenet equations* are obtained in the following way. According to (3.13) and (3.14), for each $i = 1, \ldots, d-1$, $\mathbf{f}_i(t)$ is a linear combination of the vectors $\mathbf{x}^{(1)}(t), \ldots, \mathbf{x}^{(i)}(t)$ and the coefficients that appear in this linear combination are continuously differentiable. Hence, differentiating this equation, we conclude that $\mathbf{f}_i'(t)$ can be expressed as a linear combination of the vectors $\mathbf{x}^{(1)}(t), \ldots, \mathbf{x}^{(i+1)}(t)$ and hence, also by (3.16), as a linear combination of $\mathbf{f}_1(t), \ldots, \mathbf{f}_{i+1}(t)$. Since $\mathbf{x}^{(d)}(t)$ exists for $t \in [0, 1]$, so too does $\mathbf{f}_d'(t)$. Also, it is certainly a linear combination of the vectors $\mathbf{f}_1(t), \ldots, \mathbf{f}_d(t)$.

We summarize these facts by the matrix equation,

$$(3.19) \qquad (\mathbf{f}_1'(t), \ldots, \mathbf{f}_d'(t))^T = K(t)(\mathbf{f}_1(t), \ldots, \mathbf{f}_d(t))^T,$$

where $K(t) = (k_{ij}(t))_{i,j=1,\ldots,d}$ is some lower Hessenberg matrix. Using the fact that $(\mathbf{f}_i(t), \mathbf{f}_j(t)) = \delta_{ij}$, $i, j = 1, \ldots, d$ we get from (3.19) that

$$(3.20) \qquad k_{ij}(t) = (\mathbf{f}_i'(t), \mathbf{f}_j(t)).$$

Moreover, for $i, j = 1, \ldots, d$ we have the equations

$$(3.21) \qquad 0 = \frac{d}{dt}(\mathbf{f}_i(t), \mathbf{f}_j(t)) = (\mathbf{f}_i'(t), \mathbf{f}_j(t)) + (\mathbf{f}_i(t), \mathbf{f}_j'(t)).$$

In other words,

$$(3.22) \qquad K + K^T = 0,$$

which means that K is a skew-symmetric matrix. It follows that K is necessarily tri-diagonal. The elements of K on the (upper) secondary diagonal determine the *curvatures* of $\mathbf{x}$ and are defined by the formula

$$(3.23) \qquad \kappa_i(t) := k_{i,i+1}(t)/s(t),$$

where we set $s(t) := ((\mathbf{x}^{(1)}(t), \mathbf{x}^{(1)}(t)))^{1/2}$. In terms of these curvatures of $\mathbf{x}$, the Frenet equations become

$$(3.24) \qquad s(t)^{-1}\mathbf{f}_i'(t) = -\kappa_{i-1}(t)\mathbf{f}_{i-1}(t) + \kappa_i(t)\mathbf{f}_{i+1}(t), \qquad i = 1, \ldots, d,$$

where we set $\kappa_0(t) = \kappa_d(t) = 0$.

For later use, we need to relate the curvatures $\kappa_1(t), \ldots, \kappa_{d-1}(t)$ of $\mathbf{x}$ to the elements of the lower triangular matrix in (3.16). According to formula (3.16) we have, in particular, that

$$\ell_{ii}(t) = (\mathbf{x}^{(i)}(t), \mathbf{f}_i(t)).$$

Now, differentiate both sides of the equation

$$\mathbf{x}^{(i)}(t) = \sum_{j=1}^{i} \ell_{ij}(t)\mathbf{f}_j(t)$$

and use the orthonormality of the Frenet frame to get

$$\ell_{i+1,i+1}(t) = (\mathbf{x}^{(i+1)}(t), \mathbf{f}_{i+1}(t))$$
$$= (\ell_{ii}(t)\mathbf{f}_i'(t), \mathbf{f}_{i+1}(t)).$$

Using the Frenet equations (3.24), this relation simplifies to

$$\ell_{i+1,i+1}(t) = \kappa_i(t)\ell_{ii}(t)s(t)$$

and thus we obtain the desired formula

$$(3.25) \qquad \kappa_i(t) = \frac{\ell_{i+1,i+1}(t)}{\ell_{ii}(t)\ell_{11}(t)}, \qquad i = 1, \ldots, d.$$

One additional formula is useful. This formula relates the curvatures directly to the derivatives of the curve at t. Specifically, from (3.16) we get

$$(3.26) \qquad L(t)L^T(t) = G_d(t),$$

where

$$G_d(t) := \left((\mathbf{x}^{(i)}(t), \mathbf{x}^{(j)}(t)) \right)_{i,j=1,\ldots,d}$$

is the Gram matrix of the vectors $\mathbf{x}^{(1)}(t), \ldots, \mathbf{x}^{(d)}(t)$. In other words, $L(t)$ is the Cholesky factorization of the positive definite symmetric matrix $G_d(t)$. Computing the determinant of both sides of (3.26) it follows that

$$(3.27) \qquad \ell_{11}^2(t) \cdots \ell_{ss}^2(t) = \det\ (\mathbf{x}^{(i)}(t), \mathbf{x}^{(j)}(t))_{i,j=1,\ldots,s}$$

for $s = 1, \ldots, d$ and so

$$(3.28) \qquad \ell_{ii}(t) = \left(\frac{G_i(t)}{G_{i-1}(t)} \right)^{1/2}, \qquad i = 1, \ldots, d,$$

where $G_0(t) = 1$ and $G_i(t)$ is the determinant of the Gram matrix of the vectors $\mathbf{x}^{(1)}(t), \ldots, \mathbf{x}^{(i)}(t), i = 1, \ldots, d$.

We now have more than enough formulas. What is needed next are the following definitions.

DEFINITION 3.2. *Let* $\mathbf{x} : [0, 1] \to \mathbb{R}^d$ *be a continuous curve composed of two segments*

$$\mathbf{x}(t) = \begin{cases} \mathbf{x}_-(t), & 0 \le t \le a, \\ \mathbf{x}_+(t), & a \le t \le 1, \end{cases}$$

$a \in (0, 1)$ *where each curve segment is given by a regular parameterization with* d *continuous derivatives such that the first* $d - 1$ *derivatives are linearly independent. We say* $\mathbf{x}$ *is Frenet frame continuous provided the Frenet frame and curvatures of* $\mathbf{x}$ *are continuous on* $[0, 1]$.

Also, for matrices we use the following terminology.

DEFINITION 3.3. *An* $(n+1) \times (n+1)$ *matrix* F *is called a Frenet matrix provided that it is lower triangular,*

$$F_{i0} = \delta_{i0}, \qquad i = 0, 1, \ldots, n,$$

$$F_{ii} = (F_{11})^i, \qquad i = 0, 1, \ldots, n,$$

and

$$F_{11} > 0.$$

The following result from Dyn and Micchelli [DM] gives a geometric characterization of Frenet matrices.

THEOREM 3.1. *Let* $\mathbf{x} : [0, 1] \to \mathbb{R}^d$ *be a continuous curve composed of two segments*

$$\mathbf{x}(t) = \begin{cases} \mathbf{x}_-(t), & 0 \leq t \leq a, \\ \mathbf{x}_+(t), & a \leq t \leq 1, \end{cases}$$

$a \in (0, 1)$ *where each curve segment is given by a regular parameterization with* d *continuous derivatives such that the first* $d-1$ *derivatives are linearly independent. Then* $\mathbf{x}$ *is Frenet frame continuous if and only if there is a* $(d+1) \times (d+1)$ *Frenet* F *matrix such that*

(3.29) $$D_d[\mathbf{x}_+](a) = F D_d[\mathbf{x}_-](a).$$

Proof. First suppose (3.29) holds for some Frenet matrix. Since F is, in particular, lower triangular, it follows that for each $i, i = 1, \ldots, d$, the span of the set of vectors $\{\mathbf{x}_-^{(1)}(a), \ldots, \mathbf{x}_-^{(i)}(a)\}$ and the span of $\{\mathbf{x}_+^{(1)}(a), \ldots, \mathbf{x}_+^{(i)}(a)\}$ are the same. Hence, the Gram–Schmidt process applied to each set yields the same vectors. This means that the Frenet frames of $\mathbf{x}_-$ and $\mathbf{x}_+$ are the same at a. Therefore using formula (3.16), for each curve segment at $t = a$, we get the matrix equation

(3.30) $$H L(a^-) = L(a^+)$$

where H is the $d \times d$ matrix formed from the last d rows and columns of the Frenet matrix F. In particular, (3.30) gives us the equations

(3.31) $$\ell_{ii}(a^+) = H_{ii} \ell_{ii}(a^-), \qquad i = 1, \ldots, d.$$

Since $H_{ii} = F_{ii} = (F_{11})^i$, $i = 1, \ldots, d$, we can use formula (3.25) and equation (3.31) to show that the curvatures of $\mathbf{x}_+$ and $\mathbf{x}_-$ agree at $t = a$, that is, $(\kappa_i)(a^+) = (\kappa_i)(a^-)$, $i = 1, \ldots, d-1$. Therefore, we have established the sufficiency of condition (3.29).

For the necessity, suppose $\mathbf{x}$ is Frenet frame continuous at $t = a$. Then (3.16) implies that the matrix H defined by (3.30) satisfies

(3.32) $$(\mathbf{x}_+^{(1)}(a), \ldots, \mathbf{x}_+^{(d)}(a))^T = H(\mathbf{x}_-^{(1)}(a), \ldots, \mathbf{x}_-^{(d)}(a))^T.$$

In particular, H is lower triangular, and equations (3.31) and (3.32) yield the formula

$$1 = \frac{\kappa_i(a^+)}{\kappa_i(a^-)} = \frac{\ell_{i+1,i+1}(a^+)}{\ell_{i+1,i+1}(a^-)} \frac{\ell_{ii}(a^-)}{\ell_{ii}(a^+)} \frac{\ell_{11}(a^-)}{\ell_{11}(a^+)}$$
$$= \frac{H_{i+1,i+1}}{H_{ii}H_{11}},$$

that is, $H_{ii} = (H_{11})^i, i = 1, 2, \ldots, d$, with $H_{11} > 0$. Finally, to identify the matrix F that satisfies (3.29) we set

$$F = \begin{pmatrix} 1 & 0 \cdots 0 \\ 0 & \\ \vdots & H \\ 0 & \end{pmatrix}. \qquad \square$$

In summary, we have given two instances where equation (3.3) has geometric significance, either M is a reparameterization matrix or a Frenet matrix F. For 3×3 matrices these classes are *identical*, since they both can be put in the form (3.8). Hence, a planar curve is GC^2 if and only if it is Frenet frame continuous. The distinction between these concepts begins to occur for curves in three dimensions. In fact, there are Frenet frame continuous space curves that *cannot* be reparameterized to make them GC^3.

3.3.   Projection and lifting of curves.

To gain further insight into the relationship between these two notions of visual smoothness of curves, we explore how they are affected by basic algebraic operations encountered in practical curve design. These computations typically are performed when a curve is specified in terms of control points (see Chapter 1). Thus we begin with vectors $\mathbf{c}_1, \ldots, \mathbf{c}_N$ in $\mathbb{R}^{d+1}$ and scalar valued functions $B_1(t), \ldots, B_N(t), t \in [0, 1]$, and consider the curve

$$\mathbf{x}(t) = \sum_{j=1}^{N} \mathbf{c}_j B_j(t).$$

Typically, these "blending" functions will be chosen to be piecewise polynomials with a finite number of breakpoints in $(0, 1)$. For the moment, we focus on one such breakpoint, say $t = a$ and write each blending function as two segments, viz.

$$B_j(t) = \begin{cases} (B_-)_j(t), & 0 \leq t \leq a, \\ (B_+)_j(t), & a \leq t \leq 1. \end{cases}$$

Now, to ensure that our basic connection equation (3.3) holds for *any* choice of control points, the blending functions must satisfy the connection equations

$$D_n[(B_+)_j](a) = M D_n[(B_-)_j](a), \qquad j = 1, \ldots, N.$$

The prevailing point of view is that the matrix M takes the role of *shape parameters*. Thus, a designer specifies a control polygon formed by the control points $c_1, \ldots, c_N$. The curve $\mathbf{x}(t)$ above provides the designer with a "smooth" version of the control polygon. However, it may not be exactly what is required and so the curve may be altered by changing M but *not* the control points.

To enrich the design paradigm, piecewise rational curves are formed from $\mathbf{x}(t)$ by *projection*. To this end, we write the control points $\mathbf{c}_i, i = 1, \ldots, N$ in the form $\mathbf{c}_i = (w_i \mathbf{d}_i, w_i)$, $i = 1, \ldots, N$, for some scalars $w_1, \ldots, w_N$ and form the projected curve

$$\mathbf{v}(t) := \sum_{j=1}^{N} w_j \mathbf{d}_j B_j(t) \Big/ \sum_{j=1}^{N} w_j B_j(t).$$

To explore the form of the shape parameters of $\mathbf{v}$ given that M is either a reparameterization matrix or a Frenet matrix, we formalize the notion of *projection*.

DEFINITION 3.4. *Given a curve* $\mathbf{x}(t) = (x_1(t), \ldots, x_{d+1}(t))^T, \mathbf{x} : [0, 1] \to \mathbb{R}^{d+1}$ *such that* $x_{d+1}(t) \neq 0, t \in [0, 1]$, *its projection is defined as*

$$P(\mathbf{x})(t) = (x_1(t)/x_{d+1}(t), \ldots, x_d(t)/x_{d+1}(t), 1)^T.$$

Some further terminology will be helpful. A nonsingular $(n + 1) \times (n + 1)$ matrix $M = (M_{ij})_{i,j=0,1,\ldots,n}$ such that

$$M_{i0} = M_{0i} = \delta_{0i}, \qquad i = 0, 1, \ldots, n,$$

is called a *connection* matrix. We define $C(a, M)$ to be the class of all real-valued functions f on $[0, 1]$ that are n-times continuously differentiable on $[0, a]$ and $[a, 1]$ and satisfy the connection equation

(3.33) $D_n[f_+](a) = M D_n[f_-](a).$

If M is a connection matrix then $C(a, M)$ is a linear space of continuous functions on $[0, 1]$ that contains constants. We also say a curve $\mathbf{x}$ is in $C(a, M)$ provided that each component of $\mathbf{x}$ is in $C(a, M)$. Also, whenever f and g are n-times continuously differentiable on the intervals $[0, a]$ and $[a, 1]$, for $h := f/g$ to be in $C(a, M)$ means that $g(a) \neq 0$ and

$$D_n[h_+](a) = D_n[h_-](a).$$

This avoids the cumbersome (and for our purpose here, irrelevant) problem that g may be zero somewhere in $[0, 1]$ other than at a.

With this bit of terminology, we shall prove some facts from Goldman and Micchelli [GM] that show the inherent algebraic flexibility of Frenet frame continuity. As a means of contrast, the first result we describe concerns reparameterization matrices.

THEOREM 3.2. *Let M be a connection matrix. Then M is a reparameterization matrix if and only if whenever a curve $\mathbf{x} : [0, 1] \to \mathbb{R}^{d+1}$ with $x_{d+1}(a) \neq 0$ is in $C(a, M)$, it follows that $P(\mathbf{x})$ is also in $C(a, M)$.*

This result says that the projection of a curve has the same shape parameters as the original curve if and only if the shape parameters are given by a reparameterization matrix. For Frenet frame continuous curves we will prove the following result.

THEOREM 3.3. *Let M be an $(n+1) \times (n+1)$ Frenet matrix with an associated Frenet frame continuous curve $\mathbf{x}(t) = (x_1(t), \ldots, x_{d+1}(t))$ such that $x_{d+1}(a) \neq 0$. Then there exists a Frenet matrix N, depending only on M and x_{d+1}, such that $P(\mathbf{x})$ is in $C(a, N)$.*

We begin the proofs of these results by first establishing some properties of the set $C(a, M)$.

PROPOSITION 3.2. *Let M be a connection matrix. Then $f \in C(a, M)$ implies $p \circ f \in C(a, M)$ for all polynomials p if and only if M is a reparameterization matrix.*

Proof. First, suppose M is a reparameterization matrix corresponding to an admissible reparameterization $r : [b, c] \to [b, c]$ with a $\in (b, c)$. Let f be n-times continuously differentiable on $[0, a]$ and $[a, 1]$. Then

(3.34)
$$D_n[f_- \circ r](a) = MD_n[f_-](a)$$

and $f \in C(a, M)$ if and only if

$$D_n[f_+](a) = D_n[f_- \circ r](a).$$

Using this equation, and the chain rule twice, we get

$$D_n[p \circ f_+](a) = R(p)D_n[f_+](a) = D_n[p \circ (f_- \circ r)](a).$$

In other words, indeed

$$D_n[p \circ f_+](a) = D_n[(p \circ f_-) \circ r](a) = MD_n[p \circ f_-](a),$$

where in the last equation we used (3.34) on the function $p \circ f_-$.

For the converse, we define the function

$$r(t) = \begin{cases} t, & 0 \le t \le a, \\ a + \sum_{j=1}^{n} M_{j1}(t-a)^j / j!, & a \le t \le 1. \end{cases}$$

Obviously, r is an admissible reparameterization. Moreover, we have

$$D_n[r_+](a) = (a, M_{11}, M_{21}, \ldots, M_{n1})^T$$
$$= M(a, 1, 0, \ldots, 0)^T = MD_n[r_-](a),$$

and so $r \in C(a, M)$. Therefore, by our hypothesis, we have for all polynomials p that

(3.35)
$$D_n[p \circ r_+](a) = MD_n[p \circ r_-](a) = MD_n[p](a).$$

Let R be the reparameterization matrix corresponding to the function r. Then (3.35) implies that for all polynomials p

(3.36) $$RD_n[p](a) = MD_n[p](a),$$

from which it follows that $R = M$.   □

This result requires us to check that whenever $f \in C(a, M)$ it follows that $p \circ f \in C(a, M)$ for *every* polynomial p in order to conclude that M is a reparameterization matrix. However, the proof of Proposition 3.2 shows that it suffices to check this implication for the polynomials of degree $\leq n$. Next we shall show that it is only necessary to check it for any *fixed* nonlinear polynomial.

PROPOSITION 3.3. *Let M be a connection matrix and suppose p is a polynomial of degree at least two. Then $f \in C(a, M)$ implies $p \circ f \in C(a, M)$ if and only if M is a reparameterization matrix.*

Proof. Suppose that $f \in C(a, M)$ implies that $p \circ f \in C(a, M)$ for some polynomial p of degree at least two. We define S to be the set of all polynomials q such that whenever $f \in C(a, M)$ it follows that $q \circ f \in C(a, M)$. Let us list some properties of the set S.

(i) Linear functions are in S.
(ii) S is a linear space.
(iii) If $q_1, q_2 \in S$ then $q_1 \circ q_2 \in S$.
(iv) $p \in S$.

We claim that these four conditions imply S is the set of *all* polynomials. First, notice that S contains polynomials of arbitrarily large degree. This fact follows from (iii) and (iv), which together imply that p composed with itself any number of times is in S. Next we show that if $q_m \in S$ and q_m has degree m, then all polynomials of degree $\leq m$ are also in S. This would prove the result.

Without loss of generality, we suppose that q_m is a monic polynomial. Then the polynomial

$$q_{m-1}(t) := q_m(t + 1/m) - q_m(t)$$

is also a monic polynomial and its degree is $m - 1$. Appropriately specializing properties (i), (ii), and (iii) we see that $q_{m-1} \in S$. Continuing in this fashion we construct monic polynomials $q_0, q_1, \ldots, q_m \in S$ such that the degree of q_j is $j, j = 0, 1, \ldots, m$. These polynomials certainly span all polynomials of degree $\leq m$. Hence, by property (ii), all polynomial of degree $\leq m$ are in S.   □

The next result represents the most general form of Proposition 3.3 that is available.

THEOREM 3.4. *Let M be an $(n + 1) \times (n + 1)$ connection matrix and g a nonlinear function with at least $n + 2$ continuous derivatives on $\mathbb{R}$. Then $f \in C(a, M)$ implies $g \circ f \in C(a, M)$ if and only if M is a reparameterization matrix.*

Proof. Let g be as above and choose any $f \in C(a, M)$. Since g is nonlinear there must be some constant $c \in R$ such that $g''(c) \neq 0$. For every real value of x we define the one parameter family of functions

$$f(x, t) = xf(t) + c, \qquad t \in [0, 1].$$

Because 1 and f are in $C(a, M)$ and $C(a, M)$ is a linear space, $f(x, \cdot) \in C(a, M)$ for all $x \in \mathbb{R}$. Consequently, $g \circ (f(x, \cdot)) \in C(a, M)$ for all $x \in \mathbb{R}$. Using the fact that g has $n + 2$ continuous derivatives, we conclude by differentiating both sides of the connection equation

$$D_n[g \circ f_+(x, \cdot)](a) = D_n[g \circ f_-(x, \cdot)](a)$$

with respect to x, that

$$\frac{\partial^2}{\partial x^2}(g \circ (f(x, \cdot))) \in C(a, M),$$

for all $x \in \mathbb{R}$. However, it is easily verified that

$$\frac{\partial}{\partial x^2}(g \circ (f(x, t)))\Big|_{x=0} = g''(c)f^2(t).$$

Therefore, $g''(c)f^2 \in C(a, M)$. Now let $p(t) := g''(c)t^2$ in Proposition 3.3 to conclude that M is a reparameterization matrix.   □

COROLLARY 3.1. *Let M be a connection matrix. Then $f, g \in C(a, M)$ implies $fg \in C(a, M)$ if and only if M is a reparameterization matrix.*

Proof. Suppose that M is a connection matrix such that whenever $f, g \in C(a, M)$, it follows that $fg \in C(a, M)$. In particular, if $f \in C(a, M)$ then $f^2 = f \cdot f \in C(a, M)$. Hence by Proposition 3.3 with $p(t) := t^2$, we conclude that M is a reparameterization matrix.

The converse follows from the simple fact that

(3.37) $$(fg) \circ r = (f \circ r)(g \circ r).$$

Specifically, if $M = R(r)$ then $f, g \in C(a, M)$ if and only if

(3.38) $$D_n[f_+](a) = D_n[f_- \circ r](a)$$

and

(3.39) $$D_n[g_+](a) = D_n[g_- \circ r](a).$$

This implies, by Leibniz's rule, that

$$D_n[(fg)_+](a) = D_n[(f_- \circ r)(g_- \circ r)](a)$$

and so (3.37) gives

$$D_n[(fg)_+](a) = D_n[(fg)_- \circ r](a).$$

In other words, $fg \in C(a, M)$.   □

COROLLARY 3.2. *Let M be a connection matrix. Then $f, g \in C(a, M)$, $g(a) \neq 0$ implies that $f/g \in C(a, M)$ if and only if M is a reparameterization matrix.*

Proof. Suppose that M is a connection matrix such that whenever $f, g \in C(a, M)$, it follows that $fg \in C(a, M)$. Choose any $f \in C(a, M)$ with $f(a) \neq 0$. Since $1 \in C(a, M)$ we conclude that $1/f \in C(a, M)$ and hence $f^2 = f/(1/f) \in C(a, M)$. When $f(a) = 0$ we set $h_\epsilon(t) = f(t) + \epsilon$ where $\epsilon \neq 0$. Then $h_\epsilon(a) = \epsilon \neq 0$ and $h_\epsilon \in C(a, M)$. Hence by what was just proved, $h_\epsilon^2 \in C(a, M)$. Letting $\epsilon \to 0$ we conclude even in this case that $f^2 \in C(a, M)$. Thus we can apply Proposition 3.3, again with $p(t) = t^2$, to conclude that M is a reparameterization matrix.

The converse statement follows in a manner that parallels the converse of the proof of Corollary 3.1. Here we use the identity

$$(f/g) \circ r = (f \circ r)/(g \circ r). \qquad \square$$

Proof of Theorem 3.2. We are now ready to prove Theorem 3.2. In fact, its proof is now an immediate consequence of Corollary 3.2 (applied to the component functions x_i and x_{d+1} of the curve $\mathbf{x}$). $\qquad \square$

Let us now turn our attention to the proof of Theorem 3.3. For this purpose, we require the following notation. Fix two functions f and g that are n-times continuously differentiable in a neighborhood of point a. By Leibniz's rule, there is a matrix $L = L(g)$ such that

$$(3.40) \qquad\qquad D_n[fg](a) = L(g)D_n[f](a).$$

Note that $L(g)$ is a lower triangular matrix whose elements depend only on $g(a), g^{(1)}(a), \ldots, g^{(n)}(a)$ and, in particular, $L_{ii} = g(a), i = 0, 1, \ldots, n$. Moreover, when $g(a) \neq 0$ it follows from (3.40), by choosing $f = 1/g$, that

$$(3.41) \qquad\qquad L^{-1} = L(g^{-1}).$$

PROPOSITION 3.4. *Let M be a connection matrix. Suppose $w : [0, 1] \to \mathbb{R}$ is a continuous function that is n times continuously differentiable on $[0, a]$ and $[a, 1]$ with $w(a) \neq 0$ for some $a \in (0, 1)$. Then $f \in C(a, M)$ if and only if $f/w \in C(a, T)$ where $T = T(w) := L(w_+^{-1})ML(w_-)$.*

Proof. Choose $f \in C(a, M)$. Then it follows that

$$D_n[(f/w)_+](a) = L(w_+^{-1})D_n[f_+](a)$$
$$= L(w_+^{-1})MD_n[f_-](a).$$

Also, we have the equation

$$D_n[f_-](a) = D_n[w_-(f/w)_-](a)$$
$$= L(w_-)D_n[(f/w)_-](a).$$

Substituting this equation in the one above gives us the formula

$$D_n[(f/w)_+](a) = TD_n[(f/w)_-](a),$$

which means that $f/w \in C(a, T)$.

Conversely, if $f/w \in C(a,T)$ then by the previous observation we conclude that $f \in C(a, L(w_+)TL(w_-^{-1}))$. However, by (3.41) we obtain the relations

$$L(w_+)TL(w_-^{-1}) = L(w_+)L(w_+^{-1})ML(w_-)L(w_-^{-1})$$
$$= M.$$

That is, $f \in (a, M)$. □

COROLLARY 3.3. *Let M be a Frenet matrix and w any function in $C(a, M)$ with $w(a) \neq 0$. Then there exists a Frenet matrix T such that whenever $f \in C(a, M)$ it follows that $f/w \in C(a,T)$.*

Proof. From Proposition 3.4, the matrix

$$T := L(w_+^{-1})ML(w_-)$$

has the desired property provided we show that it is a Frenet matrix. To this end, we recall that lower triangular matrices form a group under matrix multiplication and hence T is lower triangular. Also, for $i = 0, 1, \ldots, n$,

$$T_{ii} = w_+^{-1}(a)M_{ii}w_-(a) = M_{ii} = (M_{11})^i$$
$$= T_{11}^i.$$

Therefore, it remains to show that the first column of T is $(1, 0, 0, \ldots, 0)^T$. First, note that since $w \in C(a, M)$ we conclude that

(3.42) $$D_n[w_+](a) = MD_n[w_-](a)$$

and, by Leibniz's rule, the first column of $L(w_-)$ is $D_n[w_+](a)$. Hence, by (3.42) the first column of $ML(w_-)$ is $D_n[w_+](a)$. Consequently, the first column of T is $L(w_+^{-1})D_n[w_+](a) = D_n[(w^{-1}w)_+](a) = D_n[1](a) = (1, 0, 0, \ldots, 0)^T$.   □

Proof of Theorem 3.3. Corollary 3.3 immediately implies Theorem 3.3. We merely apply it to the component functions x_i and $w = x_{d+1}$ of the curve $\mathbf{x}$.   □

Before changing the subject to the construction of blending functions that yield Frenet frame continuous curves, we make one final comment concerning algebraic manipulation of curves and visual smoothness. This observation has to do with the notion of *lifting*, which is also a typical algebraic operation on curves. Given a curve in the form

$$\mathbf{x}(t) = (x_1(t), \ldots, x_d(t), 1)^T,$$

we lift it by multiplying each of its components by some fixed functions w, viz.

$$L(\mathbf{x})(t) := (w(t)x_1(t), \ldots, w(t)x_d(t), w(t))^T.$$

We ask the question, do Theorems 3.2 and Theorems 3.3 hold for lifting? The answer is yes for Theorem 3.2 but surprisingly, it is no for Theorem 3.3.

THEOREM 3.5. *Let M be a Frenet matrix and $w \in C(a, M)$ with $w(a) \neq 0$. If there exists a Frenet matrix T such that whenever $f \in C(a, M)$ it follows that $wf \in C(a, T)$, then $M = T$ and M is a reparameterization matrix.*

The proof is based on the following fact.

LEMMA 3.1. *Let R and T be any $(n+1) \times (n+1)$ matrices. If $C(a, R) \supseteq C(a, T)$ then $R = T$.*

Proof. For every $f \in C(a, T)$, we have both equations

$$D_n[f_+](a) = T D_n[f_-](a)$$

and

$$D_n[f_+](a) = R D_n[f_-](a),$$

from which it follows that

(3.43) $$T D_n[f_-](a) = R D_n[f_-](a).$$

For each $\mathbf{x} = (x_0, \ldots, x_n)^T \in \mathbb{R}^{n+1}$, define the function

$$
h(t) = \begin{cases}
\displaystyle\sum_{j=0}^{n} x_j (t-a)^j / j!, & 0 \le t \le a, \\
\displaystyle\sum_{j=0}^{n} (T\mathbf{x})_j (t-a)^j / j!, & a \le t \le 1.
\end{cases}
$$

Then

$$D_n[h_+](a) = T\mathbf{x} = T D_n[h_-](a),$$

and so $h \in C(a, T)$. Hence, choosing $f = h$ in (3.43) gives $T\mathbf{x} = R\mathbf{x}$ for all $\mathbf{x} \in \mathbb{R}^{n+1}$. $\square$

Proof of Theorem 3.5. Let M and w be as in the hypothesis of Theorem 3.5. Choose an $f \in C(a, M)$. Since 1 is also in $C(a, M)$, we conclude that both w and wf lie in $C(a, T)$. But T is a Frenet matrix and so by Corollary 3.3 there is a Frenet matrix R (depending only on w and T) such that whenever $g \in C(a, T)$, it follows that $g/w \in C(a, R)$. Choosing $g = wf$, we conclude that $f = wf/w \in C(a, R)$. Thus we have shown $C(a, M) \subseteq C(a, R)$. We can now use Lemma 3.1 to conclude that $M = R$. But $1 \in C(a, T)$ and so by the definition of R, $1/w \in C(a, R) = C(a, M)$. In summary, we have shown for any $w \in C(a, M)$ with $w(a) \ne 0$ that $1/w \in C(a, M)$. We claim that this implies that M is a reparameterization matrix. To see this, choose any $f \in C(a, M)$ and $\epsilon \ne 0$ such that $f(a) \ne \pm\epsilon$. Then $f \pm \epsilon \in C(a, M)$ and so by the above observation $(f+\epsilon)^{-1}$ and $(f - \epsilon)^{-1}$ are both in $C(a, M)$. Hence, their difference

$$\frac{1}{f-\epsilon} - \frac{1}{f+\epsilon} = \frac{2\epsilon}{f^2 - \epsilon}$$

is in $C(a, M)$, as well. The function on the right-hand side of this equation is also nonzero at $t = a$ and so its reciprocal is also in $C(a, M)$. In conclusion, we have proved that whenever $f \in C(a, M)$ it follows that $p \circ f \in C(a, M)$, where $p(t) := (t^2 - \epsilon^2)/2\epsilon$. We now apply Proposition 3.3 to conclude that M is a reparameterization matrix. $\square$

3.4. Connection matrices and piecewise polynomial curves.

This finishes our discussion of various ideas centering on the basic connection equation (3.3). Our next goal in this chapter is the construction of linear spaces of piecewise polynomials determined by connection matrices. This motivates us to introduce the following linear spaces.

DEFINITION 3.5. *Let π_n be the space of all polynomials of degree $\leq n$. Given points $t_0, \ldots, t_k$, with $-\infty < t_0 < t_1 < \cdots < t_k < \infty$, we set $I_0 = (-\infty, t_0), I_i = (t_{i-1}, t_i), i = 1, \ldots, k$ and $I_{k+1} = (t_k, \infty)$. Also let $A^0, A^1, \ldots, A^k$ be $n \times n$ nonsingular matrices. We introduce the space of functions on $\mathbb{R}$*

$$\mathcal{S}_{n,k} = \left\{ f : f|_{I_i} \in \pi_n, D_{n-1}[f_+](t_i) = A^i D_{n-1}[f_-](t_i), \ i = 0, 1, \ldots, k \right\}.$$

Observe that when each A^i is the identity matrix

$$\mathcal{S}_{n,k} = \left\{ f : f|_{I_i} \in \pi_n, i = 0, 1, \ldots, k, f \in C^{n-1}(\mathbb{R}) \right\},$$

which is the space of spline functions of degree $\leq n$ with knots at $t_0, \ldots, t_k$.

We begin with an elementary observation about $\mathcal{S}_{n,k}$.

LEMMA 3.2.

$$\dim \mathcal{S}_{n,k} = n + k + 2.$$

Proof. We construct a basis for $\mathcal{S}_{n,k}$ in the following way. The space $\mathcal{S}_{n,k}$ does not generally contain even a single polynomial. However, there are analogues of polynomials in $\mathcal{S}_{n,k}$. We obtain them by a process of *extension*. Start with any polynomial p in π_n on the interval I_0. We extend it as a polynomial to I_1 so that the extension satisfies the connection equation at t_0 and the leading coefficients of the two polynomials are the *same*. This is done across each knot until we determine a unique function $f \in \mathcal{S}_{n,k}$ such that

$$f(t) = p(t), \qquad t \in I_0$$
$$f^{(n)}(t_j^+) = f^{(n)}(t_j^-), \qquad j = 0, 1, \ldots, k.$$

When each $A^j, j = 0, 1, \ldots, k$, is the identity matrix $f = p$. In the general case, we let $f_0, f_1, \ldots, f_n$ be the extensions into $\mathcal{S}_{n,k}$ of the monomials $1, t, \ldots, t^n$.

We need $k+1$ more functions. These functions play the roles of the truncated power functions for spline functions. For each knot $t_j, j = 0, 1, \ldots, k$ let f_{j+n+1} be the unique function in $\mathcal{S}_{n,k}$ defined by the properties

$$f_{j+n+1}(t) = 0, \ t < t_j,$$

and

$$f_{j+n+1}^{(n)}(t_\ell^+) - f_{j+n+1}^{(n)}(t_\ell^-) = \delta_{\ell j}, \qquad \ell = 0, 1, \ldots, k.$$

When each $A^j, j = 0, 1, \ldots, k$, is the identity matrix $f_{j+n+1}(t) = \frac{1}{n!}(t - t_j)_+^n$ where $t_+^n := (\max(0, t))^n$.

We claim that these functions are linearly independent on $\mathbb{R}$ and span $\mathcal{S}_{n,k}$. To prove this, it suffices to notice that on $I_0, f_j, \ j = 0, 1, \ldots, n$, are monomials and $f_{j+n+1}, j = 0, 1, \ldots, k$, are zero on I_0. Also, in a neighborhood of any knot $t_j, j = 0, 1, \ldots, k$, all the functions $f_{\ell+n+1}, \ell = 0, 1, \ldots, k, \ell \neq j$ have a continuous nth derivative while the jump in the nth derivative of f_{j+n+1} at t_j is one. The fact that they span $\mathcal{S}_{n,k}$ now follows by induction on k. □

We wish to construct a basis of *locally supported* functions for $\mathcal{S}_{n,k}$ that are variation diminishing. When the matrices $A^0, A^1, \ldots, A^k$ were nonsingular and *totally positive*, this was done in Dyn and Micchelli [DM]. We shall describe this result next. In view of the importance of Frenet matrices we will always assume here that $A^0, A^1, \ldots, A^k$ are lower triangular matrices. Nevertheless, the results of Dyn and Micchelli [DM] that we are about to present do not require this hypothesis.

Let $\mathcal{A}_n$ denote the class of all totally positive $n \times n$ nonsingular lower triangular matrices. First we wish to bound the number of zeros of a function f in $\mathcal{S}_{n,k}$ when $\{A^i\}_{i=0}^k \subseteq \mathcal{A}_n$. As we will see, this information will allow us to construct locally supported basis functions for $\mathcal{S}_{n,k}$ that are variation diminishing. Obviously a function in $\mathcal{S}_{n,k}$ may vanish on intervals, and so we must be quite explicit as to how we "count" zeros. Here, as in Dyn and Micchelli [DM], we use the zero counting convention of Goodman [Go]. Thus we say $f(a)^+ = 1, 0, -1$, if for a sufficiently small positive number ϵ, f is positive, zero, or negative on the interval $(a, a + \epsilon)$. Also, we set $f(a)^- = h(0)^+$, where $h(t) := f(a - t), t \in \mathbb{R}$. Therefore, when $f(a)^+ f(a)^- \neq 0$, f is nonzero in a neighborhood of a and there are nonnegative integers $\ell, r \leq n$, such that

$$f(a^-) = \cdots = f^{(\ell-1)}(a^-) = f(a^+) = \cdots = f^{(r-1)}(a^+) = 0,$$
$$f^{(\ell)}(a^-) f^{(r)}(a^+) \neq 0.$$

Set $q = \max(\ell, r)$. We say that f has a *point zero* of multiplicity m where

$$m = \begin{cases} q, & f(a)^- f(a)^+ (-1)^q > 0, \\ q+1, & f(a)^- f(a)^+ (-1)^q < 0. \end{cases}$$

In other words, we always have $f(a)^- f(a)^+ = (-1)^m$.

As an example, suppose that $f(a^-) = f(a^+) = 0$ while $f'(a^-) f'(a^+) \neq 0$. That is, $\ell = r = 1$. There are two cases that can occur depending on the sign of $f(a)^- f(a)^+$. Typical instances of this are depicted in Fig. 3.2.

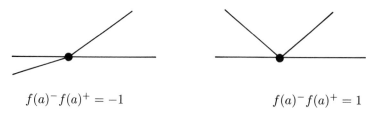

$$f(a)^- f(a)^+ = -1 \qquad\qquad f(a)^- f(a)^+ = 1$$

FIG. 3.2. *Zero counting convention.*

Since $q = 1$, in this case the multiplicity of the zero of f at a is one in the first case, while it is two in the second. If $\ell = 1$ and $r = 2$ then $q = 2$; the possibilities are shown in Fig. 3.3. Since $q = 2$, in the first case depicted above $m = 3$ while in the second case $m = 2$.

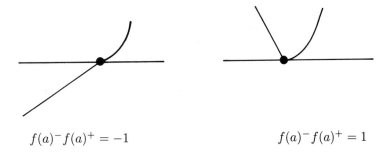

$$f(a)^- f(a)^+ = -1 \qquad\qquad\qquad f(a)^- f(a)^+ = 1$$

FIG. 3.3.   *Zero counting convention.*

Generally, we count a zero interval of f in the following way. The first case we consider is that $a < b$, $f(a)^- f(b)^+ \neq 0$, and $f(x) = 0$ for $a < x < b$. Obviously in this case, a and b correspond to knots of f and we say that f has a zero interval of multiplicity $n + 2 + k$ where k equals the number of knots in (a, b). When $f(x) = 0$, for all $x < t_j$, while $f(t_j)^+ \neq 0$ we say that $(-\infty, t_j)$ is an interval zero of multiplicity $n + 1 + j$. Finally, when f vanishes on (t_j, ∞) with $f(t_j)^- \neq 0$, we say that (t_j, ∞) is a zero interval of multiplicity $n + 1 + k - j$. The total number of zeros of f on the interval I is denoted by $Z(f|I)$. Also, recalling our notation for counting sign changes of a vector (see Chapter 1) we define

$$S_n^-(t, f) = S^-(D_n[f](t))$$

and

$$S_n^+(t, f) = S^+(D_n[f](t))$$

for values of t distinct from a knot. At a knot there are corresponding left and right values of these quantities. Our main result about zeros of $f \in S_{n,k}$ is covered by our next result, taken from Dyn and Micchelli [DM].

THEOREM 3.6 *Let* $\{A^i\}_{i=0}^k \subset A_n$. *Then for every* $f \in S_{n,k} \backslash \{0\}$

(3.44) $$Z(f|\mathbb{R}) \leq n + k + 1.$$

The proof of this result is lengthy. We describe it, piece by piece, with several auxiliary facts, the first of which gives an upper bound on the multiplicity of a zero of f at an interior knot.

LEMMA 3.3.   *Suppose that* $\{A^i\}_{i=0}^k \subset A_n$. *Let* $f \in S_{n,k}$ *be of exact polynomial degree* p, q *in* (t_{i-1}, t_i) *and* (t_i, t_{i+1}), *respectively, but not identically zero in either* *interval. Let m_i be the multiplicity of the zero of f at t_i. Then*

(3.45) $$m_i \leq S_p^+(t_i^-, f) - S_q^-(t_i^+, f) + 1.$$

Proof. Suppose $f(t_i^-) = \cdots = f^{(\ell-1)}(t_i^-) = 0 \neq f^{(\ell)}(t_i^-)$ and $f(t_i^+) = \cdots = f^{(r-1)}(t_i^+) = 0 \neq f^{(r)}(t_i^+)$ for some nonnegative integers $\ell, r \leq n$ with $\ell \leq p$ and $r \leq q$.

Since each A^i is a nonsingular lower triangular matrix, we see that, in fact, $\ell = r$. There are two cases to consider. First, suppose $m_i < n$ so that both ℓ and $r < n$. Thus, $f^{(\ell)}(t_i^-)$ and $f^{(\ell)}(t_i^+)$ have exactly the same sign, since the $r + 1$st diagonal element of A^i is positive. This means that $f(t_i)^- f(t_i)^+ = (-1)^\ell$ and so $m_i = \ell = r$, in this case. Now by the total positivity of A^i we have

$$\begin{aligned}
S_{n-1}^-(t_i^+, f) &:= S^-(D_{n-1}[f](t_i^+)) \\
&= S^-(A^i D_{n-1}[f](t_i^-)) \\
&\leq S^-(D_{n-1}[f](t_i^-)) := S_{n-1}^-(t_i^-, f),
\end{aligned}$$

that is,

(3.46) $$S_{n-1}^-(t_i^+, f) \leq S_{n-1}^-(t_i^-, f).$$

Next we observe that

(3.47) $$S_q^-(t_i^+, f) \leq S_{n-1}^-(t_i^+, f) + 1,$$

since

$$\begin{aligned}
S_q^-(t_i^+, f) &:= S^-(D_q[f](t_i^+)) \\
&\leq S^-(D_n[f](t_i^+)) \\
&\leq S^-(D_{n-1}[f](t_i^+)) + 1 \\
&= S_{n-1}^-(t_i^+, f) + 1.
\end{aligned}$$

Finally, we need that

(3.48) $$S_{n-1}^-(t_i^-, f) \leq S_p^+(t_i^-, f) - m_i,$$

which comes from the inequalities

$$\begin{aligned}
S_{n-1}^-(t_i^-, f) &:= S^-(D_{n-1}[f](t_i^-)) \\
&\leq S^-(D_p[f](t_i^-)) \\
&\leq S^+(D_p[f](t_i^-)) - \ell \\
&:= S_p^+(t_i^-, f) - m_i.
\end{aligned}$$

Now, combining (3.46)–(3.48), we get

$$\begin{aligned}
S_p^+(t_i^-, f) &\geq m_i + S_{n-1}^-(t_i^-, f) \geq m_i + S_{n-1}^-(t_i^+, f) \\
&\geq m_i - 1 + S_q^-(t_i^+, f).
\end{aligned}$$

The second case, when $m_i \geq n$, is much easier to prove. In fact in this case $S_q^-(t_i^+, f) = 0$ and $S^+(t_i^-, f) = n$.   □

For the next result we need the Budan–Fourier theorem for polynomials.

LEMMA 3.4. *Let f be a polynomial of exact degree n. Then*

$$(3.49) \qquad Z(f|(a,b)) \leq S_n^-(a^+, f) - S_n^+(b^-, f).$$

We can even allow $a = -\infty$ or $b = \infty$ above with the convention that $S_n^-(a^+, f) = n$ for $a = -\infty$ and $S_n^+(b^-, f) = 0$ for $b = \infty$.

Lemma 3.4 is a very useful result and is central to the proof of Theorem 3.6. For this reason, it is worth a brief digression to provide its proof.

Proof. For any constants $x_0, \ldots, x_n$ we have

$$S^-(x_0, \ldots, x_n) + S^+(x_0, \ldots, (-1)^n x_n) = n.$$

Thus, Lemma 3.4 has the equivalent form

$$\begin{aligned} Z(f|(a,b)) \leq n &- S^+(f(a), -f^{(1)}(a), \ldots, (-1)^n f^{(n)}(a)) \\ &- S^+(f(b), f^{(1)}(b), \ldots, f^{(n)}(b)). \end{aligned}$$

It is this version that we will prove, by induction on n. The linear case $n = 1$ is straightforward. Assume that the result is valid for all polynomials of degree $< n$. Since f has exact degree n, we can find a $\delta_0 > 0$, sufficiently small, such that for all $\epsilon \in (0, \delta_0)$, $f^{(i)}(b - \epsilon) f^{(i)}(a + \epsilon) \neq 0, i = 0, 1, \ldots, n,$

$$\begin{aligned} S^+&(f(b), f^{(1)}(b), \ldots, f^{(n)}(b)) \\ &= S^-(f(b - \epsilon), f^{(1)}(b - \epsilon), \ldots, f^{(n)}(b - \epsilon)) \end{aligned}$$

and

$$\begin{aligned} S^+&(f(a), -f^{(1)}(a), \ldots, (-1)^n f^{(n)}(a)) \\ &= S^-(f(a + \epsilon), -f^{(1)}(a + \epsilon), \ldots, (-1)^n f^{(n)}(a + \epsilon)). \end{aligned}$$

Now choose an interval $(c, d) = (a + \epsilon, b - \epsilon)$ so that the above equations hold and, in addition, all the zeros of f on (a, b) are in (c, d). By induction

$$\begin{aligned} Z(f^{(1)}|(c,d)) \leq n - 1 &- S^-(f^{(1)}(c), -f^{(2)}(c), \ldots, (-1)^n f^{(n)}(c)) \\ &- S^-(f^{(1)}(d), f^{(2)}(d), \ldots, f^{(n)}(d)), \end{aligned}$$

and, by Rolle's theorem,

$$\begin{aligned} Z(f|(c,d)) \\ \leq 1 + Z(f^{(1)}|(c,d)) &- S^-(f(c), -f^{(1)}(c)) \\ &- S^-(f(d), f^{(1)}(d)). \end{aligned}$$

Combining these inequalities proves the result.   □

It is worth mentioning that Rolle's theorem only guarantees a zero of $f^{(1)}$ between two consecutive zeros of f. However, if $S^-(f(c), -f^{(1)}(c)) = 1$ then we have an additional zero between c and the first zero of $f^{(1)}$ in (c, d). It is this simple improvement of Rolle's theorem that is embodied in the Budan–Fourier lemma (see Fig. 3.4).

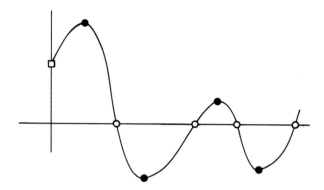

FIG. 3.4.   *Extra zero for the derivative.* $S^-(f(0), -f^{(1)}(0)) = 1$.

The next result extends Lemma 3.4 to functions in $\mathcal{S}_{n,k}$ that do not vanish on intervals.

LEMMA 3.5. *Suppose* $\{A^i\}_{i=0}^k \subset \mathcal{A}_n$ *and let* $f \in \mathcal{S}_{n,k}$ *have no zero intervals in* (a, b). *Then*

$$(3.50) \qquad Z(f|(a,b)) \le S_{p_a}^-(a^+, f) - S_{p_b}^+(b^-, f) + u(a, b),$$

where $u(a, b) := \#\{i : a < t_i < b\}$ *and* p_a, p_b *are the exact degree of* f *just to the right of* a *and left of* b, *respectively.*

Proof. When $k = -1$, that is, (a, b) contains no knot in its interior, this result is just Lemma 3.4. Otherwise, we suppose that for some i, j with $0 \le i \le j \le k$ that

$$t_{i-1} \le a < t_i < \cdots < t_j < b \le t_{j+1}.$$

Let p_m denote the exact degree of f in the interval (t_m, t_{m+1}) for $m = -1, 0, \ldots, k$ with the convention that $t_{-1} = -\infty$ and $t_{k+1} = \infty$. Then Lemma 3.4, appropriately specialized to the cases at hand, gives

$$(3.51) \qquad Z(f|(a, t_i)) \le S_{p_{i-1}}^-(a^+, f) - S_{p_{i-1}}^+(t_i^-, f),$$

$$(3.52) \qquad Z(f|(t_r, t_{r+1})) \le S_{p_r}^-(t_r^+, f) - S_{p_r}^+(t_{r+1}^-, f),$$

and

$$(3.53) \qquad Z(f|(t_j, b)) \le S_{p_j}^-(t_j^+, f) - S_{p_j}^+(b^-, f).$$

Also, we need the fact that, by Lemma 3.2, the multiplicity m_r of the zero of f at t_r satisfies

$$(3.54) \qquad m_r \le S_{p_{r-1}}^+(t_r^-, f) - S_{p_r}^-(t_r^+, f) + 1,$$

for $r = i, \ldots, j$. Since

$$Z(f|(a,b))$$
$$= Z(f|(a,t_i)) + \sum_{r=i}^{j-1} Z(f|(t_r, t_{r+1}))$$
$$+ Z(f|(t_j,b)) + \sum_{r=i}^{j} m_r,$$

we conclude, from inequalities (3.51)–(3.54), that

$$Z(f|(a,b))$$
$$\le S_{p_{i-1}}^-(a^+, f) - S_{p_{i-1}}^+(t_i^-, f)$$
$$+ \sum_{r=i}^{j-1} (S_{p_r}^-(t_r^+, f) - S_{p_r}^+(t_{r+1}^-, f))$$
$$+ S_{p_j}^-(t_j^+, f) - S_{p_j}^+(b^-, f)$$
$$+ \sum_{r=i}^{j} (S_{p_{r-1}}^+(t_r^-, f) - S_{p_r}^-(t_r^+, f) + 1)$$
$$= S_{p_{i-1}}^-(a^+, f) - S_{p_j}^+(b^-, f) + j - i + 1. \qquad \square$$

As a preparation for the proof of Theorem 3.6 we list several special cases of this lemma in the next corollary.

COROLLARY 3.4. *Suppose* $\{A^i\}_{i=0}^k \subset \mathcal{A}_n$. *Let* $f \in \mathcal{S}_{n,k}$ *and* i, j *integers with* $0 \le i < j \le k$.

(i) *If* f *has no zero intervals in* (t_i, t_j) *and* $f(t_i)^- = f(t_j)^+ = 0$ *then* $Z(f|(t_i,t_j)) \le j - i - n - 1$.

(ii) *If* f *has no zero intervals in* $(-\infty, t_i)$ *and* $f(t_i)^+ = 0$ *then* $Z(f|(-\infty, t_i)) \le i$.

(iii) *If* f *has no zero intervals in* (t_i, ∞) *and* $f(t_i)^- = 0$ *then* $Z(f|(t_i, \infty)) \le k - i$.

(iv) *If* f *has no zero intervals in* $\mathbb{R}$ *then* $Z(f|\mathbb{R}) \le n + k + 1$.

Proof. All these assertions follow from Lemma 3.5. We consider each in turn. For (i), we apply Lemma 3.5 to the interval (t_i, t_j) keeping in mind that $S_{p_i}^-(t_i^+, f) = 0$, when $f(t_i)^- = 0$, while $S_{p_{j-1}}^+(t_j^-, f) = n$ when $f(t_j)^+ = 0$, and $u(t_i, t_j) = j - i - 1$. For (ii), we choose an a such that all the zeros of f in $(-\infty, t_i)$ lie in (a, t_i). Then, $S_{p_a}^-(a^+, f) = n$ and, as above, $S_{p_{i-1}}^+(t_i^-, f) = n$ and $u(a, t_i) = i$. Therefore, by applying Lemma 3.5 to the interval (a, t_i) the result is obtained. Case (iii) is similar. Choose b such that f has all its zeros $< b$. Then $S_{p_b}^+(b^-, f) = S_{p_i}^-(t_i^+, f) = 0$ and $u(t_i, b) = k - i$. Finally for the last case, we choose a and b so that all the zeros of f are in (a, b). Then $S_{p_a}^-(a^+, f) = n, S_{p_b}^+(b^-, f) = 0$ and $u(a, b) = k + 1$. $\square$

Proof of Theorem 3.6. If f has no zero intervals in $\mathbb{R}$ the result is given by Corollary 3.4, part (iv). Next, suppose f has no zero intervals in (t_0, t_k) but vanishes identically in either $(-\infty, t_0)$ or (t_k, ∞). Three possibilities occur depending on whether f vanishes on $(-\infty, t_0)$, (t_k, ∞), or both. When f vanishes identically on $(-\infty, t_0)$, but has no zero interval in (t_0, ∞), we can use Corollary 3.4, part (iii), to get

$$Z(f|\mathbb{R}) = n + 1 + Z(f|(t_0, \infty))$$
$$\leq n + k + 1.$$

When f vanishes identically on (t_k, ∞), but has no zero interval in $(-\infty, t_k)$, we can use Corollary 3.4, part (ii), to get once again

$$Z(f|\mathbb{R}) = n + 1 + Z(f|(-\infty, t_k))$$
$$\leq n + k + 1.$$

If f vanishes identically outside of (t_0, t_k), but has no zero intervals in (t_0, t_k), then we use Corollary 3.4, part (i), to get

$$Z(f|\mathbb{R}) = 2n + 2 + Z(f|(t_0, t_k))$$
$$\leq 2n + 2 + k - n - 1 = n + k + 1.$$

There remains the case when f does indeed have a zero intervals in (t_0, t_k). Let the zero intervals be $(a_0, b_0), \ldots, (a_\nu, b_\nu)$ with $\nu \geq 0$, $t_0 \leq a_0$, $b_0 \leq t_k$, $b_\ell < a_{\ell+1}$, $\ell = 0, 1, \ldots, \nu - 1$, and $a_\ell < b_\ell$, $\ell = 0, 1, \ldots, \nu$. Each a_ℓ, b_ℓ, $\ell = 0, 1, \ldots, \nu$ corresponds to some knot of f. First suppose that f is not identically zero outside either $(-\infty, t_0)$ or (t_k, ∞). Then we have

$$Z(f|(-\infty, a_0) \leq \#\{m : t_m < a_0\},$$
$$Z(f|(b_\ell, a_{\ell+1})) \leq \#\{m : b_\ell < t_m < a_{\ell+1}\} - n, \qquad \ell = 1, \ldots, \nu,$$

and

$$Z(f|(b_\ell, \infty)) \leq \#\{m : b_\nu < t_m\}.$$

These inequalities follow from Corollary 3.4, parts (ii), (i), and (iii), respectively. Our zero counting convention leads us to the formula

$$Z(f|(a_\ell, b_\ell)) = n + 2 + \#\{m : a_\ell < t_m < b_\ell\}, \qquad \ell = 0, 1, \ldots, \nu.$$

Therefore, by adding up the bounds above we obtain the inequality

$$Z(f|\mathbb{R}) \leq \#\{m : t_m \notin \{a_0, \ldots, a_\nu, b_0, \ldots, b_\nu\}\}$$
$$+ (\nu + 1)(n + 2) - \nu n$$
$$= k + 1 - 2(\nu + 1) + (\nu + 1)(n + 2) - \nu n$$
$$= n + k + 1.$$

Next, suppose f vanishes on $(-\infty, t_0)$ while $f(t_0)^+ f(t_k)^+ \neq 0$. All the previous bounds remain valid except for $Z(f|(-\infty, a_0))$. This quantity is bounded by using the fact that

$$Z(f|(-\infty, t_0)) = n + 1,$$

and the inequality

$$Z(f|(t_0, a_0)) \leq \#\{m : t_0 < t_m < a_0\} - n,$$

obtained from Corollary 3.4, part (i). Consequently, taking account of the knot t_0, we still have, even in this case, the inequality

$$Z(f|(-\infty, a_0)) \leq \#\{m : t_m < a_0\}.$$

A similar argument takes care of the case when f vanishes identically on (t_k, ∞) and $f(t_k)^- f(t_0)^- \neq 0$ and the case when f identically vanishes outside of (t_0, t_k). $\square$

3.5. B-spline basis.

Our next proposition provides the existence of a basis of locally supported functions for the space $\mathcal{S}_{n,k}$. For this purpose we require the following lemma.

LEMMA 3.6. Let $\{A^i\}_{i=0}^n \subset \mathcal{A}_n$ and suppose $t_0 < \cdots < t_{n+1}$. Then there exists a unique function $S(t) = S(t|t_0, \ldots, t_{n+1})$ such that
(i) $S \in \mathcal{S}_{n,n+1}$,
(ii) $S(t) > 0, t_0 < t < t_{n+1}$, and zero otherwise,
(iii) $S^{(n)}(t_0^+) S^{(n)}(t_{n+1}^-) \neq 0, S^{(n)}(t_0^+) > 0$, and
(iv)

$$\int_{\mathbb{R}} S(t)dt = \frac{1}{(n+1)!}.$$

Proof. Let $f_{n+1}, \ldots, f_{2n+1}$ be the functions in $\mathcal{S}_{n,n+1}$ (constructed in the proof of Lemma 3.2) defined so that for $j = 0, 1, \ldots, n$

$$f_{n+1+j}(t) = 0, \qquad t < t_j$$

and

$$f_{n+1+j}^{(n)}(t_\ell^+) - f_{n+1+j}^{(n)}(t_\ell^-) = \delta_{j\ell}$$

for $\ell = 0, 1, \ldots, n+1$. Obviously there are nonzero constants $c_0, \ldots, c_n$ such that the function

$$f(t) := \sum_{j=0}^n c_j f_{n+1+j}(t)$$

satisfies $f^{(\ell)}(t_{n+1}^+) = 0, \ell = 0, 1, \ldots, n$. This function is necessarily zero on $(-\infty, t_0]$ and (t_{n+1}, ∞). Therefore, according to Theorem 3.6 with $(k = n + 1)$ it follows that

$$Z(f|\mathbb{R}) \leq 2n + 2.$$

However, we also have by the construction of f that $Z(f|\mathbb{R}) \geq 2n + 2$, since f vanishes on $(-\infty, t_0]$ and (t_{n+1}, ∞). Hence we conclude f has no further zeros in (t_0, t_{n+1}). This means that $f^{(n)}(t_0^+)f^{(n)}(t_{n+1}^-) \neq 0$ and that f has no zero in (t_0, t_{n+1}). Therefore, for some constant c, $S := cf$ satisfies all our requirements.

To prove the uniqueness of S we suppose that there are two functions S_1 and S_2 that satisfy (i)–(iv). Then, in view of (iv), the difference $g := S_1 - S_2$ has at least one zero in (t_0, t_{n+1}). However, by (ii) it also has $2n + 2$ zeros, as it vanishes on $(-\infty, t_0)$ and (t_{n+1}, ∞). This gives at least $2n + 3$. But (i) and Theorem 3.6 (with $k = n + 1$) allow only $2n + 2$.   □

Note that when each $A^i, i = 0, 1, \ldots, n+1$, is the identity matrix the function S in Lemma 3.6 is the B-spline of Curry and Schoenberg [CS]. Moreover, if $[t_0, \ldots, t_{n+1}]f$ is the $n + 1$st divided difference of f at $t_0, \ldots, t_{n+1}$ then

$$\int_\mathbb{R} S(t)f^{(n+1)}(t)dt = [t_0, \ldots, t_{n+1}]f$$

and

$$S(t) = \frac{1}{n!}[t_0, \ldots, t_{n+1}](\cdot - t)_+^n, \qquad t \in \mathbb{R},$$

is the function described in Lemma 3.6, since properties (i)–(iv) can be checked directly from this formula.

Lemma 3.6 establishes the existence of geometrically continuous B-splines corresponding to any family of totally positive lower triangular matrices. Our intention in the remainder of this chapter is to develop properties of the B-splines that parallel the special case when all the connection matrices are the identity matrix.

It is worthwhile observing at this juncture that B-splines exist even if the connection matrices are not totally positive. In fact, when the knots are at the integers and the same connection matrix is used at each knot, the corresponding B-splines are studied in [DEM]. In this generality the B-splines are not necessarily positive and therefore have unexpected shapes. For instance, the functions in Fig. 3.5 are all B-splines of degree three.

We now consider piecewise polynomials on $\mathbb{R}$ corresponding to an *infinite* partition of $\mathbb{R}$ into bounded intervals. Thus we are given an infinite sequence $\{t_i\}_{i\in\mathbb{Z}}$ with $\lim_{i\to\pm\infty} t_i = \pm\infty$ and $t_i < t_{i+1}, i \in \mathbb{Z}$, a corresponding family of matrices $\{A^i\}_{i\in\mathbb{Z}} \subset \mathcal{A}_n$ and the space

$$\mathcal{S}_n = \left\{ f : f|_{(t_i, t_{i+1})} \in \pi_n, \ D_{n-1}[f_+](t_i) = A^i D_{n-1}[f_-](t_i), \ i \in \mathbb{Z} \right\}.$$

For each $i \in \mathbb{Z}$, we let $S_i(t) = S(t|t_i, \ldots, t_{i+n+1}), t \in \mathbb{R}$, be the function constructed in Lemma 3.6 corresponding to the knots $t_i, \ldots, t_{i+n+1}$ and matrices $A^i, \ldots, A^{i+n+1}$.

PROPOSITION 3.5. *Suppose* $\{A^i\}_{i\in\mathbb{Z}} \subset \mathcal{A}_n$. *Then* $f \in \mathcal{S}_n$ *if and only if it can be expressed as a unique linear combination of the functions* $\{S_i\}_{i\in\mathbb{Z}}$, *that is,*

(3.55) $$f(t) = \sum_{i\in\mathbb{Z}} c_i S_i(t), \qquad t \in \mathbb{R}.$$

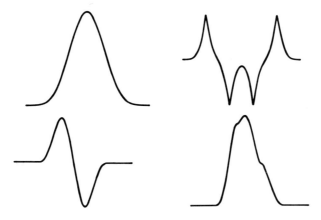

FIG. 3.5. *Geometrically continuous B-splines.*

Proof. First we prove that for each i, the functions $S_{i-n}, \ldots, S_i$ are linearly independent on the interval (t_i, t_{i+1}). To this end, we consider a typical linear combination of these functions, viz.

$$g(t) = \sum_{j=i-n}^{i} c_j S_j(t), \qquad t \in \mathbb{R}.$$

Then $g \in \mathcal{S}_{n,2n+1}(t_{i-n}, \ldots, t_{i+n+1})$ and therefore, by Theorem 3.6, when g is not identically zero, we obtain the inequality

$$Z(g|\mathbb{R}) \leq n + 2n + 1 = 3n + 1.$$

Now, g vanishes to the left of t_{i-n} and to the right of t_{i+n+1}. This gives us at least $2n + 2$ zeros for g. If it vanishes identically on (t_i, t_{i+1}) we would get $n + 2$ more zeros for a total of at least $3n + 4$ zeros. This number exceeds the bound above unless g is identically zero on $\mathbb{R}$. Thus we have established the linear independence of the functions $S_{i-n}, \ldots, S_i$ on (t_i, t_{i+1}).

On the interval (t_i, t_{i+1}) these $n + 1$ functions are polynomials of degree $\leq n$ and hence must span all such polynomials. In particular, for $f \in \mathcal{S}$ there are unique constants $c_{i-n}, \ldots, c_i$ such that

$$f(t) = \sum_{j=i-n}^{i} c_j S_j(t), \qquad t \in (t_i, t_{i+1}).$$

Let us now consider the function

$$g_i(t) := f(t) - \sum_{j=i-n}^{i} c_j S_j(t),$$

which has been arranged to be zero on (t_i, t_{i+1}). In other words,

$$g_i^{(\ell)}(t_{i+1}^-) = 0, \qquad \ell = 0, 1, \ldots, n-1,$$

from which it follows that $g_i^{(\ell)}(t_{i+1}^+) = 0$, $\ell = 0, 1, \ldots, n-1$. In view of Lemma 3.6, part (iii), we can choose a unique constant c_{i+1} such that the function

$$g_{i+1}(t) := g_i(t) - c_{i+1} S_{i+1}(t)$$

satisfies the equations $g_{i+1}^{(\ell)}(t_{i+1}^+) = 0$, $\ell = 0, 1, \ldots, n$. This means that g vanishes identically on the next interval (t_{i+1}, t_{i+2}). In other words, we have

$$f(t) = \sum_{j=i-n}^{i+1} c_j S_j(t), \qquad t \in (t_i, t_{i+2}).$$

Proceeding to the right of t_{i+2} in this way past each knot gives us unique constants $c_{i-n}, c_{i-n+1}, \ldots$ such that

$$f(t) = \sum_{j=i-n}^{\infty} c_j S_j(t), \qquad t > t_i.$$

Similarly, we proceed to the left of t_i, one knot at a time to finally obtain the desired (unique) representation of f. $\qquad \square$

We remark that the functions $\{S_i\}_{i \in \mathbb{Z}}$ can be used to generate a basis for the space $\mathcal{S}_{n,k}$. Clearly, every $f \in \mathcal{S}_{n,k}$ is in $\mathcal{S}_n$ and hence can be written as

$$f(t) = \sum_{j \in \mathbb{Z}} c_j S_j(t), \qquad t \in \mathbb{R}.$$

If we restrict this representation to some interval (a, b), with $t_{-1} < a < t_0 < t_k < b < t_{k+1}$, we have the formula

$$f(t) = \sum_{j=-n-1}^{k} c_j S_j(t), \qquad t \in (a, b).$$

Thus the $n + k + 2$ functions $S_{-n-1}, \ldots, S_k$ give a locally supported basis for $\mathcal{S}_{n,k}$.

To proceed further, we introduce the subclass $\mathcal{B}_n$ of all matrices A in $\mathcal{A}_n$ with the property that $A_{j0} = A_{0j} = \delta_{0j}, j = 0, 1, \ldots, n-1$, that is, A is a connection matrix. Whenever $\{A_j\}_{j \in \mathbb{Z}} \subset \mathcal{B}_n$, we can be sure that all functions in $\mathcal{S}_n$ are continuous on $\mathbb{R}$ and that constants belong to $\mathcal{S}_n$. Thus, from Proposition 3.5 when $\{A^i\}_{i \in \mathbb{Z}} \subseteq \mathcal{B}_n$, there are unique constants $\{a_i\}_{i \in \mathbb{Z}}$ such that

$$\sum_{i \in \mathbb{Z}} a_i S_i(t) = 1, \qquad t \in \mathbb{R}.$$

We want to show that $a_i > 0$ for all $i \in \mathbb{Z}$. For this purpose, we restrict the above equation to the interval (t_{i+n}, t_{i+n+1}). Then the function

$$F(t) = \sum_{j=i}^{i+n} a_j S_j(t)$$

has the property that $F(t) = 1$ for $t \in (t_{i+n}, t_{i+n+1})$. Hence, it follows that $F^{(\ell)}(t_{i+n}^+) = 0, \ell = 1, \ldots, n-1$. Set

$$h(t) = \begin{cases} F'(t), & t < t_{i+n}, \\ 0, & t \geq t_{i+n}. \end{cases}$$

For each $j \in \mathbb{Z}$, let $\hat{A}^j$ be the $(n-1) \times (n-1)$ submatrix of A^j obtained by deleting its first row and column. Also, denote by $\mathcal{S}_{n-1}$ the space of piecewise polynomials corresponding to the family $\hat{\mathcal{A}} := \{\hat{A}^j\}_{j \in \mathbb{Z}}$ on the same partition $\{t_j\}_{j \in \mathbb{Z}}$. Clearly, $\hat{\mathcal{A}} \subseteq \mathcal{A}_{n-1}$ because $\mathcal{A} \subseteq \mathcal{B}_n$ and therefore, by Theorem 3.6, $Z(h|\mathbb{R}) \leq 2n$ (here $k = n$ and n is replaced by $n-1$). However, as h vanishes to the left of t_i and the right of t_{i+n} it has, by our zero counting convention, at least $2n$ zeros. Thus, F' doesn't vanish in (t_i, t_{i+n}) and so F must be strictly positive there. However, since $F^{(n)}(t_i^+) = a_i S_i^{(n)}(t_i^+)$ and $S_i^{(n)}(t_i^+) > 0$ we get by (iii) of Lemma 3.6 that $a_i > 0$, $i \in \mathbb{Z}$.

We now introduce the functions $M_i := a_i S_i, i \in \mathbb{Z}$, which are everywhere positive on (t_i, t_{i+n+1}), zero otherwise, and satisfy the equation

$$\sum_{i \in \mathbb{Z}} M_i(t) = 1, \qquad t \in \mathbb{R}.$$

3.6.   Dual functionals.

In Barry, Dyn, Goldman, and Micchelli [BDGM], a family of linear functionals $\{L_i\}_{i \in \mathbb{Z}}$ were constructed that are "dual" to the set of functions $\{M_i\}_{i \in \mathbb{Z}}$ in the sense that

$$L_i(M_j) = \delta_{ij}, \qquad i, j \in \mathbb{Z}.$$

The basic ingredient in this construction is a certain *finite* dimensional linear space of piecewise polynomials. To explain this idea we consider a family of $(n+1) \times (n+1)$ matrices $\{C^i\}_{i \in \mathbb{Z}} \subset \mathcal{A}_{n+1}$ and define

$$\mathcal{P} = \mathcal{P}^C = \{f : f|_{[t_i, t_{i+1}]} \in \pi_n, D_n[f_+](t_i) = C^i D_n[f_-](t_i), i \in \mathbb{Z}\}.$$

Note that at each knot we impose $n+1$ connection conditions on the derivatives of a function $\mathcal{P}$. In contrast, for the space $\mathcal{S}_n$ only n are imposed. When each C^i is the identity matrix, $\mathcal{P}$ becomes the space π_n. Generally, since each C^i is nonsingular we can begin on any interval between consecutive knots with some polynomial p and extend it *uniquely* to all of $\mathbb{R}$ as an element of $\mathcal{P}$. In particular, $\dim \mathcal{P} = n+1$ and moreover any nontrivial element of $\mathcal{P}$ cannot vanish on an interval. The analogue of Theorem 3.6 for $\mathcal{P}$ is given next.

THEOREM 3.7. *Suppose that* $\{C^i\}_{i\in\mathbb{Z}} \subset \mathcal{A}_{n+1}$ *and* $\{t_i\}_{i\in\mathbb{Z}}$ *is a partition of* $\mathbb{R}$. *Then for any* $f \in \mathcal{P}\backslash\{0\}$ *we have*

$$(3.56) \qquad\qquad Z(f|\mathbb{R}) \leq n.$$

Proof. The proof of this result parallels the proof of Theorem 3.6, but it is much simpler. First, keep in mind that a function $f \in \mathcal{P}\backslash\{0\}$ *only has point zeros*, which are counted as before. Let m_i be the multiplicity of the zero of f at t_i. With the same notation used in Lemma 3.2, we see that (3.46) and (3.48) become

$$(3.57) \qquad\qquad S_n^-(t_i^+, f) \leq S_n^-(t_i^-, f)$$

and

$$(3.58) \qquad\qquad S_n^-(t_i^-, f) \leq S_p^+(t_i^-, f) - m_i.$$

(The only difference here is $n - 1$ is replaced by n). Therefore, (3.47) becomes the inequality

$$(3.59) \qquad\qquad S_q^-(t_i^+, f) \leq S_n^-(t_i^+, f).$$

This means, in the present circumstance, that

$$m_i \leq S_p^+(t_i^-, f) - S_q^-(t_i^+, f), \qquad i \in \mathbb{Z}.$$

Hence, if we apply the Budan–Fourier Lemma 3.4 to f on intervals between successive knots in a given interval (a, b), as in Lemma 3.5, we simply get

$$(3.60) \qquad\qquad Z(f|(a, b)) \leq S_{p_a}^-(a^+, f) - S_{p_b}^+(b^-, f).$$

This implies that f has at most a finite number of zeros on any interval (a, b), and hence a finite number of zeros on $\mathbb{R}$. Now, choose (a, b) large enough to include all the zeros of f in its interior. Then $S_{p_a}^-(a^+, f) = n$ and $S_{p_b}^+(b^-, f) = 0$ and so (3.60) gives the desired bound (3.56). $\qquad\square$

We use this result in the following way. We return to our space $\mathcal{S}_n$ determined by a partition $\{t_i\}_{i\in\mathbb{Z}}$ and $n \times n$ matrices $\{A^i\}_{i\in\mathbb{Z}} \subset \mathcal{A}_n$. Define $(n+1) \times (n+1)$ matrices

$$(3.61) \qquad\qquad C_i = \begin{bmatrix} & & & 0 \\ & A^i & & \vdots \\ & & & 0 \\ 0 \cdots 0 & & 1 \end{bmatrix},$$

so that $\mathcal{P}^C \subseteq \mathcal{S}_n$ and every element in $\mathcal{P}^C$ has the same leading coefficient on every interval (t_i, t_{i+1}). We also introduce the $(n + 1) \times (n + 1)$ matrices

$$R^T = \begin{bmatrix} 0 & & \cdots & & 0 & 1 \\ & & & & & \\ \vdots & & & & \cdot & 0 \\ & & & & & \\ 0 & & & \cdot & & \\ 0 & (-1)^{n-1} & & & & \vdots \\ (-1)^n & 0 & & \cdots & & 0 \end{bmatrix}$$

and

$$(3.62) \qquad\qquad E_i^T := R^{-1} C_i^{-1} R.$$

We claim that each E_i is in $\mathcal{A}_{n+1}$. In fact, since C_i is in $\mathcal{A}_{n+1}$, this observation is a consequence of the following standard fact.

LEMMA 3.7. *Let C be an $(n+1) \times (n+1)$ nonsingular totally positive matrix. Then the matrix B defined by*

$$B^T = R^{-1} C^{-1} R$$

is also totally positive

Proof. For every $0 \le i_1 < \cdots < i_p \le n$ and $0 \le j_1 < \cdots < j_p \le n$ we have

$$B^T \left(\begin{matrix} j_1, \ldots, j_p \\ i_1, \ldots, i_p \end{matrix} \right) = (-1)^{\sum_{m=1}^{p}(i_m + j_m)} C^{-1} \left(\begin{matrix} n - i_p, \ldots, n - i_1 \\ n - j_p, \ldots, n - j_1 \end{matrix} \right)$$

$$= \frac{C \left(\begin{matrix} i_1', \ldots, i_{n+1-p}' \\ j_1', \ldots, j_{n+1-p}' \end{matrix} \right)}{C \left(\begin{matrix} 0, 1, \ldots, n \\ 0, 1, \ldots, n \end{matrix} \right)},$$

where $\{i_1', \ldots, i_{n+1-p}'\}, \{j_1', \ldots, j_{n+1-p}'\}$ are the complementary indices to $\{n - i_p, \ldots, n - i_1\}, \{n - j_p, \ldots, n - j_1\}$ relative to $\{0, 1, \ldots, n\}$, respectively. In the last equation, we used the formula relating the minors of an inverse of a matrix to the minors of the matrix itself, cf. Karlin [K, p. 5].   □

Note that $(E_i)_{nj} = \delta_{nj}, j = 0, 1, \ldots, n$ and so any piecewise polynomial in $P^{\mathcal{E}}$ has the same leading coefficient on every knot interval (t_i, t_{i+1}). We specify a family of functions $\{N_i\}_{i \in \mathbb{Z}}$ by the following conditions:

(i) $N_i \in P^{\mathcal{E}}, \ \mathcal{E} := \{E_i\}_{i \in \mathbb{Z}}$,
(ii) $N_i(t_j) = 0, j = i + 1, \ldots, i + n$, *and*
(iii) $N_i^{(n)}(t) = n!(-1)^n, \ t \in \mathbb{R}$.

The existence of nontrivial functions satisfying (i) and (ii) is clear since $\dim P^{\mathcal{E}} = n$. Theorem 3.7 implies that each N_i has *exactly* n zeros given by (ii) and no more. Therefore, N_i is of exact degree n and can be normalized to satisfy (iii). Note that when each A^i is the identity matrix, N_i is given explicitly by

$$(3.63) \qquad N_i(t) := (t_{i+1} - t) \cdots (t_{i+n} - t), \qquad t \in \mathbb{R}.$$

For the general case, we follow Barry, Dyn, Goldman, and Micchelli [BDGM] and define for every $x \in \mathbb{R}, i \in \mathbb{Z}$ and function f, which is n times continuously differentiable in some neighborhood of x, the linear functional

$$(3.64) \qquad L_i(x)(f) := \sum_{r=0}^{n} \frac{(-1)^{n-r}}{n!} N_i^{(n-r)}(x) f^{(r)}(x).$$

First we shall show that for any f in $\mathcal{S}_n$, $L_i(x)(f)$ is a *continuous* function of x on the interval (t_i, t_{i+n+1}). Clearly $L_i(x)f$ is defined and continuous except at the knots $\{t_{i+1}, \ldots, t_{i+n}\}$. Now, choose $x = t_j$ for some $j \in \{i+1, \ldots, i+n\}$ and observe that

$$L_i(x^+)(f) = \frac{1}{n!} \sum_{r=0}^{n} (-1)^{n-r} N_i^{(n-r)}(t_j^+) f^{(r)}(t_j^+)$$

$$= \frac{1}{n!} (D_n[f_+](t_j), RD_n[(N_i)_+](t_j)).$$

Since $N_i(t_j) = 0$ this equation becomes

$$L_i(x^+)(f) = \frac{1}{n!} (C_i D_n[f_-](t_j), RE_i D_n[(N_i)_-](t_j))$$

$$= \frac{1}{n!} (D_n[f_-](t_j), C_i^T RE_i D_n[(N_i)_-](t_j)).$$

However, (3.62) gives us the formula

$$C_i^T RE_i = C_i^T RR^T C_i^{-T} R^{-T}.$$

Moreover, by the definition of R we see that

$$R^T = R^{-1}$$

and so $C_i^T RE_i = R$. Using this equation above confirms that $L_i(x^+)(f) = L_i(x^-)(f)$. Thus $L_i(x)f$ is indeed a continuous piecewise polynomial on (t_i, t_{i+n+1}).

We claim that it is, in fact, a *constant* there. To see this, we pick any $x \in (t_i, t_{i+n+1}) \backslash \{t_{i+1}, \ldots, t_{i+n}\}$ and compute the derivative of $L_i(x)f$ with respect to x, viz.

$$\frac{d}{dx}(L_i(x)(f))$$

$$= \sum_{r=0}^{n} \frac{(-1)^{n-r}}{n!} N_i^{(n+1-r)}(x) f^{(r)}(x)$$

$$+ \sum_{r=0}^{n} \frac{(-1)^{n-r}}{n!} N_i^{(n-r)}(x) f^{(r+1)}(x)$$

$$= \frac{(-1)^n}{n!} N_i^{(n+1)}(x) f(x) + \frac{1}{n!} N_i(x) f^{(n+1)}(x)$$

$$+ \frac{1}{n!} \sum_{r=0}^{n-1} (-1)^{n-r} \left\{ N_i^{(n-r)}(x) f^{(r+1)}(x) - N_i^{(n-r)}(x) f^{(r+1)}(x) \right\}$$

$$= 0.$$

Therefore $L_i(x)f$ is indeed a constant, independent of x, on (t_i, t_{i+n+1}). Our principal observation about these linear functions comes next.

THEOREM 3.8. *Suppose that* $\{A^i\}_{i \in \mathbb{Z}} \subset \mathcal{B}_n$ *and* $\{t_i\}_{i \in \mathbb{Z}}$ *is a partition of* $\mathbb{R}$. *Then* $L_i(x)(f)$ *is constant independent of* x *in* (t_i, t_{i+n+1}) *and*

(3.65) $$L_i(x)(M_j) = \delta_{ij}, \qquad i, j \in \mathbb{Z}.$$

Proof. To prove equation (3.65), we fix an $i \in \mathbb{Z}$. If $j \leq i - n - 1$ or $j \geq i + n + 1$ then $L_i(x)(M_j) = 0$, because M_j vanishes on (t_i, t_{i+n+1}). For $i - n \leq j < i$, choose any $x \in (t_{i+n}, t_{i+n+1})$. Again, $L_i(x)(M_j) = 0$, since M_j vanishes on (t_{i+n}, t_{i+n+1}). Finally, for $i < j \leq i + n$, we choose an $x \in (t_i, t_{i+1})$ and conclude, as before, that $L_i(x)(M_j) = 0$. Since $L_i(x)(f)$ is *independent* of $x \in (t_i, t_{i+n+1})$ when $f \in \mathcal{S}_n$ we have proved (3.65), as long as $j \neq i$.

It remains to demonstrate that $L_i(x)(M_i) = 1$. Here we use the fact that $M_i, i \in \mathbb{Z}$, were chosen to form a partition of unity, that is,

(3.66) $$1 = \sum_{j \in \mathbb{Z}} M_i(t), \qquad t \in \mathbb{R}.$$

Since $L_i(x)(1) = 1$ for $x \in (t_i, t_{i+n+1})$, we get, by applying $L_i(x)$ to both sides of (3.66) and using what we have already proved, that

$$1 = L_i(x)(1) = \sum_{j \in \mathbb{Z}} L_i(x)(M_j) = L_i(x)(M_i). \qquad \square$$

Our next result, from Barry, Dyn, Goldman, and Micchelli [BDGM], ties the families of functions $\{N_i\}_{i \in \mathbb{Z}}$ and $\{M_i\}_{i \in \mathbb{Z}}$ together by a *binomial identity*. To give meaning to this statement we need the correct interpretation of the function $(x - y)^n$ relative to our set of matrices $\{A^i\}_{i \in \mathbb{Z}}$.

Recall the fact that the functions $\{N_i\}_{i \in \mathbb{Z}}$ were constructed in $\mathcal{P}^{\mathcal{E}}$ relative to the family $\mathcal{E} = \{E_i\}_{i \in \mathbb{Z}}$ given by (3.62). In the next result, the family $\mathcal{C} = \{C_i\}_{i \in \mathbb{Z}}$ of matrices defined by (3.61) and the corresponding space $\mathcal{P}^{\mathcal{C}}$ will also appear. Recall the fact, mentioned earlier, that $\mathcal{P}^{\mathcal{C}} \subset \mathcal{S}_n$.

PROPOSITION 3.6. *Suppose that* $\{A^i\}_{i \in \mathbb{Z}} \subset \mathcal{A}_n$. *There exists a unique function* $g(x, y), x, y \in \mathbb{R}$ *such that* $g(\cdot, y) \in \mathcal{P}^{\mathcal{C}}$ *for each* $y \in \mathbb{R}$. *Moreover, whenever for some* $i \in \mathbb{Z}$, *both* $x, y \in (t_i, t_{i+1})$ *then* $g(x, y) = (x - y)^n$.

For any $k \in \mathbb{Z}$ *and* $i = 0, 1, \ldots, n$, *let* $p_i(\cdot, t_k), q_i(\cdot, t_k)$ *be the unique functions such that* $p_i(\cdot, t_k) \in \mathcal{P}^{\mathcal{C}}, q_i(\cdot, t_k) \in \mathcal{P}^{\mathcal{E}}$ *and*

$$p_i(t, t_k) = q_i(t, t_k) = \frac{(t - t_k)^i}{i!}, \qquad t \in (t_k, t_{k+1}),$$

$i = 0, 1, \ldots, n$. *Then the function* g *can be expressed in the form*

(3.67) $$g(x, y) = n! \sum_{i=0}^{n} (-1)^{n-i} p_i(x, t_k) q_{n-i}(y, t_k).$$

Note that the left-hand side of (3.67) is independent of t_k. Also, in the special case that each $A^i, i \in \mathbb{Z}$, is an identity matrix, we have

$$p_i(x) = q_i(x) = \frac{(x - t_k)^i}{i!}, \qquad x \in \mathbb{R}$$
$$g(x, y) = (x - y)^n,$$

and so (3.67) is just the binomial theorem.

Proof. The functions $p_i(\cdot, t_k), q_i(\cdot, t_k)$, $i = 0, 1, \ldots, n$, are well-defined because of the previously mentioned fact that every polynomial of degree $\leq n$ has a unique extension from any knot interval to all of $\mathbb{R}$ in either $\mathcal{P}^{\mathcal{C}}$ or $\mathcal{P}^{\mathcal{E}}$. The existence of $g(x, y)$ is not clear since we want it to reduce to $(x - y)^n$ whenever x, y are in the same knot interval. We establish its existence in the following way. To this end, we first note that there is at most one function $g(x, y)$, $x, y \in \mathbb{R}$, with the desired properties. To see this, suppose that there is another such function, say $\tilde{g}(x, y)$. Now fix any knot interval $(t_i, t_{i+1}), i \in \mathbb{Z}$, and a $y \in (t_i, t_{i+1})$. For $x \in (t_i, t_{i+1})$, $\tilde{g}(x, y)$ and $g(x, y)$ agree. Since they are both in $\mathcal{P}^{\mathcal{C}}$ as functions of x, they must agree everywhere on $\mathbb{R}$. But $i \in \mathbb{Z}$ was arbitrarily chosen and so g and $\tilde{g}$ are the same.

Next for any integer k, the right-hand side of (3.67) is in $\mathcal{P}^{\mathcal{C}}$ as a function of x for any fixed y. When x, y are in (t_k, t_{k+1}), it reduces to $(x - y)^n$ by the binomial theorem. The essence of the proof is to show that the right-hand side of (3.67) agrees with $(x - y)^n$ whenever x, y are in *any* knot interval (t_i, t_{i+1}) $i \in \mathbb{Z}$. Then, by what we said above, it follows that this function is *independent* of k and has all the properties expected of g.

Call the right-hand side of (3.67) $H(x, y)$. The proof of our claim is made by induction. We suppose for some $r \in \mathbb{Z}$ that $H(x, y)$ reduces to $(x - y)^n$ when $x, y \in (t_{r-1}, t_r)$. We will show that $H(x, y) = (x - y)^n$ when $x, y \in (t_r, t_{r+1})$. This would mean, by induction, that for any $x, y \in (t_r, t_{r+1})$ with $r \geq k$ it would follow that $H(x, y) = (x - y)^n$. Similarly, moving to the left we will also show under the same condition above that $H(x, y) = (x - y)^n$, when $x, y \in (t_{r-2}, t_{r-1})$.

We begin the proof of these two assertions with a forward induction step. For $i \in \mathbb{Z}$, let

$$\mathbf{P}_i(x) := (p_0(x, t_k), \ldots, p_n(x, t_k))^T, \qquad x \in (t_{i-1}, t_i)$$

and

$$\mathbf{Q}_i(x) := (q_0(x, t_k), \ldots, q_n(x, t_k))^T, \qquad x \in (t_{i-1}, t_i).$$

We also need the monomials

$$v_{li}(x) = \frac{(x - t_i)^l}{k!}, \qquad l = 0, 1, \ldots, n$$

and the vector

$$\mathbf{V}_i(x) = (v_{0i}(x), \ldots, v_{ni}(x))^T.$$

Finally, define the matrices

$$(D(P)(t))_{jl} := p_j^{(l)}(t, t_k), \qquad j, l = 0, 1, \ldots, n$$

and

$$(D(Q)(t))_{jl} := q_j^{(l)}(t, t_k), \qquad j, l = 0, 1, \ldots, n.$$

With this notation it follows that for all $\ell \in \mathbb{Z}$

(3.68) $$D(P_+)(t_\ell) = D(P_-)(t_\ell)C_\ell^T$$

and

(3.69) $$D(Q_+)(t_\ell) = D(Q_-)(t_\ell)E_\ell^T,$$

since $p_i \in \mathcal{P}^\mathcal{C}$ and $q_i \in \mathcal{P}^\mathcal{E}, i = 0, 1, \ldots, n$.

Now, suppose for $x, y \in (t_{r-1}, t_r)$

$$(x - y)^n = n! \sum_{i=0}^{n} (-1)^{n-i} p_i(x, t_k) q_{n-i}(y, t_k)$$

$$= n!(P_r(x), R\mathbf{Q}_r(y)).$$

By Taylor's theorem we also have the equations

$$\mathbf{P}_r(x) = D(P_-)(t_r)\mathbf{V}_r(x), \qquad x \in (t_{r-1}, t_r),$$

and

$$\mathbf{Q}_r(y) = D(Q_-)(t_r)\mathbf{V}_r(y), \qquad y \in (t_{r-1}, t_r).$$

Substituting these equations into the previous one gives us the formulas

$$(x - y)^n = n!(D(P_-)(t_r)\mathbf{V}_r(x), RD(Q_-)(t_r)\mathbf{V}_r(y))$$

$$= n!(\mathbf{V}_r(x), \{D(P_-)(t_r)^T RD(Q_-)(t_r)\} \mathbf{V}_r(y)).$$

Since the binomial theorem says that

$$(x - y)^n = n!(\mathbf{V}_r(x), R\mathbf{V}_r(y))$$

we conclude that

(3.70) $$H(x, y) = (x - y)^n, \text{ for } x, y \in (t_{r-1}, t_r)$$

if and only if

(3.71) $$D(P_-)(t_r)^T RD(Q_-)(t_r) = R.$$

We could just as easily expand $\mathbf{P}_r$ and $\mathbf{Q}_r$ on (t_{r-1}, t_r) in a monomial basis about t_{r-1} (we will need to do this for our backward induction step). If we proceed in this way, we get also that

(3.72) $$D(P_+)(t_{r-1})^T RD(Q_+)(t_{r-1}) = R$$

as a condition ensuring that (3.70) holds.

Call the matrix on the left-hand side of equation (3.71) F_r. Suppose that $x, y \in (t_r, t_{r+1})$, then Taylor's theorem, (3.68), and (3.69) give

(3.73)
$$\begin{aligned}
\mathbf{P}_{r+1}(x) &= D(P_+)(t_r)\mathbf{V}_r(x), \\
&= D(P_-)(t_r)C_r^T\mathbf{V}_r(x),
\end{aligned}$$

and similarly,

(3.74)
$$\mathbf{Q}_{r+1}(y) = D(Q_-)(t_r)E_r^T\mathbf{V}_r(y).$$

Hence, using equations (3.71)–(3.74) we conclude that

$$\begin{aligned}
H(x, y) &= n!\, (\mathbf{P}_{r+1}(x), R\mathbf{Q}_{r+1}(y)) \\
&= n!(D(P_-)(t_r)C_r^T\mathbf{V}_r(x), RD(Q_-)(t_r)R^{-1}C_r^{-1}R\mathbf{V}_r(y)) \\
&= n!(\mathbf{V}_r(x), C_rF_rR^{-1}C_r^{-1}R\mathbf{V}_r(y)) \\
&= n!(\mathbf{V}_r(x), R\mathbf{V}_r(y)) \\
&= (x - y)^n,
\end{aligned}$$

where in the last equation we again used the binomial theorem. This computation advances the induction hypothesis in the forward direction.

To step backwards from the interval (t_{r-1}, t_r) we now suppose $x, y \in (t_{r-2}, t_{r-1})$ and now proceed to compute $H(x, y)$, knowing that (3.72) holds. Specifically, we have the equations

$$\begin{aligned}
H(x, y) &= n!(\mathbf{P}_{r-1}(x), R\mathbf{Q}_{r-1}(y)) \\
&= n!(D(P_-)(t_{r-1})\mathbf{V}_{r-1}(x), RD(Q_-)(t_{r-1})\mathbf{V}_{r-1}(y)) \\
&= n!(\mathbf{V}_{r-1}(x), D(P_-)(t_{r-1})^T RD(Q_-)(t_{r-1})\mathbf{V}_{r-1}(y)).
\end{aligned}$$

According to (3.72), (3.68)–(3.69), and (3.62) we obtain

$$C_{r-1}D(P_-)(t_{r-1})^T RD(Q_-)(t_{r-1})R^{-1}C_{r-1}^{-1}R = R,$$

which is an equivalent way of saying that

$$D(P_-)(t_{r-1})^T RD(Q_-)(t_{r-1}) = R.$$

Hence, the above expression for H simplifies to

$$\begin{aligned}
H(x, y) &= n!(\mathbf{V}_{r-1}(x), R\mathbf{V}_{r-1}(y)) \\
&= (x - y)^n. \qquad \square
\end{aligned}$$

THEOREM 3.9. *Suppose that $\{A^i\}_{i\in\mathbb{Z}} \subseteq \mathcal{B}_n$ and $\{t_i\}_{i\in\mathbb{Z}}$ is a partition of $\mathbb{R}$. Then*

$$g(x, y) = \sum_{i\in\mathbb{Z}} M_i(x)N_i(y), \qquad x, y \in \mathbb{R}.$$

Proof. As we already pointed out, $\mathcal{P}^\mathcal{C}$ is a subspace of $\mathcal{S}_n$. Hence, by Proposition 3.5 we can express $g(x, y)$ in the form

$$(3.75) \qquad g(x, y) = \sum_{i \in \mathbb{Z}} c_i(y) M_i(x), \qquad x, y \in \mathbb{R},$$

where necessarily each $c_i \in \mathcal{P}^\mathcal{E}$. This can easily be seen from formula (3.67) of Proposition 3.6. Now, pick any knot interval (t_r, t_{r+1}). By Proposition 3.6, (3.75) specializes to the formula

$$(3.76) \qquad (x - y)^n = \sum_{i \in \mathbb{Z}} c_i(y) M_i(x), \qquad x, y \in (t_r, t_{r+1}).$$

Apply the linear functional $L_r(y)$ to both sides of this equation, as a function of x. Using definition (3.64) of $L_i(y)$, we get on the left $N_i(y)$ and on the right, by Theorem 3.8, we get $c_r(y)$. Thus c_r and N_r agree on the interval (t_r, t_{r+1}). Since both are in $\mathcal{P}^\mathcal{E}$ they agree everywhere.   □

When each A^i is the identity matrix, Theorem 3.9 reduces to a well-known identity of Marsden, cf. Schumaker [Sch]

$$(3.77) \qquad (x - y)^n = \sum_{i \in \mathbb{Z}} (t_{i+1} - y) \cdots (t_{i+n} - y) M_i(x).$$

3.7. Knot insertion and variation diminishing property of the B-spline basis.

We shall now show that our basis functions $\{M_i\}_{i \in \mathbb{Z}}$ are variation diminishing (see Chapters 1 and 2 for a definition of this concept). We approach the proof of this claim by using a *knot insertion* formula for the space $\mathcal{S}_n$, taken from Barry, Goldman, and Micchelli [BGM].

The technique of knot insertion has already been mentioned in Chapter 2. In that chapter, we studied cardinal splines and inserted knots at all half integers. Here we begin with the space $\mathcal{S}_n$ and insert one knot at a time. Specifically, we pick any point $\hat{t} \notin \{t_i\}_{i \in \mathbb{Z}}$ where for definiteness we suppose that $\hat{t} \in (t_q, t_{q+1})$, for some $q \in \mathbb{Z}$ and define the new partition $\{\hat{t}_i\}_{i \in \mathbb{Z}}$ as

$$\hat{t}_j = \begin{cases} t_j, & j \leq q, \\ \hat{t}, & j = q + 1, \\ t_{j-1}, & j \geq q + 2. \end{cases}$$

We have to decide what matrix to associate with the new knot $\hat{t}$. Since we want the new space $\hat{\mathcal{S}}$ on the new partition to contain $\mathcal{S}$ we add the $n \times n$ identity matrix to our family $\{A^i\}_{i \in \mathbb{Z}}$. Specifically, set

$$\hat{A}^j = \begin{cases} A^j, & j \leq q, \\ I, & j = q + 1, \\ A^{j-1}, & j \geq q + 2, \end{cases}$$

which ensures that $\mathcal{S} \subset \hat{\mathcal{S}}$. Also, with this definition we let

$$\hat{E}_j = \begin{cases} E_j, & j \le q, \\ I, & j = q+1, \\ E_{j-1}, & j \ge q+2, \end{cases}$$

and so $\mathcal{P}^{\mathcal{E}} = \mathcal{P}^{\hat{\mathcal{E}}}$.

Let us now relate the new functions $\{\hat{N}_i\}_{i\in\mathbb{Z}}, \{\hat{M}_i\}_{i\in\mathbb{Z}}$, and dual functionals $\{\hat{L}_j\}_{j\in\mathbb{Z}}$ corresponding to the space $\hat{\mathcal{S}}$ to the original families of functions $\{N_i\}_{i\in\mathbb{Z}}, \{M_i\}_{i\in\mathbb{Z}}$ and dual functionals $\{L_i\}_{i\in\mathbb{Z}}$. For this purpose, we require the following lemma.

LEMMA 3.7. *For any* $x \notin \{t_{i+1}, \ldots, t_{i+n-1}\}$

$$N_i(x) - N_{i-1}(x) \neq 0.$$

Proof. Set $Y(t) = N_i(t) - N_{i-1}(t)$. Suppose to the contrary that $Y(x) = 0$ for some $x \notin \{t_{i+1}, \ldots, t_{i+n-1}\}$. Then Y has n zeros, since it obviously vanishes at $t_{i+1}, \ldots, t_{i+n-1}$. Also, Y is a polynomial of degree $n-1$ on each of the knot intervals (t_k, t_{k+1}), $k \in \mathbb{Z}$. Let $\tilde{E}_i$ denote the $n \times n$ submatrix of E_i obtained by deleting its last row and column. Then

$$Y \in \mathcal{P}^{\tilde{\mathcal{E}}}$$

where $\tilde{\mathcal{E}} := \{\tilde{E}_i\}_{i\in\mathbb{Z}}$ and Theorem 3.7 implies that

$$Z(Y|\mathbb{R}) \le n - 1.$$

This contradiction proves the result.   □

PROPOSITION 3.7. *Let* $\hat{t} \in (t_q, t_{q+1})$ *and suppose that* $\{A^i\}_{i\in\mathbb{Z}} \subset \mathcal{B}_n$. *Then*

$$\hat{N}_j(t) = \begin{cases} N_j(t), & j \le q-n, \\ \dfrac{N_j(\hat{t})N_{j-1}(t) - N_{j-1}(\hat{t})N_j(t)}{N_j(\hat{t}) - N_{j-1}(\hat{t})}, & q-n+1 \le j \le q, \\ N_{j-1}(t), & j \ge q+1. \end{cases}$$

Proof. Only the middle formula needs to be proved. Denote the right-hand side of the above equation by $U(t)$. Then, clearly

$$U \in \mathcal{P}^{\hat{\mathcal{E}}}, \qquad U^{(n)}(t) = n!(-1)^n$$

since both N_{j-1} and N_j satisfy these conditions. Moreover, U vanishes at $\hat{t}$ and $t_j, \ldots, t_{j+n-1}$. But these points are precisely the zeros of $\hat{N}_j$.   □

THEOREM 3.10. *Suppose that* $\{A^i\}_{i\in\mathbb{Z}} \subset \mathcal{B}_n$. *Then for* $x \in (t_j, t_{j+n})$ *we have*

$$\hat{L}_j(x) = \begin{cases} L_j(x), & j \le q-n, \\ \dfrac{N_j(\hat{t})L_{j-1}(x) - N_{j-1}(\hat{t})L_j(x)}{N_j(\hat{t}) - N_{j-1}(\hat{t})}, & q-n+1 \le j \le q, \\ L_{j-1}(x), & j \ge q+1. \end{cases}$$

Proof. The proof is an immediate consequence of Proposition 3.7 and definition (3.64) of the linear functionals $L_i(x)$.   □

The next result is the knot insertion formula for the pair $\mathcal{S}_n$ and $\hat{\mathcal{S}}_n$.

THEOREM 3.11. *Suppose that* $\{A^i\}_{i\in\mathbb{Z}} \subset \mathcal{B}_n$ *and* $\{t_i\}_{i\in\mathbb{Z}}$ *is a partition of* $\mathbb{R}$. *Then, for any* $f \in \mathcal{S}_n$ *given as*

$$f = \sum_{j\in\mathbb{Z}} c_j M_j,$$

we have

$$f = \sum_{j\in\mathbb{Z}} \hat{c}_j \hat{M}_j,$$

where

$$(3.78) \qquad \hat{c}_j = \begin{cases} c_j, & j \leq q - n, \\ \dfrac{N_j(\hat{t})c_{j-1} - N_{j-1}(\hat{t})c_j}{N_j(\hat{t}) - N_{j-1}(\hat{t})}, & q - n + 1 \leq j \leq q, \\ c_{j-1}, & j \geq q + 1. \end{cases}$$

Proof. To prove the result we merely apply the dual functionals $\{\hat{L}_j\}_{j\in\mathbb{Z}}$ to both sides of the formula

$$\sum_{j\in\mathbb{Z}} c_j M_j = \sum_{j\in\mathbb{Z}} \hat{c}_j \hat{M}_j$$

and use Theorem 3.10.   □

Theorem 3.11 allows us to prove the variation diminishing property of the basis $\{M_i\}_{i\in\mathbb{Z}}$.

THEOREM 3.12. *Suppose that* $\{A^i\}_{i\in\mathbb{Z}} \subset \mathcal{B}_n$ *and* $\{t_i\}_{i\in\mathbb{Z}}$ *is a partition of* $\mathbb{R}$. *Then*

$$S^- \left(\sum_{j\in\mathbb{Z}} c_j M_j \right) \leq S^- \left(\{c_j\}_{j\in\mathbb{Z}} \right).$$

Proof. First we point out that the knot insertion formula (3.78) implies that each $\hat{c}_j$ is a *convex* combination of c_j and c_{j-1}. To prove this claim we observe that

$$\operatorname{sgn} N_i(t) = \begin{cases} 1, & t < t_{i+1}, \\ (-1)^r, & t_{i+r} < t < t_{i+r+1}, \\ (-1)^n, & t_{i+n} < t, \end{cases}$$

which follows from the fact that N_i vanishes only at $t_{i+1}, \dots, t_{i+n}$ and $\lim_{t\to-\infty} N_i(t) = \infty$. Hence, for any integer j with $q - n + 1 \leq j \leq q$, we have $\operatorname{sgn} N_j(\hat{t}) \operatorname{sgn} N_{j-1}(\hat{t}) = -1$.

Using this fact, it follows that

$$(3.79) \qquad S^-(\{\hat{c}_j\}_{j\in\mathbb{Z}}) \leq S^-(\{c_j\}_{j\in\mathbb{Z}}).$$

Now we add knots successively and form sequences of partitions $\{t_j^k\}_{j\in\mathbb{Z}}, k = 1, 2, \ldots$, remembering to add at each new knot the identity matrix to our family of matrices. The knots are added in each interval (t_ℓ, t_m) so that their successive differences go to zero as $k \to \infty$. Thus we have

$$f := \sum_{j\in\mathbb{Z}} c_j M_j = \sum_{j\in\mathbb{Z}} c_j^k M_j^k, \qquad k = 1, 2, \ldots,$$

and from (3.79) we get

(3.80) $S^-(\{c_j^k\}_{j\in\mathbb{Z}}) \leq S^-(\{c_j\}_{j\in\mathbb{Z}}).$

Let $r = S^-(f)$ and suppose that $y_1 < \cdots < y_{r+1}$ are the points at which f alternates in sign. For definiteness, we may as well assume that sgn $f(y_\ell) = (-1)^\ell$, $\ell = 1, \ldots, r + 1$. At each point y_ℓ, there are most $n + 1$ integers i such that $M_i^k(y_\ell) \neq 0$. Call this set J_ℓ^k and choose k sufficiently large so that $J_1^k < \cdots < J_{r+1}^k$. Since the M_i^k, $i \in \mathbb{Z}$, are nonnegative and sum to one, there must be an $i_\ell^k \in J_\ell^k$, $\ell = 1, \ldots, r + 1$ such that

$$(-1)^\ell c_{i_\ell^k}^k \geq (-1)^\ell f(y_\ell), \qquad \ell = 1, \ldots, r + 1.$$

Hence, $S^-(\{c_i^k\}_{j\in\mathbb{Z}}) \geq r$ and from (3.80) we get $S^-(f) \leq S^-(\{c_j\}_{j\in\mathbb{Z}})$. $\square$

References

[B₁]   B. A. BARSKY, *The Beta-Spline: A Local Representation Based on Shape Parameters and Fundamental Geometric Measures*, Ph.D. thesis, Department of Computer Sciences, University of Utah, December 1981.

[B₂]   B. A. BARSKY, *Computer Graphics and Geometric Modeling Using Beta-Splines*, Springer-Verlag, Berlin, 1988.

[BDGM] P. J. BARRY, N. DYN, R. N. GOLDMAN, AND C. A. MICCHELLI, *Identities for piecewise polynomial spaces determined by connection matrices*, Aequationes Math., 42(1991), 123–136.

[BGM]   P. J. BARRY, R. N. GOLDMAN, AND C. A. MICCHELLI, *Knot insertion algorithms for piecewise polynomial spaces determined by connection matrices*, Adv. Comp. Math., 1(1993), 139–171.

[CS]   H. B. CURRY AND I. J. SCHOENBERG, *On Pólya frequency functions IV: The fundamental spline functions and their limits*, J. d'Anal. Math., 17(1966), 71–107.

[DEM]   N. DYN, A. EDELMAN, AND C. A. MICCHELLI, *On locally supported basis functions for the representation of geometrically continuous curves*, Analysis, 7(1987), 313–341.

[DM]   N. DYN AND C. A. MICCHELLI, *Piecewise polynomial spaces and geometric continuity of curves*, Numer. Math., 54(1988), 319–337.

[GM]   R. N. GOLDMAN AND C. A. MICCHELLI, *Algebraic aspects of geometric continuity*, in Mathematical Methods in Computer Aided Geometric Design, T. Lyche and L.L. Schumaker, eds., Academic Press, Boston, 1989, 353–371.

[Go]   T. N. T. GOODMAN, *Properties of beta-splines*, J. Approx. Theory, 44(1985), 132–153.

[Gr]   J. A. GREGORY, *Geometric continuity*, in Mathematical Methods in Computer Aided Geometric Design, T. Lyche and L. L. Schumaker, eds., Academic Press, Boston, 1989, 313–332.

[K]   S. KARLIN, *Total Positivity*, Stanford University Press, Stanford, CA, 1968.

[Kn]   D. E. KNUTH, *The Art of Computer Programming*, Vol. 1, *Fundamental Algorithms*, Second ed., Addison-Wesley Publishing Company, Reading, MA, 1973.

[M]   J. R. MANNING, *Continuity conditions for spline curves*, Comput. J., 17(1974), 181–186.

[S]   M. A. SABIN, *Spline curves*, report VTO/M S/154, British Aircraft Corporation, Weybridge, England, 1969.

[Sch]   L. L. SCHUMAKER, *Spline Functions: Basic Theory*, John Wiley and Sons, New York, 1981.

[Sp]   M. SPIVAK, *Differential Geometry*, Publish or Perish, Inc., Boston, 1975.

CHAPTER 4

Geometric Methods for Piecewise Polynomial Surfaces

4.0. Introduction.

This chapter explores the powerful idea of generating multivariate smooth piecewise polynomials as the volume of "slices" of polyhedra. The traditional finite element approach to the problem of constructing smooth piecewise polynomials puts down a grid on the domain of interest, specifies a given smoothness, and then finds, by ad hoc methods, a desirable (compactly supported) B-spline. Here we are guided by the geometric principle that polyhedral B-splines can be constructed as volumes. To make this a viable alternate approach to constructing smooth piecewise polynomials, we must develop analytic and computational tools to study these volume functions. This is done via recurrence formulas that enable us to delineate their smoothness properties, describe their knot regions, and provide efficient methods to compute with them. Three specific B-splines are studied: the multivariate B-spline, the truncated power, and the cube spline associated respectively with the simplex, positive orthant, and the cube. First, various formulas and properties of the multivariate B-spline are derived. Then using this function as a starting point we develop analogous facts about the truncated power and cube spline.

We do not discuss how one builds spline spaces from multivariate B-splines. The interested reader can consult the papers [DM1], [DM3], [H] as well as [DMS], [G], [GL], [S], and reference [Se] from Chapter 5 for information on this important topic. At the end of the chapter we include some correspondence by I.J. Schoenberg and H.B. Curry that describes some of the historical development of B-splines.

4.1. B-splines as volumes.

Many years ago Alfred Cavaretta mentioned to me in passing that I.J. Schoenberg had a way to interpret the univariate B-spline *geometrically* as the volume of the intersection of a simplex and a hyperplane. At the time, this appeared to be no more than a curiosity and I made no effort to look into Curry and Schoenberg [CS], the original source, for a precise description of the result. Only years later, in the spring of 1978 to be precise, when I was visiting the Mathematics Research Center at the University of Wisconsin by invitation of Carl de Boor, did I appreciate the value of their observation.

It was my first day in Madison. I had just driven from Yorktown, New York and arrived at the WARF building on the University of Wisconsin campus. I went to Carl's office, which was filled with visitors. Besides Carl himself, there were B. Sendov from Sofia, Bulgaria and Dick Askey from the Mathematics Department. They were engaged in a lively conversation about some new multivariate interpolation scheme discovered by P. Kergin who was just then completing his Ph.D. in Mathematics at the University of Toronto. My first reaction to Kergin's result was a sense of confusion. But I was going to be around Madison until the end of the summer and so there was plenty of time to clarify the matter. Things did become clear, mostly because of Curry and Schoenberg's geometric interpretation of the B-spline. What we present in this chapter are the details that have since followed. They took much longer than a summer to discover and represent the collective efforts of many people to understand this elegant way of studying multivariate splines.

I think the appropriate way to start this chapter on multivariate piecewise polynomials is to describe Schoenberg's geometric construction of the B-spline. Let us recall the definition of the univariate B-spline of degree $n - 1$ with knots at $t_0, \ldots, t_n$, viz.

$$(4.1) \qquad S(t|t_0, \ldots, t_n) = \frac{1}{(n-1)!} [t_0, \ldots, t_n](\cdot - t)_+^{n-1}, \qquad t \in \mathbb{R}.$$

Here $[t_0, \ldots, t_n]g$ denotes the divided difference of a function g at $t_0, \ldots, t_n$, which is given explicitly by

$$[t_0, \ldots, t_n]g = \sum_{j=0}^n \frac{g(t_j)}{\Pi_{i \neq j}(t_j - t_i)}$$

when $t_0, \ldots, t_n$ are distinct. We mentioned in Chapter 3 that S is the kernel for the integral representation of the divided difference operator. Thus we have for any $g \in C^n(\mathbb{R})$ the formula

$$(4.2) \qquad [t_0, \ldots, t_n]g = \int_{\mathbb{R}} S(t|t_0, \ldots, t_n)g^{(n)}(t)dt.$$

There is another useful formula for the divided difference. It says that

$$(4.3) \qquad [t_0, \ldots, t_n]g = \int_{\Delta^n} g^{(n)}((\mathbf{v}, \mathbf{t}))dm(\mathbf{v}),$$

where $(\mathbf{v}, \mathbf{t}) = \sum_{j=0}^n v_j t_j$ is the standard inner product on $\mathbb{R}^{n+1}$ of the vectors $\mathbf{v} = (v_0, \ldots, v_n)^T, \mathbf{t} = (t_0, \ldots, t_n)^T$,

$$\Delta^n = \left\{ \mathbf{v} = (v_0, \ldots, v_n)^T : v_j \geq 0, j = 0, \ldots, n, \sum_{j=0}^n v_j = 1 \right\}$$

is the standard n-simplex, and $dm(\mathbf{t}) = dt_1 \cdots dt_n$ is Lebesgue measure on Δ^n.

One way to prove this well-known formula when $t_0, \ldots, t_n$ are distinct (the general case would follow by continuity) is to use the divided difference recurrence formula

$$(4.4) \qquad [t_0, \ldots, t_n]g = \frac{[t_1, \ldots, t_n]g - [t_0, \ldots, t_{n-1}]g}{t_n - t_0}$$

and show, by integrating relative to dt_n in (4.3), that the integral satisfies the same recurrence formula.

Now, we combine formulas (4.2) and (4.3) to eliminate the divided difference and at the same time replace $g^{(n)}$ by g. Thus for any $g \in C(\mathbb{R})$ we obtain the equation

$$(4.5) \qquad \int_{\Delta^n} g((\mathbf{v}, \mathbf{t})) dm(\mathbf{v}) = \int_{\mathbb{R}} S(t|t_0, \ldots, t_n) g(t) dt.$$

This formula is the crucial observation. To interpret it geometrically, we choose any $n + 1$ vectors $\mathbf{v}^0, \mathbf{v}^1, \ldots, \mathbf{v}^n$ in $\mathbb{R}^n$ such that the first component of $\mathbf{v}^i$ is $t_i, i = 0, 1, \ldots, n$ (see Fig. 4.1). We express this symbolically by saying that

$$(4.6) \qquad \mathbf{v}^i|_{\mathbb{R}^1} = t_i, \qquad i = 0, 1, \ldots, n.$$

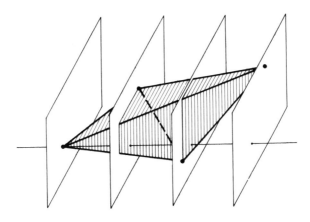

FIG. 4.1. *Lifting knots.*

(Generally we use $\mathbf{v}|_{\mathbb{R}^d}$ for the vector in $\mathbb{R}^d$ composed of the first d components of the vector $\mathbf{v}$.) We also require that the vectors $\mathbf{v}^0, \mathbf{v}^1, \ldots, \mathbf{v}^n$ are chosen so that the simplex

$$(4.7) \qquad [V] := \left\{ \sum_{j=0}^{n} u_j \mathbf{v}^j : \sum_{j=0}^{n} u_j = 1, \; u_\ell \geq 0, \qquad \ell = 0, 1, \ldots, n \right\},$$

where $V = \{\mathbf{v}^0, \ldots, \mathbf{v}^n\}$, has *nonzero* n-dimensional volume. There are many choices of $\mathbf{v}^0, \ldots, \mathbf{v}^n$ with this property, as long as the points $t_0, \ldots, t_n$ are not all the same. For instance, if we have $t_0 \leq t_1 \leq \cdots \leq t_n$ with $t_0 \neq t_n$ then the vectors

$$
\begin{aligned}
\mathbf{v}^0 &= (t_0, 0, 0, \ldots, 0)^T \\
\mathbf{v}^1 &= (t_1, 1, 0, \ldots, 0)^T
\end{aligned}
$$

(4.8)

$$
\vdots
$$

$$
\begin{aligned}
\mathbf{v}^{n-1} &= (t_{n-1}, 0, 0, \ldots, 1)^T \\
\mathbf{v}^n &= (t_n, 0, 0, \ldots, 0)^T
\end{aligned}
$$

will do, and in this case

$$
\mathrm{vol}_n[V] = (t_n - t_0)/n! \ .
$$

We recall that the volume of $[V]$ in (4.7) is given by

(4.9)
$$
\mathrm{vol}_n[V] = \frac{1}{n!} |\det(\mathbf{v}^0, \ldots, \mathbf{v}^n)|,
$$

where

(4.10)
$$
\det(\mathbf{v}^0, \ldots, \mathbf{v}^n) :=
\begin{vmatrix}
1 & (\mathbf{v}^0)_1 & \cdots & (\mathbf{v}^0)_n \\
1 & (\mathbf{v}^1)_1 & \cdots & (\mathbf{v}^1)_n \\
\vdots & \vdots & & \vdots \\
1 & (\mathbf{v}^n)_1 & \cdots & (\mathbf{v}^n)_n
\end{vmatrix}.
$$

We consider the linear mapping $T : \mathbb{R}^{n+1} \to \mathbb{R}^n$ given by

$$
\mathbf{y} = T(\mathbf{u}) = \sum_{j=0}^{n} u_j \mathbf{v}^j,
$$

where $\mathbf{u} = (u_0, \ldots, u_n)^T$. T maps Δ^n bijectively onto $[V]$ and the first component of $T(\mathbf{u})$ is $(\mathbf{u}, \mathbf{t})$. Thus, by a change of variables in (4.5), we get

$$
\int_{\mathbb{R}} S(t|t_0, \ldots, t_n) g(t) dt = (n! \mathrm{vol}_n[V])^{-1} \int_{[V]} g(y_1) dy_1 \cdots dy_n
$$

$$
= \int_{\mathbb{R}} \frac{\mathrm{vol}_{n-1}\{\mathbf{y} \in [V] : \mathbf{y}|_{\mathbb{R}^1} = t\}}{n! \mathrm{vol}_n[V]} g(t) dt,
$$

and we obtain the formula

(4.11)
$$
S(t|t_0, \ldots, t_n) = \frac{\mathrm{vol}_{n-1}\{\mathbf{y} \in [V] : \mathbf{y}|_{\mathbb{R}^1} = t\}}{n! \mathrm{vol}_n[V]}
$$

at all common points of continuity of these functions. This formula, from Curry and Schoenberg [CS], expresses the B-spline as the volume of a "slice" of a simplex (see Fig. 4.2).

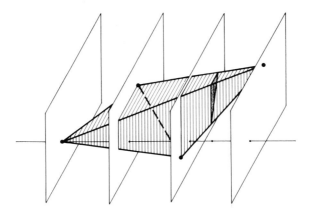

FIG. 4.2.   *The value of a univariate quadratic B-spline by slicing.*

In (4.11), we use the convention that for $n = 1$ the numerator in (4.11) is the characteristic function of the interval $[t_0, t_n]$. Also we want to point out the remarkable fact that (4.11) holds for *any* vectors $\mathbf{v}^0, \ldots, \mathbf{v}^n$ such that $\mathbf{v}^i|_{\mathbb{R}} = t^i, i = 0, 1, \ldots, n$, and $\text{vol}_n[V] > 0$.

We mention two applications of formula (4.11).   The first provides a lower bound for S.   To this end, we go back to the special choice of vectors (4.8) and notice that a typical $\mathbf{y} = \sum_{j=0}^{n} u_j \mathbf{v}^j \in [V]$ with $u_j \geq 0$, $j = 0, 1, \ldots, n, \sum_{j=0}^{n} u_j = 1$, has the property that $\mathbf{y}|_{\mathbb{R}^1} = t$ if and only if

$$\sum_{j=1}^{n-1} u_j(t_n - t_j) \leq t_n - t$$

and

$$\sum_{j=1}^{n-1} u_j(t_j - t_0) \leq t - t_0.$$

Therefore

$$S(t|t_0, \ldots, t_n) = \frac{1}{t_n - t_0} \text{vol}_{n-1} P,$$

where

$$P := \left\{ (u_1, \ldots, u_{n-1}) : \sum_{j=1}^{n-1} u_j \frac{t_n - t_j}{t_n - t} \leq 1, \quad \sum_{j=1}^{n-1} u_j \frac{t_j - t_0}{t - t_0} \leq 1, \right.$$

$$\left. u_\ell \geq 0, \quad \ell = 1, \ldots, n - 1 \right\}.$$

To estimate the volume of P we introduce the quantities

$$a := \min(t - t_0, t_n - t)$$

and

$$b := \max_{1 \le j \le n-1} (t_n - t_j, t_j - t_0).$$

Notice that the simplex

$$\left\{ (u_1, \ldots, u_{n-1}) : \sum_{j=1}^{n-1} u_j \le \frac{a}{b}, \quad u_\ell \ge 0, \quad \ell = 1, \ldots, n-1 \right\}$$

is contained in P. Therefore, we get the lower bound

$$S(t|t_0, \ldots, t_n) \ge \frac{1}{(n-1)!(t_n - t_0)} \left(\frac{\min(t - t_0, t_n - t)}{\max\limits_{1 \le j \le n} (t_n - t_j, t_j - t_0)} \right)^{n-1}$$

and, in particular, S is positive on (t_0, t_n).

Curry and Schoenberg [CS] had a nice application of formula (4.11). The following result comes from their paper.

PROPOSITION 4.1. *The function*

$$\log S(t|t_0, \ldots, t_n)$$

is concave on the open interval (t_0, t_n).

Proof. The proof uses the *Brunn–Minkowski inequality,* which states that for any two nonempty convex sets A_1 and A_2 in $\mathbb{R}^{n-1}$

$$\text{vol}_{n-1}(tA_1 + (1 - t)A_2)$$

$$\ge (t(\text{vol}_{n-1} A_1)^{1/(n-1)} + (1 - t)(\text{vol}_{n-1} A_2)^{1/(n-1)})^{n-1},$$

cf. Milman and Schecktman [MS, p. 134] for a proof of the result. We apply this result to the family of convex subsets of $\mathbb{R}^{n-1}$ defined by

$$A(x) := \{ \mathbf{y} : \mathbf{y} \in [V], \; \mathbf{y}|_{\mathbb{R}^1} = x \}, \qquad x \in (t_0, t_n),$$

and use the fact that for every $x_1, x_2 \in (t_0, t_n)$ and $t \in [0, 1]$, $A(tx_1 + (1-t)x_2) = tA(x_1) + (1 - t)A(x_2)$ to conclude that the function

$$\log S^{1/(n-1)}(x) = \frac{1}{n - 1} \log S(x)$$

is indeed concave.   □

We are not aware of any other application of formula (4.11). However, Schoenberg clearly had something else in mind. In a letter written to P. Davis of Brown University in 1965, Schoenberg describes a method to construct a *bivariate* B-spline (see the end of the chapter for a copy of this letter and other historical notes pertaining to the multivariate B-splines). His sketch of the quadratic case (below) clearly shows that his surface has cross sections, near a vertex, that are univariate B-splines.

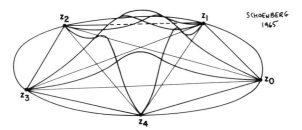

A SKETCH OF THE SPLINE FUNCTION z=M(x,y; z_0, z_1, z_2, z_3, z_4)

FIG. 4.3.   *Schoenberg's sketch of the bivariate quadratic B-spline.*

4.2.   Multivariate B-splines: Smoothness and recursions.

In his letter to Davis, Schoenberg still seemed to be fascinated by formula (4.3), albeit for analytic functions g and complex points $t_0, \ldots, t_n$ and therefore he was restricted to two variables. It was in de Boor [deB] that the multivariate B-spline was first defined. The definition came at the end of his survey paper on univariate B-spline and it follows next.

Given a set of vectors $X = \{\mathbf{x}^0, \ldots, \mathbf{x}^n\}$ in $\mathbb{R}^d$ with $n \geq d$, let $[X]$ denote its convex hull and suppose that

$$\mathrm{vol}_d[X] > 0.$$

This condition can be also expressed by saying that the $(n+1) \times (d+1)$ matrix

$$M := \begin{pmatrix} 1 & (\mathbf{x}^0)_1 & \cdots & (\mathbf{x}^0)_d \\ & \vdots & & \\ 1 & (\mathbf{x}^n)_1 & \cdots & (\mathbf{x}^n)_d \end{pmatrix}$$

has rank $d+1$. In other words, the vectors $(1, \mathbf{x}^0)^T, \ldots, (1, \mathbf{x}^n)^T$ span $\mathbb{R}^{d+1}$. When $n = d$ and M has rank $d+1$ we sometimes say that the points $\mathbf{x}^0, \ldots, \mathbf{x}^n$ are *affinely independent*.

Choose any $\mathbf{v}^0, \ldots, \mathbf{v}^n \in \mathbb{R}^n$ such that

(4.12) $$\mathbf{v}^0|_{\mathbb{R}^d} = \mathbf{x}^0, \ldots, \mathbf{v}^n|_{\mathbb{R}^d} = \mathbf{x}^n$$

and the simplex $[V]$, where $V = \{\mathbf{v}^0, \ldots, \mathbf{v}^n\}$, has positive volume. The *multivariate B-spline* is defined by the formula

(4.13) $$S(\mathbf{x}|X) = \frac{1}{n!} \frac{\mathrm{vol}_{n-d}\{\mathbf{y} : \mathbf{y} \in [V], \mathbf{y}|_{\mathbb{R}^d} = \mathbf{x}\}}{\mathrm{vol}_n[V]}.$$

We use the convention that when $n = d$ the numerator above is the characteristic function of the simplex $[X]$. Note that the multivariate B-spline is unchanged by any reordering of the elements in X.

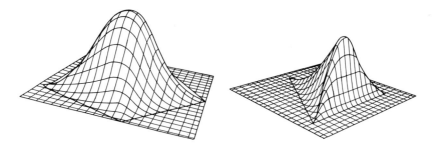

FIG. 4.4. *Computer generated bivariate quadratic B-splines.*

At first sight, it seems very difficult to make use of definition (4.13). For instance, the following result from Micchelli [Mic2] on the smoothness and piecewise polynomial nature of the multivariate B-spline seems to be inaccessible by means of (4.13). To formulate the result, we require some terminology.

We say that the set $X = \{\mathbf{x}^0, \ldots, \mathbf{x}^n\}$ is in *general position* provided that for every subset $Y \subseteq X$ of cardinality $d+1$ it follow that $\mathrm{vol}_d[Y] > 0$. Another way of describing this concept is to say that X is in general position whenever every subset of $d+1$ vectors in X is affinely independent.

Recall that the dimension of a convex set is defined to be one less than the maximum number of affinely independent vectors it contains. We use $\langle Y \rangle$ to denote the affine space generated by affine linear combinations of the vectors in Y, that is, if $Y = \{\mathbf{v}^1, \ldots, \mathbf{v}^m\}$ then

$$\langle Y \rangle := \left\{ \sum_{j=1}^{m} u_j \mathbf{v}^j : \sum_{j=1}^{m} u_j = 1 \right\}.$$

Also, we call $\langle Y \rangle$ an X-plane whenever Y is a subset of X with $\dim \langle Y \rangle = d-1$. When X is in general position *every* $Y \subseteq X$ with $\#Y = d$ determines an X-plane that contains Y, namely $\langle Y \rangle$.

We use the following standard multivariate notation for derivatives of functions,

$$(D^{\mathbf{k}} f)(\mathbf{x}) := \frac{\partial^{|\mathbf{k}|_1}}{\partial x_1^{k_1} \cdots \partial x_d^{k_d}} f(\mathbf{x}),$$

where $\mathbf{x} = (x_1, \ldots, x_d)^T$, $\mathbf{k} = (k_1, \ldots, k_d)^T \in \mathbb{Z}_+^d$, $|\mathbf{k}|_1 = k_1 + \cdots + k_d$, and

$$C^{\ell}(\mathbb{R}^d) := \{f : D^{\mathbf{k}} f \in C(\mathbb{R}^d), \qquad \mathbf{k} = (k_1, \ldots, k_d)^T, \qquad |\mathbf{k}|_1 \leq \ell\}.$$

Finally, we use $\Pi_n(\mathbb{R}^d)$ for the set of all polynomials of total degree $\leq n$. A typical element p in $\Pi_n(\mathbb{R}^d)$ has the form

$$p(\mathbf{x}) = \sum_{|\mathbf{k}|_1 \leq n} c_{\mathbf{k}} \mathbf{x}^{\mathbf{k}}$$

for some real scalars $c_{\mathbf{k}}$, $|\mathbf{k}|_1 \leq n$, and the dimension of $\Pi_n(\mathbb{R}^d)$ is $\binom{n+d}{d}$.

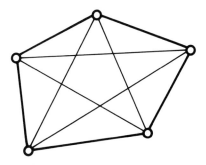

FIG. 4.5. *Knot regions for a bivariate quadratic B-spline.*

THEOREM 4.1. *Let* $X = \{\mathbf{x}^0, \ldots, \mathbf{x}^n\} \subset \mathbb{R}^d$ *be in general position. Then* $S(\cdot|X)$ *is in* $C^{n-d-1}(\mathbb{R}^d)$, *is positive in the interior of* $[X]$, *and on any region in* $\mathbb{R}^d \setminus \cup \{\langle Y \rangle : Y \subseteq X, \#Y = d\}$, *is a polynomial of total degree* $\leq n - d$.

Before we prove this result we note the following consequence of Theorem 4.1. Let X be any set in $\mathbb{R}^d$ that is in general position with, $\#X = n+1$. Then every line $\mathbf{x} + t\mathbf{y}, t \in \mathbb{R}$ must intersect at least $n - d + 2$ X-planes. The reason for this is that, as a function of t, $S(\mathbf{x} + t\mathbf{y}|X)$ is a compactly supported spline of degree $n - d$ with knots at points t where $\mathbf{x} + t\mathbf{y}$ intersects an X-plane. Thus it must have at least $n - d + 2$ knots.

The first step in the proof of Theorem 4.1 is to revisit Schoenberg's proof of (4.11). He began with (4.5) and then derived (4.11). Reversing this argument, we conclude that the multivariate B-spline defined by (4.13) has the property that for all $f \in C(\mathbb{R})$

$$(4.14) \qquad \int_{\mathbb{R}^d} S(\mathbf{x}|X)f(\mathbf{x})d\mathbf{x} = \int_{\Delta^n} f(X\mathbf{t})dm(\mathbf{t}),$$

where $X\mathbf{t} := \sum_{j=0}^{n} t_j \mathbf{x}^j$, $\mathbf{t} = (t_0, \ldots, t_n)^T$. Here we use "double duty" notation and let X signify both the set of vectors $\{\mathbf{x}^0, \ldots, \mathbf{x}^n\}$ and the $d \times (n+1)$ matrix X whose columns are $\mathbf{x}^0, \ldots, \mathbf{x}^n$. By the way, equation (4.14) demonstrates that equation (4.13) is *independent* of the choice of vectors $\mathbf{v}^0, \ldots, \mathbf{v}^n \in \mathbb{R}^n$ that satisfy (4.12).

Equation (4.14) gives us a way to extend the meaning of the multivariate B-spline. Specifically, we can think of the multivariate B-spline as the linear functional

$$(4.15) \qquad \int_X f := \int_{\Delta^n} f(X\mathbf{t})dm(\mathbf{t}),$$

where $X = \{\mathbf{x}^0, \ldots, \mathbf{x}^n\}$. One advantage of this point of view is that no hypothesis is required on the set $\{\mathbf{x}^0, \ldots, \mathbf{x}^n\}$. If, in fact, $\mathrm{vol}_d[X] > 0$ then the distribution (4.15) corresponds to an integrable function that is zero outside of

[X]. Below we will derive several equations involving the multivariate B-spline and related distributions. They are to be understood as equalities among distributions, unless otherwise noted.

We begin to build the proof of Theorem 4.1. For this purpose, we require the next lemma.

LEMMA 4.1. *Suppose that* $\text{vol}_d[X] > 0$, $X \subset \mathbb{R}^d$, $\#X < \infty$. *Given any* $\mathbf{x} \in [X]$ *there is a* $Y \subseteq X$ *with* $\#Y = d+1$ *and* $\text{vol}_d[Y] > 0$ *such that* $\mathbf{x} \in [Y]$.
  Proof. Let

$$C := \{\mathbf{t} = (t_0, \ldots, t_n)^T : \mathbf{t} \in \Delta^n, \mathbf{x} = X\mathbf{t}\}.$$

Clearly C is a nonempty compact convex set and hence has an extreme point $\mathbf{u} = (u_0, \ldots, u_n)^T$ (that is, is not a convex combination of two distinct points of C). If the set $I := \{j : u_j > 0, 0 \leq j \leq n\}$ has more than $d+1$ elements we can find a $\mathbf{c} = (c_0, \ldots, c_n)^T \neq 0$ such that

$$\sum_{j=0}^{n} c_j \mathbf{x}^j = 0, \qquad \sum_{j=0}^{n} c_j = 0,$$

and $c_j = 0$, if $j \notin I$. Let $\mathbf{u}(\epsilon) = \mathbf{u} + \epsilon\mathbf{c}$ and note that there exists an $\epsilon_0 > 0$ such that for all $\epsilon, |\epsilon| < \epsilon_0$, $\mathbf{u}(\epsilon) \in C$. This contradicts the fact that $\mathbf{u}$ is an extreme point of C. Thus we conclude that $\#I \leq d+1$. Now, define m to be the least positive integer such that $\mathbf{x} \in [V]$ for some $V \subseteq X$ with $\#V = m$. So far we have shown that $1 \leq m \leq d+1$

Let us reorder the vectors in X so that $V = \{\mathbf{x}^0, \ldots, \mathbf{x}^{m-1}\}$. The argument used above also shows that the vectors $(1, \mathbf{x}^0)^T, \ldots, (1, \mathbf{x}^{m-1})^T$ are linearly independent. Since the vectors $(1, \mathbf{x}^0)^T, \ldots, (1, \mathbf{x}^n)^T$ span $\mathbb{R}^{d+1}$, we may extend the vectors $(1, \mathbf{x}^0)^T, \ldots, (1, \mathbf{x}^{m-1})^T$ to a basis $(1, \mathbf{x}^0)^T, \ldots, (1, \mathbf{x}^{m-1})^T, (1, \mathbf{x}^{i_1})^T, \ldots, (1, \mathbf{x}^{i_{d+1-m}})^T$ of $\mathbb{R}^{d+1}$. Hence $Y := \{\mathbf{x}^0, \ldots, \mathbf{x}^{m-1}, \mathbf{x}^{i_1}, \ldots, \mathbf{x}^{i_{d+1-m}}\}$ is the set we seek (see Fig. 4.6). □

PROPOSITION 4.2 *Given any finite set* $X = \{\mathbf{x}^0, \ldots, \mathbf{x}^n\} \subset \mathbb{R}^d$ *with* $\text{vol}_d[X] > 0$ *then*

(4.16) $S(\mathbf{x}|X) > 0, \qquad \mathbf{x} \in [X]^0 := \text{interior of } [X],$

and $\log S(\mathbf{x}|X)$ *is concave on* $[X]^0$.

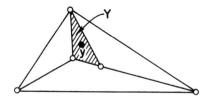

FIG. 4.6. *Proper triangle.*

Proof. The proof of (4.16) is similar to the univariate case. Choose $\mathbf{x} \in [X]^0$ and any set $Y \subseteq X$ such that $\#Y = d + 1$, $\mathbf{x} \in [Y]$, and $\mathrm{vol}_d[Y] > 0$. By reordering the vectors of X, we may as well assume that $Y = \{\mathbf{x}^0, \dots, \mathbf{x}^d\}$. Our choice of vectors in (4.13) are

$$\mathbf{v}^0 = \left(\mathbf{x}^0, 0, 0, \dots, 0\right)^T, \dots, \mathbf{v}^d = \left(\mathbf{x}^d, 0, 0, \dots, 0\right)^T$$
$$\mathbf{v}^{d+1} = \left(\mathbf{x}^{d+1}, 1, 0, \dots, 0\right)^T, \dots, \mathbf{v}^n = (\mathbf{x}^n, 0, 0, \dots, 1)^T.$$

For this choice, it is easy to see that $n!\mathrm{vol}_n[\mathbf{v}^0, \dots, \mathbf{v}^n] = d!\mathrm{vol}_d[Y]$. Let $\lambda_0(\mathbf{x}), \dots, \lambda_d(\mathbf{x})$ be the barycentric components of $\mathbf{x}$ relative to the simplex $[Y]$ and consider following the convex subset of $\mathbb{R}^{n-d}$

$$(4.17) \quad V(\mathbf{x}|Y) = \left\{ \mathbf{t} = (t_{d+1}, \dots, t_n)^T : \sum_{j=d+1}^n t_j \lambda_\ell(\mathbf{x}^j) \leq \lambda_\ell(\mathbf{x}), \right.$$
$$\left. \ell = 0, 1, \dots, d, \ \mathbf{t} \geq \mathbf{0} \right\}.$$

A typical element $\mathbf{y} = \sum_{j=0}^n t_j \mathbf{v}^j$ in $[\mathbf{v}^0, \dots, \mathbf{v}^n]$ satisfies $\mathbf{y}|_{\mathbb{R}^d} = \mathbf{x}$ if and only if

$$\lambda_\ell(\mathbf{x}) \geq \sum_{j=d+1}^n t_j \lambda_\ell(\mathbf{x}^j), \qquad \ell = 0, 1, \dots, d,$$

and so

$$(4.18) \quad S(\mathbf{x}|X) = \frac{\mathrm{vol}_{n-d} V(\mathbf{x}|Y)}{d!\mathrm{vol}_d(Y)}.$$

Therefore, for all $\mathbf{x}$ in the *interior* of some $[Y]$, $Y \subseteq X, \#Y = d + 1$ we have $S(x|X) > 0$ because the numbers $a := \min\{\lambda_j(\mathbf{x}) : j = 0, 1, \dots, d\}$ and $b := \max\{\lambda_j(\mathbf{x}^i) : j = 0, 1, \dots, d, i = d + 1, \dots, n\}$ are positive and the set

$$\left\{ \mathbf{t} = (t_{d+1}, \dots, t_n)^T : \sum_{i=d+1}^n t_i \leq \frac{a}{b}, \ \mathbf{t} \geq \mathbf{0} \right\}$$

is contained in $V(\mathbf{x}|Y)$.

The points that lie in the interior of some $[Y]$ are dense in $[X]^0$ (since the only points excluded must lie on an X-plane). Call the numerator in (4.13) $H(\mathbf{x}|X)$. Then, as in the proof of Proposition 4.1, the Brunn–Minkowski inequality implies $H^{1/(n-d)}(t\mathbf{u} + (1-t)\mathbf{v}) \geq tH^{1/(n-d)}(\mathbf{u}) + (1-t)H^{1/(n-d)}(\mathbf{v})$, for all $\mathbf{u}, \mathbf{v} \in [X]$ and $t \in [0, 1]$, from which it follows that $\log S$ is concave on $[X]^0$. Thus for any $\mathbf{x} \in [X]^0$ we choose $\mathbf{u}, \mathbf{v} \in [X]^0$ such that $\mathbf{x} = \frac{1}{2}(\mathbf{u} + \mathbf{v})$ and both $\mathbf{u}$ and $\mathbf{v}$ are in the interior of some $[Y], Y \subseteq X$ with $\#Y = d + 1$. Then

$$\log S(\mathbf{x}) \geq \frac{1}{2} \log S(\mathbf{u}) + \frac{1}{2} \log S(\mathbf{v}),$$

which proves S is positive on $[X]^0$ (see Fig. 4.7).   □

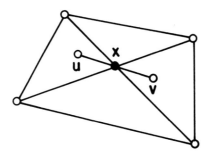

FIG. 4.7. *Exceptional point.*

LEMMA 4.2. *Suppose that* $X = \{\mathbf{x}^0, \ldots, \mathbf{x}^n\} \subset \mathbb{R}^d$ *and* $\mathrm{vol}_d[X\backslash\{\mathbf{x}^j|\}] > 0$, $j = 0, 1, \ldots, n$, *then* $S(\cdot|X)$ *is continuous on* $\mathbb{R}^d$.

Proof. By Proposition 4.2 it suffices to show that $S(\cdot|X)$ is continuous at any boundary point of $[X]$, since a concave (or convex) function is continuous in the interior of its domain, cf. Roberts and Varberg [RV].

To this end, suppose $\mathbf{y} \in \partial[X] :=$ boundary of $[X]$. We claim that there exists a $Y \subseteq X$ with $\#Y = d + 1$ and $\mathrm{vol}_d[Y] > 0$ such that $\mathbf{y}$ lies on some common supporting hyperplane of $[Y]$ and $[X]$.

The proof of this fact is a consequence of Lemma 4.1. To describe it in detail we recall some standard terminology, cf. Nemhauser and Wolsey [NW] or Schryer [S]. A convex *polyhedron* P is defined as the intersection of finitely many affine half-spaces. A convex polyhedron P in $\mathbb{R}^d$ is bounded if and only if it is of the form $[X]$ for some $X \subset \mathbb{R}^d$, $\#X < \infty$, [Sch, Cor. 7.1c, p. 89] and in this case P is called a *polytope*. The hyperplane $H = \{(\mathbf{v}, \mathbf{x}) = b : \mathbf{x} \in \mathbb{R}^d\}$ supports a convex polyhedron P, if $\max\{(\mathbf{v}, \mathbf{x}) : \mathbf{x} \in P\} = b$. A face F is the intersection of the supporting hyperplane H and the polyhedron P and, in this case, we say H represents F. A polyhedron P is of dimension k if the maximum number of affinely independent points in P is $k + 1$ and it is called full dimensional provided that $\dim P = d$. A face F of P is called a facet of P, if $\dim F = \dim P - 1$. According to [NW, Thm. 3.5, p. 91], facets of a polyhedron P determine it in the following sense. For each facet, there is a supporting hyperplane $(\mathbf{v}^i, \mathbf{x}) = b_i$, $\mathbf{x} \in \mathbb{R}^d$ representing it and P is represented by its facets

$$P = \{\mathbf{x} \in \mathbb{R}^d : (\mathbf{v}^i, \mathbf{x}) \le b_i, \qquad i = 1, \ldots, t\}.$$

Since $[X]$ is a full-dimensional convex polytope, there is a facet F of $[X]$ containing $\mathbf{y}$. Let H be the hyperplane that represents F. Clearly, $F = [X']$ where $X' := X \cap H$ and, since $[X]$ is full dimensional, $\#X' < \#X$. We identify $\mathbf{y}$ and F with a vector space of dimension $d - 1$. Hence by Lemma 4.1, there is a set $Y' \subseteq X'$ with $\#Y' = d - 1$ such that $\mathbf{y} \in [Y']$ and $\mathrm{vol}_{d-1}[Y'] > 0$. Since $[X]$ is full dimensional, there is an $\mathbf{x} \in X\backslash H$. Now, set $Y := \{\mathbf{x}\} \cup Y'$ and observe

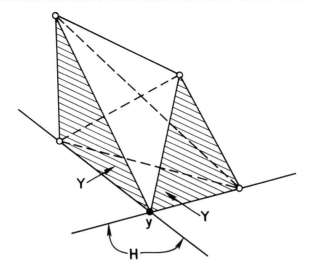

FIG. 4.8. *Common supporting hyperplane.*

that not only is $\text{vol}_d Y > 0$ but also H is a common supporting hyperplane of $[Y]$ and $[X]$.

By reordering the elements of X we suppose that $Y = \{\mathbf{x}^0, \ldots, \mathbf{x}^d\}$ and also let $\lambda_0(\mathbf{x}), \ldots, \lambda_d(\mathbf{x})$ be the barycentric components of $\mathbf{x}$ relative to $[Y]$. Then, for some $m \in \{0, 1, \ldots, d\}$, $\lambda_m(\mathbf{x}) = 0$ is the equation of the hyperplane H. Thus $\lambda_m(\mathbf{x}) \geq 0$ for all $\mathbf{x} \in [X]$ and in particular

$$\lambda_m(\mathbf{x}^\ell) \geq 0, \qquad \ell = d+1, \ldots, n.$$

Moreover, for some $k \in \{d+1, \ldots, n\}$, $\lambda_m(\mathbf{x}^k) > 0$, otherwise we would contradict our hypothesis says that $\text{vol}_d[X \backslash \{\mathbf{x}^j\}] > 0$, $j = 0, 1, \ldots, n$. Now, according to (4.17), if $\mathbf{t} = (t_{d+1}, \ldots, t_n)^T \in V(\mathbf{x}|Y)$ for some $\mathbf{x} \in [X]$, we have $\sum_{j=d+1}^n t_j \leq 1$ and $t_k \lambda_m(\mathbf{x}^k) \leq \lambda_m(\mathbf{x})$. Hence, for all $\mathbf{x} \in [X]$ we obtain the inequality

$$S(\mathbf{x}|X) \leq (d!(n-d-1)! \lambda_m(\mathbf{x}^k) \text{vol}_d[Y])^{-1} \lambda_m(\mathbf{x}).$$

This means, for $\mathbf{x} = \mathbf{y}$, that $S(\mathbf{y}|X) = 0$ and also, for any sequence $\mathbf{y}_j \in [X]$, $j = 0, 1, 2, \ldots$, such that $\lim_{j \to \infty} \mathbf{y}_j = \mathbf{y}$, it follows that $\lim_{j \to \infty} S(\mathbf{y}_j|X) = S(\mathbf{y}|X) = 0$. Since $S(\mathbf{x}|X) = 0$, for $\mathbf{x} \notin [X]^0$, the result follows. □

The next proposition gives a formula for the derivative of the multivariate B-spline. To describe it, we recall the notation for a directional derivative of f

$$(D_\mathbf{y} f)(\mathbf{x}) := \sum_{j=1}^d y_j \frac{\partial f(\mathbf{x})}{\partial x_j},$$

where $\mathbf{y} = (y_1, \ldots, y_d)^T$ and $\mathbf{x} = (x_1, \ldots, x_d)^T$.

PROPOSITION 4.3. *Suppose that* $X := \{\mathbf{x}^0, \ldots, \mathbf{x}^n\} \subset \mathbb{R}^d$. *Given any* $\mathbf{y} \in \mathbb{R}^d$ *such that for some constants* $c_0, \ldots, c_n$

$$(4.19) \qquad \mathbf{y} = \sum_{j=0}^n c_j \mathbf{x}^j, \qquad \sum_{j=0}^n c_j = 0,$$

we have

$$(4.20) \qquad D_{\mathbf{y}} S(\cdot | X) = \sum_{j=0}^n c_j S(\cdot | X_j),$$

where

$$X_j := X \backslash \{\mathbf{x}^j\}, \qquad j = 0, 1, \ldots, n.$$

Proof. By the definition (4.14) of the multivariate B-spline, equation (4.20) means that

$$(4.21) \qquad \int_X D_{\mathbf{y}} f = -\sum_{j=0}^n c_j \int_{X_j} f$$

for all $f \in C^1([X])$. It suffices to prove (4.21) for a class of functions that are dense in $C^1([X])$. For example, the set of functions $\{f_{\mathbf{u}} : \mathbf{u} \in \mathbb{R}^d\}$ where $f_{\mathbf{u}}(\mathbf{x}) = g((\mathbf{u}, \mathbf{x})), \mathbf{x} \in \mathbb{R}^d$, and $g(t) = e^t, t \in \mathbb{R}$ will suffice for this purpose. We compute the left side of (4.21) when $f = f_{\mathbf{u}}$

$$\int_X D_{\mathbf{y}} f_{\mathbf{u}} = (\mathbf{u}, \mathbf{y}) \int_{\Delta^n} g'\left(\sum_{j=0}^n t_j (\mathbf{u}, \mathbf{x}^j)\right) dm(\mathbf{t})$$

$$= (\mathbf{u}, \mathbf{y}) \int_{\Delta^n} g^{(n)}\left(\sum_{j=0}^n t_j (\mathbf{u}, \mathbf{x}^j)\right) dm(\mathbf{t}),$$

which, by (4.3), equals

$$(\mathbf{u}, \mathbf{y}) \left[(\mathbf{u}, \mathbf{x}^0), \ldots, (\mathbf{u}, \mathbf{x}^n)\right] g.$$

Similarly, the right-hand side of (4.21) for $f = f_{\mathbf{u}}$ can be evaluated as

$$-\sum_{j=0}^n c_j \left[(\mathbf{u}, \mathbf{x}^0), \ldots, (\mathbf{u}, \mathbf{x}^{j-1}), (\mathbf{u}, \mathbf{x}^{j+1}), \ldots, (\mathbf{u}, \mathbf{x}^n)\right] g.$$

Both sides of this equation are continuous functions of $\mathbf{u}$. Thus, it suffices to prove they are equal when $(\mathbf{u}, \mathbf{x}^0), \ldots, (\mathbf{u}, \mathbf{x}^n)$ are distinct. To this end, we use the recurrence formula (4.4) (with t_j replaced by $(\mathbf{u}, \mathbf{x}^j)$) and (4.19) to obtain

$$
(\mathbf{u}, \mathbf{y}) \left[(\mathbf{u}, \mathbf{x}^0), \dots, (\mathbf{u}, \mathbf{x}^n) \right] g
$$

$$
= \sum_{j=0}^{n} c_j ((\mathbf{u}, \mathbf{x}^j) - (\mathbf{u}, \mathbf{x}^0)) \left[(\mathbf{u}, \mathbf{x}^0), \dots, (\mathbf{u}, \mathbf{x}^n) \right] g
$$

$$
= \sum_{j=0}^{n} c_j \left(\left[(\mathbf{u}, \mathbf{x}^1), \dots, (\mathbf{u}, \mathbf{x}^n) \right] g \right.
$$

$$
\left. - \left[(\mathbf{u}, \mathbf{x}^0), \dots, (\mathbf{u}, \mathbf{x}^{j-1}), (\mathbf{u}, \mathbf{x}^{j+1}), \dots, (\mathbf{u}, \mathbf{x}^n) \right] g \right)
$$

$$
= -\sum_{j=0}^{n} c_j \left[(\mathbf{u}, \mathbf{x}^0), \dots, (\mathbf{u}, \mathbf{x}^{j-1}), (\mathbf{u}, \mathbf{x}^{j+1}), \dots, (\mathbf{u}, \mathbf{x}^n) \right] g. \qquad \square
$$

We have now accumulated enough facts for the proof of Theorem 4.1. In fact, we will prove the following result, which subsumes Theorem 4.1.

THEOREM 4.2. *Let* $X = \{\mathbf{x}^0, \dots, \mathbf{x}^n\} \subset \mathbb{R}^d, \#X = n + 1,$ *and* $\mathrm{vol}_d[X] > 0$. *Then* $S(\cdot|X)$ *is positive in* $[X]^0$, *and on any region in*

$$
[X] \backslash \{ \langle Y \rangle : Y \subseteq X, \#Y = d \}
$$

it is a polynomial of total degree $\leq n - d$. *Moreover, if for some* $\ell \in \{0, 1, \dots, n - d - 1\}$, $\mathrm{vol}_d[X \backslash Y] > 0$ *for all* $Y \subseteq X$ *with* $\#Y \leq \ell + 1$ *then* $S(\cdot|X) \in C^\ell(\mathbb{R}^d)$.

Proof. The positivity of $S(\cdot|X)$ in $[X]^0$ has been proved in Proposition 4.2. Let us now prove it is a piecewise polynomial. We prove the result by induction on n. The case $n = d$ is valid by our definition (4.13). Suppose it is true for all X with $\#X < n$ and let X be a set with $\#X = n$. Choose any region R in $[X]^0$, not intersected by X-planes $\langle Y \rangle$ where $\#Y = d$ and $Y \subseteq X$. Note that for each of the sets $X_j = X \backslash \{\mathbf{x}^j\}, j = 0, 1, \dots, n$, we have $\#X_j = n - 1$ and any X_j-plane is an X-plane. Hence, R is not cut by X_j-planes for any $j \in \{0, 1, \dots, n\}$. Moreover, since $\mathrm{vol}_d[X] > 0$, the vectors $(1, \mathbf{x}^0)^T, \dots, (1, \mathbf{x}^n)^T$ span $\mathbb{R}^{d+1}$. Hence, every $\mathbf{y} \in \mathbb{R}^d$ can be written in the form (4.19) for some constants $c_0, \dots, c_n$. Thus, for *every* $\mathbf{y} \in \mathbb{R}^d$ we conclude, from (4.20), that on the region R, $D_{\mathbf{y}} S(\cdot|X)$ is a polynomial of degree $n - d - 1$. From this fact it easily follows that $S(\cdot|X)$ is a polynomial of degree $\leq n - d$.

The last claim is also proved by induction, this time on ℓ. The case $\ell = 0$ is covered by Lemma 4.2. Now, suppose that it is true for all $0 < \ell' < \ell$ and that $\mathrm{vol}_d[X \backslash Y] > 0$ for all $Y \subseteq X$, $\#Y \leq \ell + 1$. Then for any $j = 0, 1, \dots, n$ $\mathrm{vol}_d[X_j \backslash Y] = \mathrm{vol}_d[X \backslash (\{\mathbf{x}^j\} \cup Y)] > 0$ whenever $Y \subseteq X$ and $\#Y \leq \ell$. We conclude, by induction, that the right-hand side of (4.20) is continuous and hence so too is $D_{\mathbf{y}} S(\cdot|X)$ for all $\mathbf{y} \in \mathbb{R}^d$. $\square$

Our next result from Micchelli [Mic2] is a *knot insertion* formula for the multivariate B-spline.

THEOREM 4.3. *If* $X = \{\mathbf{x}^0, \dots, \mathbf{x}^n\}$ *and*

$$
(4.22) \qquad \mathbf{y} = \sum_{j=0}^{n} c_j \mathbf{x}^j, \qquad \sum_{j=0}^{n} c_j = 1,
$$

then

(4.23)
$$S(\cdot|X) = \sum_{j=0}^{n} c_j S(\cdot|X_j \cup \{\mathbf{y}\}).$$

We give two proofs of this result. The first proof is "geometric" and confirms Theorem 4.3 when $\mathrm{vol}_d[X] > 0$, $c_j \geq 0, j = 0, 1, \dots, n$ and $n \geq d+1$. However, it proves that (4.23) holds pointwise on $\mathbb{R}^d$. This proof depends on the following lemma.

LEMMA 4.3. *Given any set of vectors* $V = \{\mathbf{v}^0, \dots, \mathbf{v}^n\} \subset \mathbb{R}^n$ *with* $\mathrm{vol}_n[V] > 0$ *and any* $\mathbf{v} \in [V]$, *the simplices*

(4.24)
$$V_j := [\mathbf{v} \cup (V \backslash \{\mathbf{v}^j\})], \qquad j = 0, 1, \dots, n,$$

satisfy

(4.25)
$$\bigcup_{j=0}^{n} V_j = [V]$$

and

(4.26)
$$V_\ell \cap V_k = [\mathbf{v} \cup (V \backslash \{\mathbf{v}^\ell, \mathbf{v}^k\})], \qquad \ell, \ k = 0, 1, \dots, n.$$

Proof. Suppose that $\mathbf{v} = \sum_{j=0}^{n} \gamma_j \mathbf{v}^j$, with $\gamma_j \geq 0$, $j = 0, 1, \dots, n$, and $\sum_{j=0}^{n} \gamma_j = 1$. Let $\Lambda = \{j : \gamma_j > 0, j = 0, 1, \dots, n\}$ and suppose that $\mathbf{x} = \sum_{j=0}^{n} \lambda_j \mathbf{v}^j$, $\lambda_j \geq 0$, $j = 0, 1, \dots, n$, $\sum_{j=0}^{n} \lambda_j = 1$, is a typical element in $[V]$. Then, there is a $k \in \{0, 1, \dots, n\}$ with

$$\lambda_k/\gamma_k = \min\{\lambda_j/\gamma_j : j \in \Lambda\}.$$

Hence, $\alpha_j := \gamma_k^{-1}(\gamma_k \lambda_j - \gamma_j \lambda_k) \geq 0$, for $j = 0, 1, \dots, n$, $\alpha_k = 0$, and moreover,

$$\mathbf{x} = \lambda_k \gamma_k^{-1} \mathbf{v} + \sum_{j=0}^{n} \alpha_j \mathbf{v}^j.$$

This equation proves (4.25).

For (4.26), we suppose that $\mathbf{x} \in V_\ell \cap V_k$ for $\ell \neq k$. We wish to show that $\mathbf{x} \in [\mathbf{v} \cup (V \backslash \{\mathbf{v}^\ell, \mathbf{v}^k\})]$. To this end, we represent $\mathbf{x}$ as

$$\mathbf{x} = \mu_{n+1}\left(\sum_{j=0}^{n} \gamma_j \mathbf{v}^j\right) + \sum_{j=0}^{n} \mu_j \mathbf{v}^j$$

$$= \delta_{n+1}\left(\sum_{j=0}^{n} \gamma_j \mathbf{v}^j\right) + \sum_{j=0}^{n} \delta_j \mathbf{v}^j,$$

for some nonnegative constants μ_j, δ_j, $j = 0, 1, \dots, n$ that necessarily satisfy the equations $\mu_j = \lambda_j - \mu_{n+1}\gamma_j$, $\delta_j = \lambda_j - \delta_{n+1}\gamma_j$, $j = 0, 1, \dots, n+1$, $\mu_\ell = \delta_k = 0$, and

$$\sum_{j=0}^{n+1} \mu_j = \sum_{j=0}^{n+1} \delta_j = 1.$$

Hence, it follows that $\delta_j - \mu_j = (\mu_{n+1} - \delta_{n+1})\gamma_j$. In particular, if $\gamma_k = 0$ then $\mu_\ell = \mu_k = 0$ and we have $\mathbf{x} \in [\mathbf{v} \cup (V \setminus \{\mathbf{v}^\ell, \mathbf{v}^k\})]$, as desired. The same conclusion holds when $\gamma_\ell = 0$. Thus, we only need to consider the case $\gamma_k \gamma_\ell > 0$. But then the equations $-\mu_k = (\mu_{n+1} - \delta_{n+1})\gamma_k$ and $\delta_\ell = (\mu_{n+1} - \delta_{n+1})\gamma_\ell$ imply that $\delta_{n+1} = \mu_{n+1}$. Therefore, $\mu_j = \delta_j, j = 0, 1, \ldots, n$, and, in particular, we get again $\mu_\ell = \mu_k = \delta_\ell = \delta_k = 0$ (see Fig. 4.9).   □

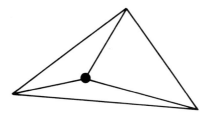

FIG. 4.9.   *Knot insertion.*

First proof of Theorem 4.3. Assume that $n \geq d + 1$. We choose $\mathbf{v}^0, \ldots, \mathbf{v}^n \in \mathbb{R}^n$ such that $\mathbf{v}^j\big|_{\mathbb{R}^d} = \mathbf{x}^j$, $j = 0, 1, \ldots, n$, and set $\mathbf{v} = \sum_{j=0}^n c_j \mathbf{v}^j$ where $c_j \geq 0$, $j = 0, 1, \ldots, n$. Then, by (4.25) and (4.26)

$$\mathrm{vol}_{n-d}\{\mathbf{u} : \mathbf{u} \in [V], \mathbf{u}\big|_{\mathbb{R}^d} = \mathbf{x}\} = \sum_{j=0}^n \mathrm{vol}_{n-d}\{\mathbf{u} : \mathbf{u} \in V_j, \mathbf{u}\big|_{\mathbb{R}^d} = \mathbf{x}\},$$

where V_j is given by (4.24). Thus,

$$(4.27) \qquad S(\mathbf{x}|X) = (\mathrm{vol}_n[V])^{-1} \sum_{j=0}^n \mathrm{vol}_n V_j S(\mathbf{x}|X_j \cup \{\mathbf{y}\}),$$

and, by formula (4.10) for the volume of a simplex, we have $\mathrm{vol}_n V_j = c_j \mathrm{vol}_n[V]$, $j = 0, 1, \ldots, n$.   □

The second proof depends only on *simple* properties of divided differences.

Second proof of Theorem 4.3. Equation (4.23) means that

$$(4.28) \qquad \int_{[X]} f = \sum_{j=0}^n c_j \int_{[X_j \cup \{\mathbf{y}\}]} f$$

for all $f \in C([X])$. As in the proof of Proposition 4.3, it suffices to prove (4.28) for all f of the form $f = f_{\mathbf{u}}$ and $\mathbf{u} \in \mathbb{R}^d$. In this case, (4.28) becomes

$$(4.29) \quad \begin{aligned} &[(\mathbf{u}, \mathbf{x}^0), \ldots, (\mathbf{u}, \mathbf{x}^n)]g \\ &= \sum_{j=0}^n c_j \left[(\mathbf{u}, \mathbf{y}), (\mathbf{u}, \mathbf{x}^0), \ldots, (\mathbf{u}, \mathbf{x}^{j-1}), (\mathbf{u}, \mathbf{x}^{j+1}), \ldots, (\mathbf{u}, \mathbf{x}^n)\right] g. \end{aligned}$$

For *any* $g \in C^n(\mathbb{R})$, both sides of this equation are continuous functions of $\mathbf{u}$ and so to prove that they are equal we may assume, without loss of generality, that $(\mathbf{u}, \mathbf{y}), (\mathbf{u}, \mathbf{x}^0), \ldots, (\mathbf{u}, \mathbf{x}^n)$ are all distinct. The divided difference on the left side of equation (4.29) is characterized as the unique linear functional that vanishes on all polynomials of degree $\leq n - 1$, is one for the function $t^n, t \in \mathbb{R}$, and is a linear combination of the value of g at $(\mathbf{u}, \mathbf{x}^0), \ldots, (\mathbf{u}, \mathbf{x}^n)$. In view of the second equation in (4.22), the right-hand side of equation (4.29) has the first two of these properties. To verify the third property we need to show that the coefficient of $g((\mathbf{u}, \mathbf{y}))$ is zero. For this purpose, we use the formula for the divided difference on distinct points for each factor in the right-hand side of (4.29) to identify the coefficient of $g((\mathbf{u}, \mathbf{y}))$ as

$$\sum_{j=0}^{n} c_j \frac{1}{\prod_{\ell \neq j}((\mathbf{u}, \mathbf{y}) - (\mathbf{u}, \mathbf{x}^\ell))}$$

$$= \frac{1}{\prod_{\ell=0}^{n}((\mathbf{u}, \mathbf{y}) - (\mathbf{u}, \mathbf{x}^\ell))} \left(\mathbf{u}, \left(\mathbf{y} - \sum_{j=0}^{n} c_j \mathbf{x}^j \right) \right) = 0. \qquad \square$$

Our next result from Micchelli [Mic2] concerns a recurrence formula for the multivariate B-spline, in the cardinality of X, which is useful for computational purposes. (It is this recurrence formula that was used to generate the surfaces in Fig. 4.4.)

THEOREM 4.4. *Suppose that* $X = \{\mathbf{x}^0, \ldots, \mathbf{x}^n\} \subset \mathbb{R}^d$ *and*

$$\text{vol}_d[X_j] > 0, \qquad j = 0, 1, \ldots, n.$$

If

$$\mathbf{x} = \sum_{j=0}^{n} c_j \mathbf{x}^j, \qquad \sum_{j=0}^{n} c_j = 1,$$

then

$$(4.30) \qquad S(\mathbf{x}|X) = \frac{1}{n-d} \sum_{j=0}^{n} c_j S(\mathbf{x}|X_j), \quad \text{a.e., } \mathbf{x} \in \mathbb{R}^d.$$

We base the proof of this formula on the following auxiliary result.

PROPOSITION 4.4. *Suppose that* $X \subset \mathbb{R}^d, \#X = n$ *and*

$$\text{vol}_d[X] > 0.$$

Then

$$(4.31) \qquad S(\mathbf{x}|\{\mathbf{x}\} \cup X) = \frac{1}{n-d} S(\mathbf{x}|X), \qquad \mathbf{x} \in \mathbb{R}^d.$$

Note that a proof of Theorem 4.4 can be constructed from Proposition 4.4 by first choosing $\mathbf{y} = \mathbf{x}$ in (4.22), and then using (4.31) to simplify the resulting expression.

Proof. We shall verify (4.31) directly from the definition (4.13) of $S(\cdot|X)$ as a ratio of volumes. But first we need a simple fact about volumes.

Let C be any set in $\mathbb{R}^{n-1}$ and $\mathbf{y}$ some vector in $\mathbb{R}^{n-1}$. The "cone" in $\mathbb{R}^n$ with base C and vertex $(\mathbf{y}, 1)^T$ is defined to be the set

$$K[C, \mathbf{y}] = \{((1-t)\mathbf{x} + t\mathbf{y}, t)^T, t \in [0,1], \ \mathbf{x} \in C\}$$

The mapping $T : C \times [0,1] \rightarrow K[C, \mathbf{y}]$ defined by $T((\mathbf{x}, t)^T) = (\mathbf{z}, t)^T$ where $\mathbf{z} = (1-t)\mathbf{x} + t\mathbf{y}$ has Jacobian $(1-t)^{n-1}$ and hence

(4.32)
$$\mathrm{vol}_n K[C, \mathbf{y}] = \frac{1}{n} \mathrm{vol}_{n-1} C.$$

In particular, we see that the volume of $K[C, \mathbf{y}]$ is *independent* of $\mathbf{y}$ (see Fig. 4.10).

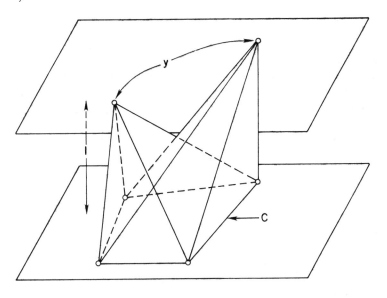

FIG. 4.10.   *Equal volumes.*

Now suppose that $X = \{\mathbf{x}^0, \ldots, \mathbf{x}^{n-1}\}$ and $\mathbf{v}^0, \ldots, \mathbf{v}^{n-1}$ are vectors in $\mathbb{R}^{n-1}$ such that $\mathbf{v}^i|_{\mathbb{R}^d} = \mathbf{x}^i, i = 0, 1, \ldots, n-1$, with $\mathrm{vol}_{n-1}\sigma > 0$, where $\sigma = [\mathbf{v}^0, \ldots, \mathbf{v}^{n-1}]$. Let $\mathbf{v}$ be any vector in $\mathbb{R}^{n-1}$ such that $\mathbf{v}|_{\mathbb{R}^d} = \mathbf{x}$. Then the cone in $\mathbb{R}^n$ with base $(\sigma, 0)$ and vertex $(\mathbf{v}, 0)^T$ is given by

$$K[\sigma, \mathbf{v}] = [\bar{\mathbf{v}}^0, \ldots, \bar{\mathbf{v}}^n] := \bar{\sigma},$$

where $\bar{\mathbf{v}}^i = (\mathbf{v}^i, 0)^T, i = 0, 1, \ldots, n-1, \bar{\mathbf{v}}^n = (\mathbf{v}, 1)^T$. Thus (4.32) gives the formula

(4.33)
$$\mathrm{vol}_n \bar{\sigma} = \frac{1}{n} \mathrm{vol}_{n-1} \bar{\sigma}.$$

(This formula also follows by appropriately specializing equations (4.9) and (4.10).) Next let $\sigma_{\mathbf{X}} := \{\mathbf{u} \in \sigma : \mathbf{u}|_{\mathbb{R}^d} = \mathbf{x}\}$ and $\bar{\sigma}_{\mathbf{X}} := \{\mathbf{u} \in \bar{\sigma} : \mathbf{u}|_{\mathbb{R}^d} = \mathbf{x}\}$. Then $\bar{\sigma}_{\mathbf{X}} = K[\sigma_{\mathbf{X}}, \mathbf{v}]$ and hence, by (4.32) (with n replaced by $n - d$), it follows that

$$(4.34) \qquad\qquad \mathrm{vol}_{n-d}\bar{\sigma}_{\mathbf{X}} = \frac{1}{n-d}\mathrm{vol}_{n-1-d}\sigma_{\mathbf{X}}.$$

Using (4.33), (4.34), and also (4.13) twice, we compute

$$S(\mathbf{x}|\{\mathbf{x}\} \cup X) = \frac{1}{n!}\frac{\mathrm{vol}_{n-d}\bar{\sigma}_{\mathbf{X}}}{\mathrm{vol}_n\bar{\sigma}}$$

$$= \frac{1}{n-d}\frac{1}{(n-1)!}\frac{\mathrm{vol}_{n-1-d}\sigma_{\mathbf{X}}}{\mathrm{vol}_{n-1}\sigma}$$

$$= \frac{1}{n-d}S(\mathbf{x}|X). \qquad \square$$

The next result gives a recurrence relation for the derivatives of S, see Micchelli [Mic1], [Mic2]. Remarkably, all derivatives of a given order satisfy the same recurrence relation. We formulate the result below only for the first derivatives. The statement for the higher derivatives is similar.

THEOREM 4.5. *Let* $\mathbf{y} \in \mathbb{R}^d$ *and* $X = \{\mathbf{x}^0, \ldots, \mathbf{x}^n\} \subset \mathbb{R}^d$ *where* $\mathrm{vol}[X_j] > 0$, $j = 0, 1, \ldots, n$. *If*

$$(4.35) \qquad\qquad \mathbf{x} = \sum_{j=0}^{n} c_j\mathbf{x}^j, \qquad \sum_{j=0}^{n} c_j = 1,$$

then

$$(n - d - 1)D_{\mathbf{y}}(S(\cdot|X))(\mathbf{x}) = \sum_{j=0}^{n} c_j D_{\mathbf{y}}(S(\cdot|X_j))(\mathbf{x}), \text{ a.e., } \mathbf{x} \in \mathbb{R}^d.$$

Proof. We use the equations in (4.35) and eliminate c_0 from (4.30). Then, differentiate both sides of (4.30) with respect to c_j, for each $j = 1, \ldots, n$, to obtain the equation

$$(n - d)D_{\mathbf{x}^j - \mathbf{x}^0}(S(\cdot|X))(\mathbf{x})$$

$$(4.36)$$

$$= \sum_{\ell=0}^{n} c_\ell D_{\mathbf{x}^j - \mathbf{x}^0}(S(\cdot|X_\ell))(\mathbf{x}) + S(\mathbf{x}|X_j) - S(\mathbf{x}|X_0).$$

Since $\mathrm{vol}_d[X] > 0$ we can find constants $a_1, \ldots, a_n$ such that

$$\mathbf{y} = \sum_{j=1}^{n} a_j(\mathbf{x}^j - \mathbf{x}^0).$$

Next, multiply both sides of (4.36) by a_j and sum from $j = 1, \ldots, n$ to obtain the equation

$$(n - d)D_{\mathbf{y}}(S(\cdot|X))(\mathbf{x}) = \sum_{\ell=0}^{n} c_\ell D_{\mathbf{y}}(S(\cdot|X_\ell))(\mathbf{x}) + \sum_{j=1}^{n} a_j(S(\mathbf{x}|X_j) - S(\mathbf{x}|X_0)).$$

The last summand above can be simplified by the derivative recurrence given in Proposition 4.3 to obtain the desired formula

$$(n-d)D_{\mathbf{y}}(S(\cdot|X))(\mathbf{x}) = \sum_{\ell=0}^{n} c_{\ell} D_{\mathbf{y}}(S(\cdot|X_{\ell}))(\mathbf{x}) + D_{\mathbf{y}}(S(\cdot|X))(\mathbf{x}). \qquad \square$$

4.3.   Multivariate B-splines: Multiple points and explicit formula.

There are two additional identities for the multivariate B-spline that we wish to describe. The first concerns the relationship of the multivariate B-spline to the *Bernstein–Bézier* polynomials. The second is a *degree-raising* formula. Both are accessible by seeing what happens to the multivariate B-spline when we repeat vectors in the set X.

We use the following notation. For every $\mathbf{k} = (k_0,\ldots,k_n)^T \in \mathbb{Z}_+^{n+1}$ and $X = \{\mathbf{x}^0,\ldots,\mathbf{x}^n\} \subset \mathbb{R}^d$ we let

$$X^{\mathbf{k}} := \left\{ \underbrace{\mathbf{x}^0,\ldots,\mathbf{x}^0}_{k_0+1}, \underbrace{\mathbf{x}^1,\ldots,\mathbf{x}^1}_{k_1+1}, \ldots, \underbrace{\mathbf{x}^n,\ldots,\mathbf{x}^n}_{k_n+1} \right\},$$

and $\binom{n}{\mathbf{k}} := \frac{n!}{\mathbf{k}!}$, $\mathbf{k}! = k_0! \cdots k_n!$. For the special case $\mathbf{k}_j = (0,\ldots,0,1,0,\ldots,0)^T$ where 1 occurs in the jth component, we use X^j for the set $X^{\mathbf{k}_j}$. Thus we have quite explicitly,

$$X^j = \{\mathbf{x}^0,\ldots,\mathbf{x}^{j-1},\mathbf{x}^j,\mathbf{x}^j,\mathbf{x}^{j+1},\ldots,\mathbf{x}^n\}, \qquad j = 0,1,\ldots,n.$$

Also, when $\#X = d+1$ and $\mathbf{k} \in \mathbb{Z}_+^{d+1}$, we set

$$\lambda^{\mathbf{k}}(\mathbf{x}) := \lambda_0^{k_0}(\mathbf{x}) \cdots \lambda_d^{k_d}(\mathbf{x}),$$

where $\lambda_0(\mathbf{x}),\ldots,\lambda_d(\mathbf{x})$ are the barycentric components of $\mathbf{x}$ relative to X.

THEOREM 4.6.   *Let* $X \subset \mathbb{R}^d, \#X = d+1$, *and* $\mathrm{vol}_d[X] > 0$. *Then, for* $\mathbf{k} \in \mathbb{Z}_+^{d+1}$ *we have*

$$(4.37) \qquad S(\mathbf{x}|X^{\mathbf{k}}) = \begin{cases} \lambda^{\mathbf{k}}(\mathbf{x})/\mathbf{k}! |\det(X)|, & \mathbf{x} \in [X], \\ 0, & \mathbf{x} \notin [X]. \end{cases}$$

When $X = \{\mathbf{e}^0, \mathbf{e}^1, \ldots, \mathbf{e}^d\}$, where $\mathbf{e}^0 = \mathbf{0}$ and $(\mathbf{e}^i)_j = \delta_{ij}$, $i,j = 1,\ldots,d$, then $[X] = \Delta^d$ and, according to Theorem 4.6, $n! S(\mathbf{x}|X^{\mathbf{k}}) = b_{\mathbf{k}}(\mathbf{x})$, the Bernstein–Bézier polynomials for Δ^d (see Chapter 1, text preceding Lemma 1.5).

The next result represents a multivariate B-spline, corresponding to a set with $n+1$ vectors, in terms of multivariate B-splines, corresponding to sets of $n+2$ vectors. In this sense, it is a "degree-raising" formula.

THEOREM 4.7.   *Let* $X \subset \mathbb{R}^d$ *and* $\#X = n+1$. *Then*

$$(4.38) \qquad\qquad S(\cdot|X) = \sum_{j=0}^{n} S(\cdot|X^j).$$

Proof of Theorems 4.6 *and* 4.7. We begin the proof of these results by setting $X = \{\mathbf{x}^0, \ldots, \mathbf{x}^n\}$ and considering the distribution $S(\cdot|X^n)$. For any $f \in C(\mathbb{R}^d)$, we have the equation

$$\int_{X^n} f = \int_{\Delta^{n+1}} f\left(\mathbf{x}^0 + \sum_{j=1}^{n-1} t_j(\mathbf{x}^j - \mathbf{x}^0) + (t_n + t_{n+1})(\mathbf{x}^n - \mathbf{x}^0)\right) dm(\mathbf{t}).$$

The mapping $G : \Delta^n \times [0,1] \rightarrow \Delta^{n+1}$ defined by $G(y_0, \ldots, y_{n+1}) = (y_0, \ldots, y_{n-1}, y_n(1 - y_{n+1}), y_n y_{n+1})^T$ has Jacobian y_n. Using this mapping to change the variables of integration in the integral above, yields the equation

$$\int_{X^n} f = \int_{\Delta^n} t_n f\left(\mathbf{x}^0 + \sum_{j=0}^{n} t_j(\mathbf{x}^j - \mathbf{x}^0)\right) dm(\mathbf{t}).$$

Appealing to the symmetry of the multivariate B-spline, under reordering of the elements of X, and repeating all vectors in X one at a time, the above computation gives, for $\mathbf{k} \in \mathbb{Z}_+^{n+1}$, the formula

$$\int_{X^{\mathbf{k}}} f = \int_{\Delta^n} \frac{\mathbf{t}^{\mathbf{k}}}{\mathbf{k}!} f(X\mathbf{t}) dm(\mathbf{t}).$$

Thus, for any polynomial h given by

$$h(\mathbf{t}) = \sum_{\mathbf{k} \in \mathbb{Z}_+^{n+1}} h_{\mathbf{k}} \frac{\mathbf{t}^{\mathbf{k}}}{\mathbf{k}!},$$

we have the relation

(4.39)
$$\int_{\Delta^n} h(\mathbf{t}) f(X\mathbf{t}) dm(\mathbf{t}) = \sum_{\mathbf{k} \in \mathbb{Z}_+^{n+1}} h_{\mathbf{k}} \int_{X^{\mathbf{k}}} f.$$

For example, the choice $h(t_0, \ldots, t_n) = t_0 + \cdots + t_n$ gives us the formula

$$\int_X f = \sum_{j=0}^{n} \int_{X^j} f,$$

because $h|_{\Delta^n} = 1$. This observation proves Theorem 4.7. Actually, the choice

$$h(t_0, \ldots, t_n) = (t_0 + \cdots + t_n)^N = \sum_{|\mathbf{k}|_1 = N} \binom{N}{\mathbf{k}} \mathbf{t}^{\mathbf{k}}$$

in (4.39) gives the equation

(4.40)
$$S(\cdot|X) = N! \sum_{|\mathbf{k}|_1 = N} S(\cdot|X^{\mathbf{k}}),$$

which generalizes (4.38). Another useful choice for h is to set $h(\mathbf{t}) = p(X\mathbf{t})$, where p is a polynomial on $\mathbb{R}^d$. Then (4.40) gives us the formula

$$(4.41) \qquad p(\cdot)S(\cdot|X) = \sum_{\mathbf{k}\in\mathbb{Z}_+^{n+1}} (D_{X\mathbf{k}}p)(0)S(\cdot|X^{\mathbf{k}}),$$

where we set $D_Y p = \prod_{\mathbf{y}\in Y} D_{\mathbf{y}} p$ for *any* set $Y \subset \mathbb{R}^d$. For example, when $\#X = d+1$ and $p(\mathbf{x}) = \lambda^{\mathbf{j}}(\mathbf{x})/\mathbf{j}!$, $\mathbf{j} \in \mathbb{Z}_+^{d+1}$, we have $p(X\mathbf{t}) = \mathbf{t}^{\mathbf{j}}/\mathbf{j}!$ for $\mathbf{t} \in \Delta^{d+1}$, and so (4.41) yields the equation

$$\frac{\lambda^{\mathbf{j}}(\mathbf{x})}{\mathbf{j}!} S(\mathbf{x}|X) = S(\mathbf{x}|X^{\mathbf{j}}).$$

But according to definition (4.13), we have

$$S(\mathbf{x}|X) = \begin{cases} \dfrac{1}{d!\,\mathrm{vol}_d[X]}, & \mathbf{x} \in [X], \\ 0, & \mathbf{x} \notin [X]. \end{cases}$$

Combining these two equations proves Theorem 4.6.   □

Our next formula is a curious identity from Micchelli [Mic2], which represents the *square* of the multivariate B-spline as a sum of products of B-splines.

THEOREM 4.8. *Suppose that* $X \subset \mathbb{R}^d$, $\#X = n+1$, *and* $\mathrm{vol}_d[X_j] > 0, j = 0, 1, \ldots, n$. *Then*

$$(4.42) \qquad S^2(\mathbf{x}|X) = \frac{1}{n-d}\sum_{j=0}^{n} S(\mathbf{x}|X_j)S(\mathbf{x}|X^j), \quad \text{a.e.,}\ \mathbf{x} \in \mathbb{R}^d.$$

Proof. From Theorem 4.4 we have almost everywhere, $\mathbf{t} = (t_0, \ldots, t_n)^T \in \Delta^n$,

$$S(X\mathbf{t}) = \frac{1}{n-d}\sum_{j=0}^{n} t_j S(X\mathbf{t}|X_j).$$

Choose any $f \in C(\mathbb{R}^d)$, multiply both sides of the above equation by $f(X\mathbf{t})$, and integrate over Δ^n. We obtain the equation

$$\int_{\mathbb{R}^d} f(\mathbf{x})S^2(\mathbf{x}|X)d\mathbf{x} = \int_{\Delta^n} f(X\mathbf{t})S(X\mathbf{t})dm(\mathbf{t})$$

$$= \frac{1}{n-d}\sum_{j=0}^{n} \int_{\Delta^n} t_j f(X\mathbf{t})S(X\mathbf{t}|X_j)dm(\mathbf{t}).$$

Specializing equation (4.39) to $\mathbf{k} = \mathbf{k}_j$, $j = 0, 1, \ldots, n$, the standard basis in $\mathbb{R}^{n+1}$, we get the desired formula

$$\int_{\mathbb{R}^d} f(\mathbf{x})S^2(\mathbf{x}|X)d\mathbf{x} = \int_{\mathbb{R}^d} f(\mathbf{x})\left(\frac{1}{n-d}\sum_{j=0}^{n} S(\mathbf{x}|X^j)S(\mathbf{x}|X_j)\right)d\mathbf{x}.   □$$

Our final identity is an explicit formula for $S(\cdot|X)$ whose statistical background is discussed in Dahmen and Micchelli [DM3]. To obtain this formula we make use of two lemmas.

LEMMA 4.4. *Let* $f \in L_0^\infty(\mathbb{R}^d)$ *(bounded functions of compact support) and suppose that for all* $\mathbf{y} \in \mathbb{R}^d$

$$\int_{\mathbb{R}^d} (1 + i\langle \mathbf{y}, \mathbf{x}\rangle)^{-d-1} f(\mathbf{x}) d\mathbf{x} = 0.$$

Then $f(\mathbf{x}) = 0$, *a.e.,* $\mathbf{x} \in \mathbb{R}^d$.

Proof. Let $g(z) = (1+z)^{-d-1}$ where $z \in \mathbb{C}$, $|z| < 1$. Then $f_\mathbf{u}(\mathbf{y}) := g(i\langle \mathbf{y}, \mathbf{u}\rangle)$, $\mathbf{y} \in \mathbb{R}^d$, has derivatives $(D^{\mathbf{j}} f_\mathbf{u})(\mathbf{y}) = (i\mathbf{u})^{|\mathbf{j}|_1} g^{(|\mathbf{j}|_1)}(0)$. Since $g^{(k)}(0) \neq 0$, $k \in \mathbb{Z}_+$, we have from our hypothesis that

$$\int_{\mathbb{R}^d} \mathbf{x}^{\mathbf{j}} f(\mathbf{x}) d\mathbf{x} = 0, \qquad \mathbf{j} \in \mathbb{Z}_+^d.$$

Therefore, since polynomials are dense in L^∞ on compact sets, we get that $f(\mathbf{x}) = 0$ a.e., $\mathbf{x} \in \mathbb{R}^d$. □

The next lemma is a *multivariate* partial fraction decomposition. To motivate what we have in mind, we begin by reviewing the univariate case. Given distinct nonzero real numbers $x_0, \ldots, x_n$, we consider the rational function

$$\frac{1}{(1 + zx_0) \cdots (1 + zx_n)}, \qquad z \in \mathbb{C}.$$

We let $\ell_0, \ldots, \ell_n$ be the Lagrange polynomials of degree n such that

$$\ell_i(y_j) = \delta_{ij}, \qquad i, j = 0, 1, \ldots, n,$$

where $y_j = -1/x_j$. They are given explicitly by the formula

$$\ell_j(x) = \frac{\omega(x) x_j}{\omega'(y_j)(1 + xx_j)}, \qquad j = 0, 1, \ldots, n,$$

where $\omega(x) := \prod_{j=0}^n (1 + xx_j)$. Consequently, we get

$$1 = \sum_{j=0}^n \ell_j(z), \qquad z \in \mathbb{C},$$

and this leads us to the partial fraction decomposition

$$\frac{1}{(1 + zx_0) \cdots (1 + zx_n)} = \sum_{j=0}^n \frac{x_j}{\omega'(y_j)(1 + zx_j)},$$

which is valid for all z, $z = iy$, $y \in \mathbb{R}$.

We need the same type of formula in the multivariate case. To this end, we consider the rational function

$$\frac{1}{(1 + \langle \mathbf{z}, \mathbf{x}^0\rangle) \cdots (1 + \langle \mathbf{z}, \mathbf{x}^n\rangle)},$$

where $\mathbf{z} = i\mathbf{y}, \mathbf{y} \in \mathbb{R}^d$. We approach the partial fraction decomposition of this function in the same way as in the univariate case, by considering a certain multivariate interpolation scheme. For this purpose, we begin with vectors $X = \{\mathbf{x}^0, \ldots, \mathbf{x}^n\}$ such that $\{\mathbf{0}, \mathbf{x}^0, \ldots, \mathbf{x}^n\}$ are in general position. For every $Y \subseteq X$ with $\#Y = d$ there is a $\mathbf{v}(= \mathbf{v}^Y)$ such that

$$1 + (\mathbf{v}^Y, \mathbf{y}) = 0, \qquad \mathbf{y} \in Y.$$

Thus,

$$(4.43) \qquad 1 + (\mathbf{v}^Y, \mathbf{x}) = \det(\{\mathbf{x}\} \cup Y)/\det(\{\mathbf{0}\} \cup Y), \qquad \mathbf{x} \in \mathbb{R}^d,$$

and also $1 + (\mathbf{v}, \mathbf{x}) \neq 0$, for all $\mathbf{x} \in X\backslash Y$, because $\{\mathbf{0}\} \cup X$ is in general position. (It is to be understood here that the ordering of the vectors in Y used in the numerator and denominator of the right-hand side of equation (4.43) is the same.) Consequently, if Y, Y' are two distinct subsets of X with cardinality d there is a $\mathbf{w} \in (X\backslash Y) \cap Y'$. Thus, $1 + (\mathbf{v}^Y, \mathbf{w}) \neq 0$, while $1 + (\mathbf{v}^{Y'}, \mathbf{w}) = 0$. This means that $\mathbf{v}^Y \neq \mathbf{v}^{Y'}$ whenever $Y \neq Y'$.

Define a polynomial l_Y of total degree $\leq n - d + 1$, by the formula

$$l_Y(\mathbf{x}) := \prod_{\mathbf{y} \in X\backslash Y} (1 + (\mathbf{x}, \mathbf{y}))\backslash \prod_{\mathbf{y} \in X\backslash Y} (1 + (\mathbf{v}^Y, \mathbf{y})).$$

Reasoning as above, we choose any two distinct sets $Y, Y' \subseteq X$ such that $\#Y = \#Y' = d$, then pick a $\mathbf{w} \in (X\backslash Y) \cap Y'$ and observe that $l_Y(\mathbf{v}^{Y'}) = 0$. In other words,

$$l_Y(\mathbf{v}^{Y'}) = \delta_{YY'}$$

and, in particular, the functions $l_Y, Y \subseteq X, \#Y = d$, are linearly independent. Since $\#\{Y : Y \subseteq X, \#Y = d\} = \binom{n+1}{d} = \dim \Pi_{n-d+1}(\mathbb{R}^d)$, we conclude that these set of functions $\{l_Y : Y \subseteq X, \#Y = d\}$, span $\Pi_{n-d+1}(\mathbb{R}^d)$. This provides us with the interpolation formula

$$p(\mathbf{z}) = \sum_{\substack{Y \subseteq X \\ \#Y = d}} p(\mathbf{v}^Y) \ell_Y(\mathbf{z}), \qquad \mathbf{z} \in \mathbb{C}^d,$$

valid for all $p \in \Pi_{n-d+1}(\mathbb{R}^d)$ (see Fig. 4.11).

In particular, we get

$$1 = \sum_{\substack{Y \subseteq X \\ \#Y = d}} \ell_Y(\mathbf{z}), \qquad \mathbf{z} \in \mathbb{C}^d.$$

Our next lemma follows immediately from this identity.

LEMMA 4.5. *Let* $X \subset \mathbb{R}^d, \#X = n + 1$, *and* $\{\mathbf{0}\} \cup X$ *be in general position. Then for every* $\mathbf{z} = i\mathbf{y}, \mathbf{y} \in \mathbb{R}^d$, *we have*

$$(4.44) \qquad \frac{1}{\prod_{\mathbf{x} \in X} (1 + (\mathbf{z}, \mathbf{x}))} = \sum_{\substack{Y \subseteq X \\ \#Y = d}} \prod_{\mathbf{y} \in X\backslash Y} \frac{\det(\{\mathbf{0}\} \cup Y)}{\det(\{\mathbf{y}\} \cup Y)} \frac{1}{\prod_{\mathbf{y} \in Y} (1 + (\mathbf{z}, \mathbf{y}))}.$$

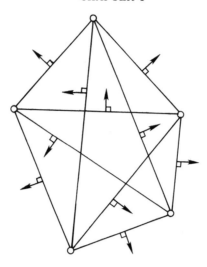

FIG. 4.11. *Ten normals determine a cube polynomial.*

We are now ready to prove the following result.

THEOREM 4.9. *Let $X \subset \mathbb{R}^d$ with $\#X = n + 1$. Suppose that $\mathbf{u} \in \mathbb{R}^d$ and $\{\mathbf{u}\} \cup X$ is in general position. Then*

$$
\begin{aligned}
&S(\mathbf{x}|X) \\
(4.45) \quad &= \frac{1}{(n-d)!} \sum_{\substack{Y \subseteq X \\ \#Y=d}} \frac{(\det(\{\mathbf{x}\} \cup Y))^{n-d}}{\displaystyle\prod_{\mathbf{y} \in X \setminus Y} \det(\{\mathbf{y}\} \cup Y)} \; sgn \det(\{\mathbf{u}\} \cup Y) \chi_{[\{\mathbf{u}\} \cup Y]}(\mathbf{x}).
\end{aligned}
$$

Proof. First, let us point out that it suffices to prove this result for $\mathbf{u} = \mathbf{0}$. Quite explicitly, if $\{\mathbf{u}\} \cup X$ is in general position then so too is the set $\{\mathbf{0}\} \cup X'$, where $X' := X - \{\mathbf{u}\}$, since

$$
\{\mathbf{0}\} \cup X' = (\{\mathbf{u}\} \cup X) - \mathbf{u}.
$$

Also, it follows from the definition of the multivariate B-spline that

$$
S(\mathbf{x}|X) = S(\mathbf{x} - \mathbf{u}|X').
$$

Thus one can apply (4.45) when $\mathbf{u} = \mathbf{0}$ to the set X' and replace $\mathbf{x}$ by $\mathbf{x} - \mathbf{u}$ to obtain the general case. Hence we assume without loss of generality that $\mathbf{u} = \mathbf{0}$ in (4.45).

Next let us consider the function $g(t) = (-1)^n(1+t)^{-1}, t \in \mathbb{R}$. Then $f_{\mathbf{z}}(\mathbf{x}) := g^{(n)}((\mathbf{z}, \mathbf{x})) = n!(1 + (\mathbf{z}, \mathbf{x}))^{-n-1}$ and for $\mathbf{z} = i\mathbf{y}, \mathbf{y} \in \mathbb{R}^d$, we have

$$
\begin{aligned}
n! \int_{\mathbb{R}^d} (1 + (\mathbf{z}, \mathbf{x}))^{-n-1} S(\mathbf{x}|X) d\mathbf{x} &= \int_{\mathbb{R}^d} g^{(n)}((\mathbf{z}, \mathbf{x})) S(\mathbf{x}|X) d\mathbf{x} \\
&= \int_{\Delta^n} g^{(n)} \left(\sum_{j=0}^n t_j (\mathbf{z}, \mathbf{x}^j) \right) dm(\mathbf{t}) \\
&= [(\mathbf{z}, \mathbf{x}^0), \ldots, (\mathbf{z}, \mathbf{x}^n)] g \\
&= \frac{1}{\prod_{\mathbf{x} \in X} (1 + (\mathbf{z}, \mathbf{x}))}.
\end{aligned}
$$

That is, we have verified the equation

$$
(4.46) \qquad \int_{\mathbb{R}^d} (1 + (\mathbf{z}, \mathbf{x}))^{-n-1} S(\mathbf{x}|X) d\mathbf{x} = \frac{1}{n!} \frac{1}{\prod_{\mathbf{x} \in X} (1 + (\mathbf{z}, \mathbf{x}))}.
$$

Specializing this formula appropriately, we have, for every $Y \subseteq X, \#Y = d$, the formula

$$
(4.47) \quad \int_{\mathbb{R}^d} (1 + (\mathbf{z}, \mathbf{x}))^{-n-1} S(\mathbf{x}|\underbrace{\{\mathbf{0}, \ldots, \mathbf{0}\}}_{n-d+1} \cup Y) d\mathbf{x} = \frac{1}{n!} \frac{1}{\prod_{\mathbf{x} \in Y} (1 + (\mathbf{z}, \mathbf{x}))}.
$$

Now set

$$
h(\mathbf{x}) = S(\mathbf{x}|X) - \sum_{\substack{Y \subseteq X \\ \#Y = d}} \prod_{\mathbf{y} \in X \backslash Y} \frac{\det(\{\mathbf{0}\} \cup Y)}{\det(\{\mathbf{y}\} \cup Y)} S(\mathbf{x}|\underbrace{\{\mathbf{0}, \ldots, \mathbf{0}\}}_{n-d+1} \cup Y).
$$

Hence from (4.46), (4.47), and Lemma 4.5, we get

$$
\int_{\mathbb{R}^d} (1 + (\mathbf{z}, \mathbf{x}))^{-n-1} h(\mathbf{x}) d\mathbf{x} = 0, \qquad \mathbf{z} = i\mathbf{y}, \ \mathbf{y} \in \mathbb{R}^d.
$$

Lemma 4.4 implies that $h(\mathbf{x}) = 0$, a.e., $\mathbf{x} \in \mathbb{R}^d$, and Theorem 4.6 gives us the equation

$$
(n-d)! S(\mathbf{x}|\underbrace{\{\mathbf{0}, \ldots, \mathbf{0}\}}_{n-d+1} \cup Y) = \frac{\lambda_0^{n-d}(\mathbf{x})}{|\det(\{\mathbf{0}\} \cup Y)|} \chi_{[\{\mathbf{0}\} \cup Y]}(\mathbf{x}),
$$

where $\lambda_0(\mathbf{x}), \ldots, \lambda_d(\mathbf{x})$ are the barycentric components of $\mathbf{x}$ relative to the simplex $[\{\mathbf{0}\} \bigcup Y]$. Specifically, we have the formula

$$
\lambda_0(\mathbf{x}) = \frac{\det(\{\mathbf{x}\} \bigcup Y)}{\det(\{\mathbf{0}\} \bigcup Y)}.
$$

Combining all of these formulas proves the result.   □

We remark that (4.45) reduces to (4.1) when $d = 1$. In fact, if $d = 1$ and $X = \{t_0, \ldots, t_n\}$ is a set of distinct positive numbers, then for the set Y consisting of one point, $Y = \{t_i\}$,

$$\prod_{y \in X \setminus Y} \det(\{y\} \cup Y) = \prod_{j \neq i}(t_j - t_i),$$

$\det(\{t\} \cup Y) = t_i - t$, $\operatorname{sgn} \det(\{0\} \cup Y) = 1$ and for $t \in \mathbb{R}_+$, $\chi_{[\{0\} \cup Y]}(t) = (t_i - t)_+^0$.

4.4. Multivariate truncated powers: Smoothness and recursion.

We have developed an extensive list of identities for the multivariate B-spline. Let us now provide similar formulas for the multivariate analog of the univariate truncated power function $t_+^{n-1} = (\max(0, t))^{n-1}, t \in \mathbb{R}$.

DEFINITION 4.1. *Let $X \subset \mathbb{R}^d$, $\#X = n$, $\mathbf{0} \notin [X]$, $\operatorname{vol}_d[X] > 0$. Then the truncated power function is defined as*

$$O(\mathbf{x}|X) = \int_0^\infty t^{-n+d-1} S(t\mathbf{x}|X) dt.$$

For $X = \{\mathbf{x}^1, \ldots, \mathbf{x}^n\}$ we introduce the convex cone generated by X,

$$\langle X \rangle_+ = \left\{ \sum_{j=1}^n t_j \mathbf{x}^j : t_\ell \geq 0, \ell = 1, \ldots, n \right\}.$$

Note that $O(\mathbf{x}|X) = 0$, if $\mathbf{x} \notin \langle X \rangle_+$.

The above definition of the multivariate truncated power demands that $n \geq d + 1$. For $n = d$, and $\operatorname{vol}_d([\{\mathbf{0}\} \cup X]) > 0$ we set

$$O(\mathbf{x}|X) = |\det(\{\mathbf{0}\} \cup X)|^{-1} \chi_{\langle X \rangle_+}(\mathbf{x}).$$

Let us first identify $O(\cdot|X)$ in the univariate case. Our requirement that $0 \notin [X]$, $X = \{x_1, \ldots, x_n\}$, means that either $x_i < 0, i = 1, \ldots, n$ or $x_i > 0, i = 1, \ldots, n$. Assume the latter case holds. Then, for $n \geq 2$

$$O(t|x_1, \ldots, x_n) = \int_0^\infty \sigma^{-n} S(t\sigma|x_1, \ldots, x_n) d\sigma$$

$$= t_+^{n-1} \int_0^\infty \sigma^{-n} S(\sigma|x_1, \ldots, x_n) d\sigma$$

$$= \frac{(-1)^{n-1}}{(n-1)!} t_+^{n-1} [x_1, \ldots, x_n] g,$$

where $g(t) = t^{-1}$. But it can easily be checked that

$$[x_1, \ldots, x_n] g = \frac{(-1)^{n-1}}{x_1 \cdots x_n}$$

and so

$$O(t|x_1, \ldots, x_n) = \frac{1}{(n-1)!} \frac{t_+^{n-1}}{x_1 \cdots x_n}.$$

This formula also holds for $n = 1$. Thus $O(\cdot|X)$ is a constant multiple of the univariate truncated power function.

In our next theorem, we summarize some of the most important properties of the truncated power function.

THEOREM 4.10. *Let* $X = \{\mathbf{x}^1, \ldots, \mathbf{x}^n\} \subset \mathbb{R}^d$ *with* $\mathbf{0} \notin [X]$. *Then*

(i) $D_{\mathbf{y}}O(\cdot|X) = \sum_{j=1}^{n} c_j O(\cdot|X_j)$,

where

$$\mathbf{y} = \sum_{j=1}^{n} c_j \mathbf{x}^j$$

and

$$X_j = X \setminus \{\mathbf{x}^j\}, \qquad j = 1, \ldots, n.$$

(ii) $$O(\mathbf{x}|X) = \frac{1}{n-d} \sum_{j=1}^{n} c_j O(\mathbf{x}|X_j),$$

where

$$\mathbf{x} = \sum_{j=1}^{n} c_j \mathbf{x}^j.$$

(iii) $O(\cdot|X) = \lim_{\epsilon \to 0+} \epsilon^{-n+d} S(\epsilon \cdot |\{\mathbf{0}\} \cup X)$.

(iv) *If every subset of* ℓ *elements in* X *is linearly independent then* $O(\cdot|X) \in C^{n-\ell-1}(\mathbb{R}^d)$. *Moreover,* $O(\cdot|X)$ *is a polynomial of degree* $\leq n - d$ *on any region not intersected by* X-*planes.*

(v) *For every* $f \in C_0(\mathbb{R}^d)$ *(continuous functions of compact support)*

(4.48) $$\int_{\mathbb{R}^d} O(\mathbf{x}|X) f(\mathbf{x}) d\mathbf{x} = \int_{\mathbb{R}^n_+} f(X\mathbf{t}) d\mathbf{t}.$$

We remark that Dahmen [Dah] used this last property (v) as the definition of the truncated power.

Proof. We begin with (iii). To this end, we first establish for any $\mathbf{y} \in \mathbb{R}^d$ the formula

$$S(\mathbf{x}|\{\mathbf{y}\} \cup X) = \int_1^\infty t^{-n+d-1} S((1-t)\mathbf{y} + t\mathbf{x}|X) dt,$$

a.e., $\mathbf{x} \in \mathbb{R}^d$.

For the proof of this formula, observe that for $\mathbf{x} \in [X]$ the ray $(1-t)\mathbf{y} + t\mathbf{x}$, $t \in [1, \infty)$ must intersect $\partial[X]$ at some value $t_0 \in [1, \infty)$. Hence for $t \geq t_0$ the integrand above vanishes and so the integral is clearly finite.

Let $f \in C(\mathbb{R}^d)$, set $X = \{\mathbf{x}^1, \ldots, \mathbf{x}^n\}$ and consider the integral

$$\int_{[\{\mathbf{y}\}\cup X]} f := \int_{\Delta^n} f\left(\mathbf{y} + \sum_{j=1}^{n} t_j(\mathbf{x}^j - \mathbf{y})\right) dm(\mathbf{t})$$

$$= \int_0^1 h^{n-1}\left(\int_{\Delta^{n-1}} f\left((1-h)\mathbf{y} + h\left(\mathbf{x}^1 + \sum_{j=2}^{n} t_j(\mathbf{x}^j - \mathbf{x}^1)\right)\right) dm(\mathbf{t})\right) dh$$

$$= \int_0^1 h^{n-1}\left(\int_{\mathbb{R}^d} f((1-h)\mathbf{y} + h\mathbf{x}) S(\mathbf{x}|X) d\mathbf{x}\right) dh$$

$$= \int_{\mathbb{R}^d} f(\mathbf{x})\left(\int_0^1 h^{n-d-1} S(h^{-1}\mathbf{x} + (1-h^{-1})\mathbf{y}|X) dh\right) d\mathbf{x}$$

$$= \int_{\mathbb{R}^d} f(\mathbf{x})\left(\int_1^\infty t^{-n+d-1} S((1-t)\mathbf{y} + t\mathbf{x}|X) dt\right) d\mathbf{x}.$$

Using the above formula, with $\mathbf{y} = \mathbf{0}$, we have for any $n > d$

$$\lim_{\epsilon \to 0+} \epsilon^{-n+d} S(\epsilon \cdot |\{\mathbf{0}\} \cup X) = \lim_{\epsilon \to 0+} \epsilon^{-n+d} \int_1^\infty t^{-n+d-1} S(t\epsilon \cdot |X) dt$$

$$= \lim_{\epsilon \to 0+} \int_\epsilon^\infty t^{-n+d-1} S(t \cdot |X) dt$$

$$= \int_0^\infty t^{-n+d-1} S(t \cdot |X) dt$$

$$= O(\cdot |X).$$

For $n = d$, we obtain, directly from the definition of the multivariate truncated power, that

$$\lim_{\epsilon \to 0+} S(\epsilon \cdot |\{\mathbf{0}\} \cup X) = \lim_{\epsilon \to 0+} |\det(\{\mathbf{0}\} \cup X)|^{-1} \chi_{[\{\mathbf{0}\}\cup X]}(\epsilon \cdot)$$

$$= |\det(\{\mathbf{0}\} \cup X)|^{-1} \chi_{\langle X\rangle_+}(\cdot).$$

Next we consider (v). Set $\tilde{X} := \{\mathbf{0}\} \cup X$. Then for $f \in C_0(\mathbb{R}^d)$ we have

$$\int_{\mathbb{R}_+^n} f(X\mathbf{t}) d\mathbf{t} = \lim_{\epsilon \to 0+} \int_{\epsilon^{-1}\Delta^n} f(\tilde{X}\mathbf{t}) dm(\mathbf{t}),$$

$$= \lim_{\epsilon \to 0+} \epsilon^{-n} \int_{\Delta^n} f(\epsilon^{-1}\tilde{X}\mathbf{t}) dm(\mathbf{t})$$

$$= \lim_{\epsilon \to 0+} \epsilon^{-n} \int_{\mathbb{R}^d} S(\mathbf{x}|\tilde{X}) f(\epsilon^{-1}\mathbf{x}) d\mathbf{x}$$

$$= \lim_{\epsilon \to 0+} \epsilon^{-n+d} \int_{\mathbb{R}^d} S(\epsilon \mathbf{x}|\{\mathbf{0}\} \cup X) f(\mathbf{x}) d\mathbf{x}$$

$$= \int_{\mathbb{R}^d} O(\mathbf{x}|X) f(\mathbf{x}) d\mathbf{x}.$$

The recurrence formula (ii) is an immediate consequence of the analogous one for the multivariate B-spline given in Theorem 4.4. Specifically, we have

$$O(\mathbf{x}|X) := \lim_{\epsilon \to 0+} \epsilon^{-n+d} S(\epsilon \mathbf{x}|\{\mathbf{0}\} \cup X).$$

To evaluate this limit we write $\mathbf{x}$ in the form

$$\mathbf{x} = \sum_{j=1}^{n} c_j \mathbf{x}^j + \left(1 - \sum_{j=1}^{n} c_j\right) \mathbf{0},$$

then apply Theorem 4.4 to the set $\{\mathbf{0}\} \cup X$ and get the formula

$$O(\mathbf{x}|X) = \lim_{\epsilon \to 0+} \frac{\epsilon^{-n+d}}{n-d} \left\{ \sum_{j=1}^{n} \epsilon c_j S(\epsilon \mathbf{x}|\{\mathbf{0}\} \cup X_j) + \left(1 - \sum_{j=1}^{n} \epsilon c_j\right) S(\epsilon \mathbf{x}|X) \right\}$$

$$= \frac{1}{n-d} \sum_{j=1}^{n} c_j O(\mathbf{x}|X_j),$$

because $\mathbf{0} \notin [X]$.

We now can use (v) to derive (i). In fact, if $f \in C_0^1(\mathbb{R}^d)$ we have for $i = 1, \ldots, n$,

$$\int_{\mathbb{R}_+^n} (D_{\mathbf{x}^i} f)(X\mathbf{t}) \, d\mathbf{t} = \int_{\mathbb{R}_+^n} \frac{d}{dt_i} (f(X\mathbf{t})) \, d\mathbf{t} = -\int_{\mathbb{R}^{n-1}} f(X_i \mathbf{t}) \, d\mathbf{t}.$$

Also, by the definition of the truncated power it follows that

$$\int_{\mathbb{R}_+^n} (D_{\mathbf{x}^i} f)(X\mathbf{t}) \, d\mathbf{t} = \int_{\mathbb{R}^d} O(\mathbf{x}|X)(D_{\mathbf{x}^i} f)(\mathbf{x}) \, d\mathbf{x}.$$

Hence, we have verified that

$$D_{\mathbf{x}^i} O(\cdot|X) = O(\cdot|X_i), \qquad i = 1, \ldots, n.$$

The general case follows by multiplying both sides of this equation by c_i and then summing the resulting expressions over $i = 1, \ldots, n$.

There remains the proof of (iv). This is similar to the proof of Theorem 4.2. We use property (i) above, the definition of $O(\cdot|X)$ in terms of the multivariate B-spline, and Lemma 4.2. $\square$

The next formula for $O(\cdot|X)$ provides a knot insertion identity for the truncated power function, similar to that given in Theorem 4.3 for $S(\cdot|X)$.

THEOREM 4.11. *Suppose that $X \subset \mathbb{R}^d$ and $X = \{\mathbf{x}^1, \ldots, \mathbf{x}^n\}$ with $\mathbf{0} \notin [X]$. If*

$$\mathbf{y} = \sum_{j=1}^{n} c_j \mathbf{x}^j,$$

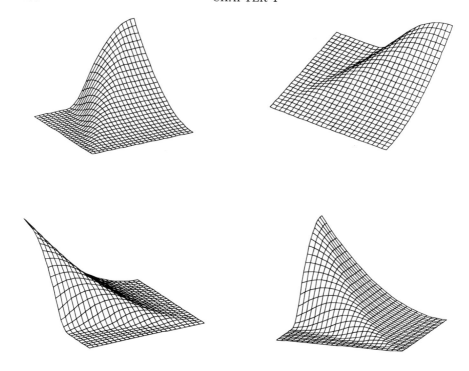

FIG. 4.12.   *A quadratic truncated power.*

and either

(4.49)
$$\sum_{j=1}^{n} c_j = 1$$

or

(4.50)
$$\sum_{j=1}^{n} c_j \neq 0, \qquad c_\ell \geq 0, \qquad \ell = 1, \ldots, n,$$

then

$$O(\cdot | X) = \sum_{j=1}^{n} c_j O(\cdot | X_j \cup \{\mathbf{y}\}).$$

Proof. We write $\mathbf{y}$ in the form

$$\mathbf{y} = \sum_{j=1}^{n} c_j \mathbf{x}^j + \left(1 - \sum_{j=1}^{n} c_j \right) \mathbf{0}$$

and use the knot insertion formula for the multivariate B-spline given in Theorem 4.3 to obtain

$$
O(\mathbf{x}|X) = \lim_{\epsilon \to 0^+} \epsilon^{-n+d} \left\{ \sum_{j=1}^{n} c_j S\left(\epsilon \mathbf{x}|\{\mathbf{0}, \mathbf{y}\} \cup X_j\right) \right.
$$

$$
\left. + \left(1 - \sum_{j=1}^{n} c_j\right) S\left(\epsilon \mathbf{x}|\{\mathbf{y}\} \cup X\right) \right\}.
$$

The second term in the bracket above is clearly zero because of (4.49). Using (4.50), we claim that $\mathbf{0} \notin [\{\mathbf{y}\} \cup X]$. To see this, we suppose that the hyperplane $\{\mathbf{x} : \mathbf{x} \in \mathbb{R}^d, (\mathbf{a}, \mathbf{x}) = b\}$ separates $\mathbf{0}$ from $[X]$. For instance, this is the case if $b > 0$ and $(\mathbf{a}, \mathbf{x}^i) > b$, $i = 1, \ldots, n$. Set $b_0 := b \min(\sum_{j=1}^{n} c_j, 1)$, then $(\mathbf{a}, \mathbf{x}) > b_0$ for all $\mathbf{x} \in [\{\mathbf{y}\} \cup X]$ and so the hyperplane $\{\mathbf{x} : \mathbf{x} \in \mathbb{R}^d, (\mathbf{a}, \mathbf{x}) = b_0\}$ separates $\mathbf{0}$ from $[\{\mathbf{y}\} \cup X]$. Thus for any $\mathbf{x}$ there is an ϵ sufficiently close to zero such that $\epsilon \mathbf{x}$ is also not in $[\{\mathbf{y}\} \cup X]$. This means that the second term in the bracket is again zero, for ϵ sufficiently small. Thus, in both cases we obtain

$$
O(\mathbf{x}|X) = \lim_{\epsilon \to 0^+} \epsilon^{-n+d} \sum_{j=1}^{n} c_j S\left(\epsilon \mathbf{x}|\{\mathbf{0}, \mathbf{y}\} \cup X_j\right),
$$

which, by definition, simplifies to

$$
O(\mathbf{x}|X) = \sum_{j=1}^{n} c_j O\left(\mathbf{x}|\{\mathbf{y}\} \cup X_j\right). \qquad \square
$$

Our next formula is the derivative recurrence relation for $O(\cdot|X)$, which is similar to that of Theorem 4.5 for $S(\cdot|X)$.

THEOREM 4.12.   For any $\mathbf{y} \in \mathbb{R}^d$ and $X = \{\mathbf{x}^1, \ldots, \mathbf{x}^n\}$ with $\mathrm{vol}_d(\{\mathbf{0}\} \cup X_j) > 0, j = 1, \ldots, n,$ we have

$$
D_{\mathbf{y}} O(\mathbf{x}|X) = \frac{1}{n-d-1} \sum_{j=1}^{n} c_j D_{\mathbf{y}} O(\mathbf{x}|X_j),
$$

where $\mathbf{x} = \sum_{j=1}^{n} c_j \mathbf{x}^j$.

Proof. We differentiate both sides of formula (ii) of Theorem 4.10 with respect to c_ℓ, $\ell = 1, \ldots, n$, and obtain the formula

$$
(4.51) \qquad D_{\mathbf{x}^\ell} O(x|X) = \frac{1}{n-d} \left\{ O(\mathbf{x}|X_\ell) + \sum_{j=1}^{n} c_j D_{\mathbf{x}^\ell} O(x|X_j) \right\}.
$$

We write $\mathbf{y}$ in the form

$$
\mathbf{y} = \sum_{j=1}^{n} a_j \mathbf{x}^j
$$

for some scalars $a_1, \ldots, a_n$, multiply both sides of (4.51) by a_ℓ, and sum both sides over $\ell = 1, \ldots, n$. This computation gives us the identity

$$(n - d)\, D\mathbf{y} O\left(\mathbf{x}|X\right) = \sum_{\ell=1}^{n} a_\ell O\left(\mathbf{x}|X_\ell\right) + \sum_{j=1}^{n} c_j D\mathbf{y} O\left(\mathbf{x}|X_j\right).$$

Next, we use formula (i) of Theorem 4.10 to conclude that

$$(n - d)\, D\mathbf{y} O\left(\mathbf{x}|X\right) = D\mathbf{y} O\left(\mathbf{x}|X\right) + \sum_{j=1}^{n} c_j D\mathbf{y} O\left(\mathbf{x}|X_j\right),$$

and so we get

$$(n - d - 1)\, D\mathbf{y} O\left(\mathbf{x}|X\right) = \sum_{j=1}^{n} c_j D\mathbf{y} O\left(\mathbf{x}|X_j\right). \qquad \square$$

Formulas for the truncated power function, obtained from repeating vectors in the set X, are based on the observation that whenever $\#X = n$,

$$(4.52) \qquad \int_{\mathbb{R}_+^{m+n}} f\left(X^{\mathbf{k}}\mathbf{r}\right) d\mathbf{r} = \frac{1}{\mathbf{k}!} \int_{\mathbb{R}_+^{n}} \mathbf{t}^{\mathbf{k}} f\left(X\mathbf{t}\right) d\mathbf{t}$$

for any $\mathbf{k} \in \mathbb{Z}_+^n$ with $|\mathbf{k}|_1 = m$ and $f \in C_0(\mathbb{R}^d)$. This formula can be proved easily by induction on n. The case $n = 1$ means that for every $\mathbf{y} \in \mathbb{R}^d$ and every $f \in C_0(\mathbb{R}^d)$

$$\int_0^\infty \cdots \int_0^\infty f\left(\mathbf{y}\left(r_0 + \cdots + r_m\right)\right) dr_0 \cdots dr_m$$
$$= \tfrac{1}{m!} \int_0^\infty t^m f\left(\mathbf{y}t\right) dt.$$

This formula follows easily from the change of variables $\mathbf{r} = (r_0, \ldots, r_m)^T = t\mathbf{x}$, $\mathbf{x} \in \Delta^m$, $t \in \mathbb{R}_+$, since $\mathrm{vol}_m \Delta^m = 1/m!$.

As with the multivariate B-spline, we use (4.52) as follows: for every $\mathbf{y} \in \mathbb{R}^d$

$$\int_{\mathbb{R}^d} \left(\left(\mathbf{y}, \mathbf{x}\right)\right)^N O\left(\mathbf{x}|X\right) f\left(\mathbf{x}\right) d\mathbf{x} = \int_{\mathbb{R}_+^n} \left(\left(\mathbf{y}, X\mathbf{t}\right)\right)^N f\left(X\mathbf{t}\right) d\mathbf{t}$$

$$= \int_{\mathbb{R}_+^n} \left(\left(X^T \mathbf{y}, \mathbf{t}\right)\right)^N f\left(X\mathbf{t}\right) d\mathbf{t}$$

$$= \sum_{|\mathbf{k}|_1 = N} \binom{N}{\mathbf{k}} \left(X^T \mathbf{y}\right)^{\mathbf{k}} \int_{\mathbb{R}_+^n} \mathbf{t}^{\mathbf{k}} f\left(X\mathbf{t}\right) d\mathbf{t}$$

$$= N! \sum_{|\mathbf{k}|_1 = N} \left(X^T \mathbf{y}\right)^{\mathbf{k}} \int_{\mathbb{R}_+^{n+m}} f\left(X^{\mathbf{k}}\mathbf{t}\right) d\mathbf{t}.$$

Thus, we get

$$\frac{((\mathbf{y}, X))^N}{N!} O\left(\mathbf{x}|X\right) = \sum_{|\mathbf{k}|_1 = N} \left(X^T \mathbf{y}\right)^{\mathbf{k}} O\left(\mathbf{x}|X^{\mathbf{k}}\right).$$

This formula is the truncated power analog of formula (4.40) for the multivariate B-spline.

To obtain the companion of Theorem 4.6 for the truncated power function, we suppose that $X = \{\mathbf{x}^1, \dots, \mathbf{x}^d\}$ and choose the polynomial $p \in \Pi_m(\mathbb{R}^d)$ defined by

$$p\left(X\mathbf{t}\right) = \frac{\mathbf{t}^{\mathbf{j}}}{\mathbf{j}!}, \quad |\mathbf{j}|_1 = m, \quad \mathbf{t} \in \mathbb{R}^d, \quad \mathbf{j} \in \mathbb{Z}_+^d.$$

Specifically, if $\mathbf{y}^1, \dots, \mathbf{y}^d$ are vectors biorthogonal relative to $\mathbf{x}^1, \dots, \mathbf{x}^d$, i.e.,

$$\left(\mathbf{x}^i, \mathbf{y}^j\right) = \delta_{ij}, \quad i, j = 1, \dots, d,$$

then for $\mathbf{j} = (j_1, \dots, j_d)^T$ we have

$$p\left(\mathbf{x}\right) = \frac{\left(\mathbf{y}^1, \mathbf{x}\right)^{j_1} \cdots \left(\mathbf{y}^d, \mathbf{x}\right)^{j_d}}{\mathbf{j}!}, \quad \mathbf{x} \in \mathbb{R}^d.$$

We compute the integer

$$\int_{\mathbb{R}_+^d} p\left(\mathbf{x}\right) O\left(\mathbf{x}|X\right) f\left(\mathbf{x}\right) d\mathbf{x} = \int_{\mathbb{R}_+^d} p\left(X\mathbf{t}\right) f\left(X\mathbf{t}\right) d\mathbf{t}$$

$$= \frac{1}{\mathbf{j}!} \int_{\mathbb{R}_+^d} \mathbf{t}^{\mathbf{j}} f\left(X\mathbf{t}\right) d\mathbf{t},$$

and so we get

$$O\left(\mathbf{x}|X^{\mathbf{j}}\right) = \frac{\left(\mathbf{y}^1, \mathbf{x}\right)^{j_1} \cdots \left(\mathbf{y}^d, \mathbf{x}\right)^{j_d}}{\mathbf{j}!} O\left(\mathbf{x}|X\right)$$

$$= \frac{\left(\mathbf{y}^1, \mathbf{x}\right)^{j_1} \cdots \left(\mathbf{y}^d, \mathbf{x}\right)^{j_d}}{\mathbf{j}! |\det\left(\{\mathbf{0}\} \cup X\right)|} \chi_{\langle X \rangle_+}\left(\mathbf{x}\right).$$

Under certain conditions on X, to be explained below, an explicit formula for the truncated power function is available. The formula we develop next is a version of Theorem 4.9 for the truncated power function. To describe it in detail, we introduce another distribution, from Dahmen and Micchelli [DM2], which is of some independent interest.

Given any finite set $X \subset \mathbb{R}^d$, with $\#X = n$, we define the distribution $E\left(\cdot|X\right)$, by requiring that

(4.53) $$E\left(\cdot|X\right)\left(f\right) := \int_{\mathbb{R}_+^n} e^{-(\mathbf{e}, \mathbf{t})} f\left(X\mathbf{t}\right) d\mathbf{t}, \quad \mathbf{e} = (1, \dots, 1)^T,$$

for all measurable functions f for which the integrand in (4.53) is absolutely integrable. When rank $X = d \leq n$, we can find an $n \times n$ nonsingular matrix V such that the $d \times n$ submatrix consisting of the first d rows of V is X. We make the change of variables $\mathbf{s} = V\mathbf{t}$ in (4.53) and obtain the formula

$$(4.54) \qquad E\left(\mathbf{x}|X\right) = |\det V|^{-1} \int_{\{\mathbf{t}:\mathbf{t}\in\mathbb{R}_+^n, X\mathbf{t}=\mathbf{x}\}} e^{-(V^{-T}\mathbf{e},\mathbf{t})} d\mathbf{t}.$$

This formula can be taken as the pointwise definition of $E\left(\cdot|X\right)$. In any case, we see that when rank $X = d$, $E\left(\cdot|X\right)$ is a nonnegative function with integral one.

When $n = d$ and rank $X = d$ we can obtain an exact expression for $E\left(\cdot|X\right)$. Specifically, the change of variables $\mathbf{s} = X\mathbf{t}$ in (4.53) gives us the equation

$$E\left(\mathbf{x}|X\right) = |\det X|^{-1} e^{-(X^{-T}\mathbf{e},\mathbf{x})} \chi_{\langle X\rangle_+}(\mathbf{x}).$$

We can rewrite this formula in a slightly different form by noting that $\det(\{\mathbf{0}\} \cup X) = \det X$ and the vector $\mathbf{v}^X = -X^{-T}\mathbf{e}$ satisfies the property that $1 + (\mathbf{v}^X, \mathbf{x}) = 0$ for all $\mathbf{x} \in X$. Thus, for any subset $Y \subseteq X$ with $\#Y = $ rank $Y = d$ we have

$$(4.55) \qquad E\left(\mathbf{x}|Y\right) = |\det Y|^{-1} e^{(\mathbf{v}^Y,\mathbf{x})} \chi_{\langle Y\rangle_+}(\mathbf{x}), \qquad \mathbf{x} \in \mathbb{R}^d.$$

Next we will build $E\left(\mathbf{x}|X\right)$ for any set X with $\#X > d$ from sums of functions of the type (4.55). Specifically, we shall prove the following result.

LEMMA 4.6. *Suppose that $X \subset \mathbb{R}^d$, $\#X = n$, rank $X = d$, and $\{\mathbf{0}\} \cup X$ is in general position. Then*

$$(4.56) \quad E\left(\mathbf{x}|X\right) = \sum_{\substack{Y \subseteq X \\ \#Y=d}} \Pi_{\mathbf{y}\in X\setminus Y} \frac{\det\left(\{\mathbf{0}\} \cup Y\right)}{\det\left(\{\mathbf{y}\} \cup Y\right)} E\left(\mathbf{x}|Y\right), \qquad a.e., \ \mathbf{x} \in \mathbb{R}^d.$$

Proof. The proof of this formula is based on the standard fact that whenever $f \in L^1(\mathbb{R}^d)$ and

$$(4.57) \qquad \int_{\mathbb{R}^d} h\left(\mathbf{x}\right) e^{i(\mathbf{y},\mathbf{x})} d\mathbf{x} = 0$$

for all $\mathbf{y} \in \mathbb{R}^d$, it follows that $h(\mathbf{x}) = 0$, a.e., $\mathbf{x} \in \mathbb{R}^d$. We will apply this condition to the function h defined to be the difference between the left-hand and right-hand side of equation (4.56). Of course, this requires that we show that (4.57) holds for this function. For this purpose, we choose $f(\mathbf{x}) = e^{-(\mathbf{z},\mathbf{x})}$, $\mathbf{x} \in \mathbb{R}^d$, and $\mathbf{z} = i\mathbf{y}, \mathbf{y} \in \mathbb{R}^d$, in equation (4.53). This gives us the formula

$$(4.58) \qquad \int_{\mathbb{R}^d} E\left(\mathbf{x}|X\right) e^{(\mathbf{z},\mathbf{x})} d\mathbf{x} = \frac{1}{\Pi_{\mathbf{x}\in X}\ (1+(\mathbf{z},\mathbf{x}))}.$$

According to our partial fraction decomposition given in Lemma 4.5, the right-hand side of (4.58) equals

$$(4.59) \qquad \sum_{\substack{Y \subseteq X \\ \#Y=d}} \Pi_{\mathbf{y}\in X\setminus Y} \frac{\det\left(\{\mathbf{0}\} \cup Y\right)}{\det\left(\{\mathbf{y}\} \cup Y\right)} \frac{1}{\Pi_{\mathbf{y}\in Y}\ (1+(\mathbf{z},\mathbf{y}))}.$$

Now we use (4.58) (with X replaced by Y) to replace the function of $\mathbf{z}$ in the sum (4.59) by the integral,

$$\int_{\mathbb{R}^d} E\left(\mathbf{x}|Y\right) e^{-(\mathbf{Z},\mathbf{X})} d\mathbf{x}.$$

This proves that h defined above satisfies (4.57) and so is identically zero.   □
    Next we consider the one parameter family of functions

$$(4.60) \qquad E_\rho\left(\mathbf{x}|X\right) := \rho^{-n} E\left(\mathbf{x}|\rho^{-1}X\right), \qquad \rho > 0,$$

under the same conditions of Lemma 4.6. Combining equation (4.55) and (4.56) with the observation that $\mathbf{v}^{\rho^{-1}Y} = \rho\mathbf{v}^Y$, $Y \subseteq X$, and (4.43), we get

$$E_\rho\left(\mathbf{x}|X\right) = \rho^{-n+d} \sum_{\substack{Y \subseteq X \\ \#Y = d}} \Pi_{\mathbf{y} \in X \backslash Y} \frac{\det\left(\{\mathbf{0}\} \cup Y\right)}{\det\left(\{\mathbf{y}\} \cup Y\right)} \, |\det Y|^{-1} e^{\rho(\mathbf{v}^Y, \mathbf{x})} \chi_{\langle Y \rangle_+}(\mathbf{x}).$$

Consequently, for all $\rho > 0$, we have the asymptotic expansion

$$E_\rho\left(\mathbf{x}|X\right) = \sum_{\ell=0}^{\infty} \rho^{\ell-n+d} p_\ell(\mathbf{x}),$$

where, for $\ell = 0, 1, 2, \ldots$,

$$p_\ell(\mathbf{x}) := \sum_{\substack{Y \subseteq X \\ \#Y = d}} \Pi_{\mathbf{y} \in X \backslash Y} \frac{\det\left(\{\mathbf{0}\} \cup Y\right)}{\det\left(\{\mathbf{y}\} \cup Y\right)} \, |\det Y|^{-1} \left(\mathbf{v}^Y, \mathbf{x}\right)^\ell \chi_{\langle Y \rangle_+}(\mathbf{x}).$$

Note that each p_ℓ is a piecewise polynomial of degree ℓ that is zero off the set $\langle X \rangle_+$. Hence for any $f \in C_0(\mathbb{R}^d)$ we have the asymptotic expansion

$$(4.61) \qquad \begin{aligned} &\int_{\mathbb{R}^d} E_\rho\left(\mathbf{x}|X\right) f(\mathbf{x}) d\mathbf{x} \\ &= \sum_{\ell=0}^{\infty} \rho^{\ell-n+d} \int_{\mathbb{R}^d} p_\ell(\mathbf{x}) f(\mathbf{x}) d\mathbf{x}. \end{aligned}$$

The left side of equation (4.61) also equals, by definitions (4.53) and (4.60), the integral

$$\int_{\mathbb{R}^n_+} e^{-\rho(\mathbf{e},\mathbf{X})} f\left(X\mathbf{t}\right) d\mathbf{t}.$$

Suppose now that $\mathbf{0} \notin [X]$, then by property (v) of Theorem 4.10 the integral above *converges* as $\rho \to 0^+$, to the integral

$$\int_{\mathbb{R}^d} O\left(\mathbf{x}|X\right) f(\mathbf{x}) d\mathbf{x}.$$

Consequently, from (4.61) we conclude that

$$p_\ell(\mathbf{x}) = 0, \quad \ell = 0, 1, \ldots, n - d - 1, \quad \text{a.e., } \mathbf{x} \in \mathbb{R}^d$$

and

$$p_{n-d}(\mathbf{x}) = O(\mathbf{x}|X), \quad \mathbf{x} \in \mathbb{R}^d.$$

We summarize these facts in the next theorem.

THEOREM 4.13. *Suppose that* $X \subset \mathbb{R}^d$, $\#X = n$, *rank* $X = d$, $\mathbf{0} \notin [X]$, *and* $\{\mathbf{0}\} \cup X$ *is in general position. Then*

$$O(\mathbf{x}|X) = \sum_{\substack{Y \subseteq X \\ \#Y = d}} \Pi_{\mathbf{y} \in X \setminus Y} \frac{\det(\{\mathbf{0}\} \cup Y)}{\det(\{\mathbf{y}\} \cup Y)} \left| \det Y \right|^{-1} (\mathbf{v}^Y, \mathbf{x})^{n-d} \chi_{\langle Y \rangle_+}(\mathbf{x}),$$

a.e., $\mathbf{x} \in \mathbb{R}^d$. *Moreover, for* $\ell = 0, 1, \ldots, n - d - 1$ *and a.e.,* $\mathbf{x} \in \mathbb{R}^d$ *we have*

$$\sum_{\substack{Y \subseteq X \\ \#Y = d}} \Pi_{\mathbf{y} \in X \setminus Y} \frac{\det(\{\mathbf{0}\} \cup Y)}{\det(\{\mathbf{y}\} \cup Y)} \left| \det Y \right|^{-1} (\mathbf{v}^Y, \mathbf{x})^\ell \chi_{\langle Y \rangle_+}(\mathbf{x}) = 0.$$

4.5. More identities for the multivariate truncated power.

Our next formula, from Dahmen and Micchelli [DM2], relates the truncated power function restricted to a hyperplane to the multivariate B-spline. We first develop the formula in a special case. Let $X \subset \mathbb{R}^d$ with $\#X = n$ and set

$$W = \{(1, \mathbf{x})^T : \mathbf{x} \in X\}.$$

We claim that

(4.62) $$O\left((t, \mathbf{x})^T | W\right) = t_+^{n-1-d} S\left(t^{-1} \mathbf{x} | X\right),$$

a.e., $(t, \mathbf{x})^T \in \mathbb{R}^{d+1}$. To prove this formula, we let $f \in C_0(\mathbb{R}^{d+1})$ and consider the integral

$$\int_{\mathbb{R}^n_+} f(W\mathbf{t}) \, d\mathbf{t}.$$

We make the change of variables $\mathbf{t} = h\mathbf{y}$, where $h \in [0, \infty)$ and $\mathbf{y} \in \Delta^{n-1}$ and obtain the equivalent expression

$$\int_0^\infty \int_{\Delta^{n-1}} h^{n-1} f((h, hX\mathbf{y})^T) \, dh \, dm(\mathbf{y})$$

$$= \int_0^\infty \int_{\mathbb{R}^d} h^{n-1} f((h, h\mathbf{x})^T) S(\mathbf{x}|X) \, d\mathbf{x} \, dh$$

$$= \int_0^\infty \int_{\mathbb{R}^d} h^{n-d-1} S(h^{-1}\mathbf{x}|X) f(h, \mathbf{x}) \, d\mathbf{x} \, dh$$

which proves (4.62).

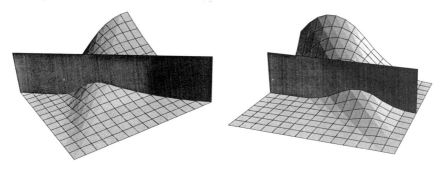

FIG. 4.13.   *A slice of a quadratic truncated power.*

The next result is a reformulation of equation (4.62) in its component free form. It states that a truncated power restricted to a hyperplane is a multivariate B-spline. To make this precise, we begin with any hyperplane $H = \{\mathbf{x} \in \mathbb{R}^d : (\mathbf{x}, \mathbf{y}) = t\}$, where $\mathbf{y} \in \mathbb{R}^d$ with $(\mathbf{y}, \mathbf{y}) = 1$ and t is a positive constant. We identify coordinates in the hyperplane H by choosing an orthonormal basis $\{\mathbf{y}^1, \ldots, \mathbf{y}^d\}$ of $\mathbb{R}^d$ with $\mathbf{y}^1 = \mathbf{y}$. Let U be the orthogonal matrix whose columns are $\mathbf{y}^1, \ldots, \mathbf{y}^d$. Then $\mathbf{x} \in H$ if and only if $\mathbf{x} = U\mathbf{w}$, where $\mathbf{w} = (t, \mathbf{u})^T$ and $\mathbf{u} \in \mathbb{R}^{d-1}$. Suppose that $(\mathbf{y}, \mathbf{x}) > 0$ for all $\mathbf{x} \in X$. This ensures that $H \cap \langle X \rangle_+$ is a nonempty bounded convex subset of $\mathbb{R}^d$. We introduce a finite set of vectors in $\mathbb{R}^{d-1}$ by setting

$$(4.63) \qquad V := \{\mathbf{v} : \exists \mathbf{x} \in X, t\,(\mathbf{y}, \mathbf{x})^{-1}\,\mathbf{x} = U\left((t, \mathbf{v})^T\right)\},$$

and claim that for $\mathbf{x} = U((t, \mathbf{u})^T), \mathbf{u} \in \mathbb{R}^{d-1}$, we have

$$(4.64) \qquad O\left(\mathbf{x}|X\right) = \Pi_{\mathbf{x} \in X}\ (\mathbf{y}, \mathbf{x})^{-1}\,t^{n-1}S\left(\mathbf{u}|V\right).$$

In other words, we obtain the following result.

THEOREM 4.14. *Let $X \subset \mathbb{R}^d$, $\#X = n$ and suppose that for some $\mathbf{y} \in \mathbb{R}^d$ with $(\mathbf{y}, \mathbf{y}) = 1$ we have $(\mathbf{y}, \mathbf{x}) > 0$ for all $\mathbf{x} \in X$. Then, for every $t > 0$ the hyperplane*

$$H = \{\mathbf{x} \in \mathbb{R}^d : (\mathbf{y}, \mathbf{x}) = t\}$$

intersects $\langle X \rangle_+$ in a nonempty bounded subset of $\mathbb{R}^d$ and, moreover,

$$(4.65) \qquad O\left(\cdot|X\right)\big|_H = \Pi_{\mathbf{x} \in X}\ (\mathbf{y}, \mathbf{x})^{-1}\,t^{n-1}S\left(\cdot|V\right)$$

where V is the subset of $\mathbb{R}^{d-1}$ defined in (4.63).

Proof. As explained above, (4.65) means that

$$(4.66) \quad O\left(U((t, \mathbf{u})^T)|X\right) = \Pi_{\mathbf{x} \in X}\ (\mathbf{y}, \mathbf{x})^{-1}\,t^{n-1}S\left(\mathbf{u}|V\right), \quad \text{a.e., } \mathbf{u} \in \mathbb{R}^{d-1}.$$

It is easy to see that for any $d \times d$ nonsingular matrix T

$$(4.67) \qquad O\left(T\mathbf{x}|X\right) = |\det T|^{-1}O\left(\mathbf{x}|T^{-1}X\right).$$

Using this formula in (4.66) with $T = U$ shows that (4.66), and hence the validity of (4.65), hinges on demonstrating that

$$(4.68) \qquad O\left((t, \mathbf{u})^T | U^T X\right) = \Pi_{\mathbf{x} \in X} \ (\mathbf{y}, \mathbf{x})^{-1} \, t^{n-1} S\left(\mathbf{u} | V\right).$$

To this end, we also need to point out that for every $X \subset \mathbb{R}^d$, $t > 0$ and an $n \times n$ diagonal matrix $D = \text{diag}\ (d_1, \dots, d_n)$ with $d_1 > 0, \dots, d_n > 0$ that

$$(4.69) \qquad O\left(\mathbf{x} | XD\right) = |\det D|^{-1} O\left(\mathbf{x} | X\right).$$

Let

$$W := \{(1, t^{-1}\mathbf{v})^T : \mathbf{v} \in V\}$$

and

$$D := \text{diag}\ (\{(\mathbf{y}, \mathbf{x}) : \mathbf{x} \in X\}).$$

Then, by (4.63), $U^T X = WD$, and so we can use (4.69) and also (4.62) (with $\mathbf{x}$ replaced by $\mathbf{u}$ and X by V and d by $d-1$) to get the formula

$$O((t, \mathbf{u})^T | U^T X) = O((t, \mathbf{u})^T | WD) = \prod_{\mathbf{x} \in X} (\mathbf{y}, \mathbf{x})^{-1} O((t, \mathbf{u})^T | W)$$

$$= t^{n-d} \prod_{\mathbf{x} \in X} (\mathbf{y}, \mathbf{x})^{-1} S(t^{-1}\mathbf{u} | t^{-1}V),$$

which reduces to (4.68).   □

The next formula, an identity for the square of the truncated power, is the analog of Theorem 4.8.

THEOREM 4.15. *Suppose that $X \subset \mathbb{R}^d$, $\#X = n$, rank $X = d$, and $\mathbf{0} \notin [X]$. Then*

$$O^2\left(\mathbf{x} | X\right) = \frac{1}{n - d} \sum_{j=0}^{n} O\left(\mathbf{x} | X_j\right) O\left(\mathbf{x} | X^j\right), \quad \text{a.e., } \mathbf{x} \in \mathbb{R}^d.$$

Proof. We use property (iii) of Theorem 4.10 and Theorem 4.8 to prove this formula. Thus we have

$$O^2\left(\mathbf{x} | X\right) = \lim_{\epsilon \to 0^+} \epsilon^{-2n+2d} S^2 \left(\epsilon \mathbf{x} | \{\mathbf{0}\} \cup X\right)$$

$$= \frac{1}{n - d} \lim_{\epsilon \to 0^+} \epsilon^{-2n+2d} \left\{ \sum_{j=1}^{n} S\left(\epsilon \mathbf{x} | \{\mathbf{0}\} \cup X_j\right) S\left(\epsilon \mathbf{x} | \{\mathbf{0}\} \cup X^j\right) \right.$$

$$\left. + S\left(\epsilon \mathbf{x} | X\right) S\left(\epsilon \mathbf{x} | \{\mathbf{0}, \mathbf{0}\} \cup X\right) \right\}.$$

Since $\mathbf{0} \notin [X]$, there is an ϵ_0 (depending on $\mathbf{x}$) such that $\epsilon_0 \mathbf{x}$ is also not in $[X]$. Hence, the second term on the right of the equation above is zero and we obtain the desired formula

$$O^2\left(\mathbf{x} | X\right) = \frac{1}{n - d} \sum_{j=1}^{n} O\left(\mathbf{x} | X_j\right) O\left(\mathbf{x} | X^j\right). \qquad □$$

Our last formula for the truncated power function is an interesting identity due to Dahmen [Dah], which expresses the multivariate B-spline in terms of the truncated power function. The background of his formula is the divided difference representation of the univariate B-spline. We recall that when $x_0, \ldots, x_n$ is an increasing sequence of real numbers, we have for all $t \in \mathbb{R}$

$$S(t|x_0, \ldots, x_n) = \frac{1}{(n-1)!} \sum_{j=0}^{n} \frac{(x_j - t)_+^{n-1}}{\Pi_{j \neq i}(x_j - x_i)}$$

$$= \frac{1}{(n-1)!} \sum_{j=0}^{n} \frac{(-1)^j (t - x_j)_+^{n-1}}{(x_j - x_0) \cdots (x_j - x_{j-1})(x_{j+1} - x_j) \cdots (x_n - x_j)}.$$

We have already pointed out that

$$O(t|x_1, \ldots, x_n) = \frac{1}{(n-1)!} \frac{t_+^{n-1}}{x_1 \cdots x_n}, \qquad t \in \mathbb{R},$$

whenever $x_i > 0, i = 1, \cdots, n$. Thus we get the equation

$$S(t|x_0, \ldots, x_n)$$

$$= \sum_{j=0}^{n} (-1)^j O(t - x_j|x_j - x_0, \ldots, x_j - x_{j-1}, x_{j+1} - x_j, \ldots, x_n - x_j).$$

Notice how we have hidden in the truncated power the fact that for a fixed $t, S(t|x_0, \ldots, x_n)$ is a rational function of $x_0, \ldots, x_n$. This notational *tour de force* leads us to the following multivariate formula.

THEOREM 4.16. *Suppose that* $X = \{\mathbf{x}^0, \ldots, \mathbf{x}^n\} \subset \mathbb{R}^d$ *and* $W := \{\mathbf{x}^i - \mathbf{x}^j : 0 \leq j < i \leq n\}$ *has the property that* $\mathbf{0} \notin [W]$. *Then*

$$S(\mathbf{x}|X) = \sum_{j=0}^{n} (-1)^j O\left(\mathbf{x} - \mathbf{x}^j | X(j)\right),$$

where $X(j) := \{\mathbf{x}^j - \mathbf{x}^0, \ldots, \mathbf{x}^j - \mathbf{x}^{j-1}, \mathbf{x}^{j+1} - \mathbf{x}^j, \ldots, \mathbf{x}^n - \mathbf{x}^0\}$.

Geometrically this formula corresponds to a "signed" decomposition of a simplex into cones (see Fig. 4.14). Its proof depends on an ancillary identity, which we present in the next lemma. Recall our notation, $D_X f$ for the differential operator $\Pi_{\mathbf{x} \in X} D_{\mathbf{x}} f$ where $D_{\mathbf{y}} f$, $\mathbf{y} \in \mathbb{R}^d$, denotes the directional derivative of f in the direction of $\mathbf{y}$.

LEMMA 4.7. *Let* $X = \{\mathbf{x}^0, \ldots, \mathbf{x}^n\} \subset \mathbb{R}^n$ *and* $W, X(j)$, $j = 0, 1, \ldots, n$ *be the sets defined in Theorem 4.16. Then for every* $f \in C^\infty(\mathbb{R}^d)$ *(functions with continuous derivatives of all orders on* $\mathbb{R}^d$) *we have*

(4.70)
$$\int_{[X]} D_W f = \sum_{j=0}^{n} (-1)^{n-j} \left(D_{W \setminus X(j)} f\right)(\mathbf{x}^j).$$

Proof. As in previous proofs, it suffices to verify equation (4.70) for functions of the form $f(\mathbf{x}) = g((\mathbf{z}, \mathbf{x}))$, $\mathbf{x} \in \mathbb{R}^d$, where $g(t) = e^t, t \in \mathbb{R}$, for all $\mathbf{z} = i\mathbf{y}, \mathbf{y} \in$

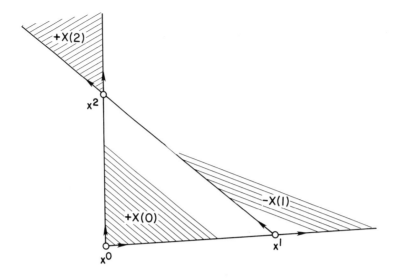

FIG. 4.14. *Signed decomposition of a simplex by cones.*

$\mathbb{R}^d$. In addition, we can assume that $(\mathbf{z}, \mathbf{x}^0), \ldots, (\mathbf{z}, \mathbf{x}^n)$ are distinct so that the left side of (4.70) becomes

$$\Pi_{\mathbf{w} \in W} \ (\mathbf{z}, \mathbf{w}) \sum_{j=0}^{n} \frac{g((\mathbf{z}, \mathbf{x}^j))}{\Pi_{j \neq \ell}((\mathbf{z}, \mathbf{x}^j) - (\mathbf{z}, \mathbf{x}^\ell))},$$

while the right side becomes

$$\sum_{j=0}^{n} (-1)^{n-j} \Pi_{\mathbf{w} \in W \setminus X(j)} \ (\mathbf{z}, \mathbf{w}) \, g\left((\mathbf{z}, \mathbf{x}^j)\right).$$

These quantities are equal because

$$\frac{\Pi_{\mathbf{w} \in W} \ (\mathbf{z}, \mathbf{w})}{\Pi_{j \neq \ell} \ ((\mathbf{z}, \mathbf{x}^j) - (\mathbf{z}, \mathbf{x}^\ell))} = \frac{\Pi_{\mathbf{w} \in W} \ (\mathbf{z}, \mathbf{w})}{(-1)^{n-j} \Pi_{\mathbf{w} \in X(j)} \ (\mathbf{z}, \mathbf{w})}$$

$$= (-1)^{n-j} \Pi_{\mathbf{w} \in W \setminus X(j)} \ (\mathbf{z}, \mathbf{w}). \qquad \square$$

To make use of this lemma we need to note that whenever $\mathbf{0} \notin [X]$

$$(4.71) \qquad \int_{\mathbb{R}^d} (D_X f)(\mathbf{x}) O(\mathbf{x}|X) \, d\mathbf{x} = (-1)^n f(\mathbf{0}).$$

To prove formula (4.71) we use part (v) of Theorem 4.10 to see that the left-hand side of equation (4.71) equals

$$\int_{\mathbb{R}^n_+} (D_X f)(X\mathbf{t}) \, d\mathbf{t} = \int_{\mathbb{R}^n_+} \frac{\partial^n}{\partial t_1 \cdots \partial t_n} (f(X\mathbf{t})) \, d\mathbf{t}$$

$$= (-1)^n f(\mathbf{0}).$$

Now, for the proof of Theorem 4.16 we choose any $h \in C_0^\infty (\mathbb{R}^d)$ and consider the function

$$f(\mathbf{x}) = \int_{\mathbb{R}^d} h(\mathbf{y} + \mathbf{x}) O(\mathbf{y}|W) \, d\mathbf{y}.$$

Differentiating under the integral sign gives the equation

$$(D_W f)(\mathbf{x}) = \int_{\mathbb{R}^d} (D_W h)(\mathbf{y} + \mathbf{x}) O(\mathbf{y}|W) \, d\mathbf{y}.$$

Thus, appropriately specializing equation (4.71) provides us with the formula

(4.72) $$(D_W f)(\mathbf{x}) = (-1)^{\#W} h(\mathbf{x}).$$

Similarly, we have

(4.73) $$\left(D_{W \setminus X(j)} f \right)(\mathbf{x}) = \int_{\mathbb{R}^d} \left(D_{W \setminus X(j)} h \right)(\mathbf{y} + \mathbf{x}) O(\mathbf{y}|W) \, d\mathbf{y}.$$

Furthermore, by specializing Theorem 4.10 part (i), it follows that for any $\mathbf{w} \in W$

(4.74) $$D_{\mathbf{w}}(O(\mathbf{x}|W)) = O(\mathbf{x}|W \setminus \{\mathbf{w}\}).$$

Using (4.74) repeatedly, we get for *any* set $R \subseteq W$

$$D_R(O(\mathbf{x}|W)) = O(\mathbf{x}|W \setminus R).$$

Hence, (4.73) becomes

$$\left(D_{W \setminus X(j)} f \right)(\mathbf{x}) = \int_{\mathbb{R}^d} \left(D_{W \setminus X(j)} h \right)(\mathbf{y} + \mathbf{x}) O(\mathbf{y}|W) \, d\mathbf{y}$$

$$= (-1)^{\#W - n} \int_{\mathbb{R}^d} h(\mathbf{y} + \mathbf{x}) D_{W \setminus X(j)}(O(\mathbf{y}|W)) \, d\mathbf{y}$$

$$= (-1)^{\#W - n} \int_{\mathbb{R}^d} h(\mathbf{y} + \mathbf{x}) O(\mathbf{y}|X(j)) \, d\mathbf{y}.$$

We now substitute these expressions for the derivatives of f into formula (4.70) of Lemma 4.7 to obtain

$$\int_{\mathbb{R}^d} h(\mathbf{x}) S(\mathbf{x}|X) \, d\mathbf{x}$$

$$= \int_{\mathbb{R}^d} h(\mathbf{x}) \sum_{j=0}^{n} (-1)^j O(\mathbf{x} - \mathbf{x}^j | X(j)) \, d\mathbf{x}.$$

From this formula it follows that

$$S(\mathbf{x}|X) = \sum_{j=0}^{n} (-1)^j O(\mathbf{x} - \mathbf{x}^j | X(j),)$$

a.e., $\mathbf{x} \in \mathbb{R}^d$.   □

If in the above proof we used the function

$$f(\mathbf{x}) = \int_{\mathbb{R}^d} h\left(\mathbf{y} - \mathbf{x}\right) O\left(\mathbf{y}|W\right) d\mathbf{y},$$

we would have

$$(D_W f)\left(\mathbf{x}\right) = h(-\mathbf{x})$$

and

$$\left(D_{W \setminus X(j)} f\right)\left(\mathbf{x}\right) = \int_{\mathbb{R}^d} h\left(\mathbf{y} - \mathbf{x}\right) O\left(\mathbf{y}|X(j)\right) d\mathbf{y}.$$

Hence it would follow that

$$S\left(\mathbf{x}|X\right) = \sum_{j=0}^{n} (-1)^{n-j} O\left(\mathbf{x}^j - \mathbf{x}|X(j)\right), \qquad \text{a.e., } \mathbf{x} \in \mathbb{R}^d,$$

and so we conclude that

$$\sum_{j=0}^{n} (-1)^{n-j} O(\mathbf{x}^j - \mathbf{x}|X(j)) = \sum_{j=0}^{n} (-1)^j O(\mathbf{x} - \mathbf{x}^j|X(j)), \quad \text{a.e, } \mathbf{x} \in \mathbb{R}^d.$$

Our last comment about the truncated power function is the simple observation that it provides the Green's function for the differential operator $D_X f$, whenever $\mathbf{0} \notin X = \{\mathbf{x}^1, \ldots, \mathbf{x}^n\}$ and rank $X = d$. To see this, we choose a nonzero vector $\mathbf{y} \in \mathbb{R}^d$ such that $\left(\mathbf{y}, \mathbf{x}^i\right) \neq 0$, $i = 1, \ldots, n$. Set $b = \min\{|\left(\mathbf{y}, \mathbf{x}^i\right)| : 1 \leq i \leq n\}$, $\epsilon_i = \text{sgn}\left(\mathbf{y}, \mathbf{x}^i\right)$, $i = 1, \ldots, n$, and $\hat{\mathbf{x}}^i := \epsilon_i \mathbf{x}^i, i = 1, \ldots, n$. Then the hyperplane $(\mathbf{y}, \mathbf{x}) = b$ separates $\mathbf{0}$ from $[\hat{X}]$, where $\hat{X} := \{\hat{\mathbf{x}}^1, \ldots, \hat{\mathbf{x}}^n\}$ and so, by formula (4.71), we have

$$\int_{\mathbb{R}^d} (D_X f)\left(\mathbf{x}\right) \left(\epsilon O\left(\mathbf{x}|\hat{X}\right)\right) d\mathbf{x} = (-1)^n f(\mathbf{0}),$$

where $\epsilon := \epsilon_1 \cdots \epsilon_n$.

4.6.   The cube spline.

We now turn our attention to the function associated with the n-cube $[0, 1]^n$. There is a useful decomposition of the cube $[0, 1]^n$ into simplices. The idea of this simplicial decomposition of the cube is to rearrange the components of a typical vector $\mathbf{y} \in [0, 1]^n$ in a nondecreasing order. As we vary over the cube each component of $\mathbf{y}$ gets a chance to be first, second, ..., last in this reordering. That is, if we let $\mathcal{P}_n$ be the set of all permutations of $\{1, 2, \ldots, n\}$, π a typical element in $\mathcal{P}_n$, and σ_π the simplex

$$\sigma_\pi := \{\mathbf{y} = (y_1, \ldots, y_n)^T : 0 \leq y_{\pi(1)} \leq \cdots \leq y_{\pi(n)} \leq 1\},$$

then the family of simplices $\{\sigma_\pi : \pi \in \mathcal{P}_n\}$ form a simplicial decomposition of the cube $[0, 1]^n$. Any two simplices in this family intersect in a common face and have equal volume.

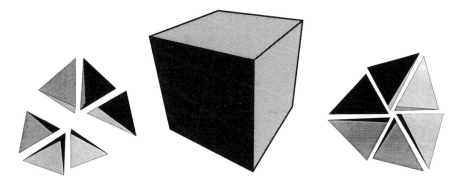

FIG. 4.15. *Triangulation of the cube.*

Recall that the multivariate B-spline is an integral over Δ^n, what we called the standard simplex. To map an element $\mathbf{t} = (t_0, t_1, \ldots, t_n)^T$ in Δ^n into σ_π we set

$$(\mathbf{t}_\pi)_{\pi(j)} := t_1 + \cdots + t_j, \qquad j = 1, \ldots, n.$$

Hence, $\mathbf{t} \to \mathbf{t}_\pi$ maps Δ^n bijectively to σ_π and, moreover, for any set of n vectors $X = \{\mathbf{x}^1, \ldots, \mathbf{x}^n\} \subset \mathbb{R}^d \backslash \{\mathbf{0}\}$ we have

$$X\mathbf{t}_\pi = X_\pi \mathbf{t}, \qquad \mathbf{t} \in \Delta^n,$$

where

$$X_\pi := \{\mathbf{0}, \mathbf{x}^{\pi(1)} + \cdots + \mathbf{x}^{\pi(n)}, \ldots, \mathbf{x}^{\pi(n-1)} + \mathbf{x}^{\pi(n)}, \mathbf{x}^{\pi(n)}\}.$$

This leads us to the following definition.

DEFINITION 4.2. *Let $X \subset \mathbb{R}^d \backslash \{\mathbf{0}\}$ and $\#X = n$. The cube spline $C(\cdot|X)$ is defined as*

$$C(\cdot|X) = \sum_{\pi \in \mathcal{P}_n} S(\cdot|X_\pi).$$

THEOREM 4.17. *Let $X \subset \mathbb{R}^d \backslash \{\mathbf{0}\}$, $\#X = n$.*
(i) *For any $f \in (\mathbb{R}^d)$*

$$\int_{\mathbb{R}^d} f(\mathbf{x}) C(\mathbf{x}|X) \, d\mathbf{x} = \int_{[0,1]^n} f(X\mathbf{t}) \, d\mathbf{t}.$$

(ii) *If $\mathbf{0} \notin [X]$ then*

$$C(\cdot|X) = \sum_{V \subseteq X} (-1)^{\#V} O\left(\cdot - \sum_{\mathbf{x} \in V} \mathbf{x}|X\right).$$

(iii) *If $\mathbf{0} \notin [X]$ and $X \subset \mathbb{Z}^d \backslash \{\mathbf{0}\}$ then*

$$O(\cdot|X) = \sum_{\mathbf{j} \in \mathbb{Z}^d} t(\mathbf{j}|X) C(\cdot - \mathbf{j}|X),$$

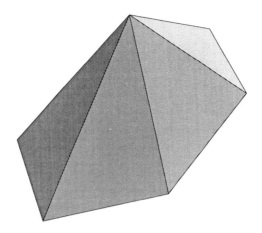

FIG. 4.16. *A linear cube spline.*

where

$$t\left(\mathbf{j}|X\right) = \#\{\mathbf{k} : \mathbf{k} \in \mathbb{Z}_+^n, \quad X\mathbf{k} = \mathbf{j}\}.$$

(iv) *If* $\mathbf{x} = \sum_{j=1}^n a_j \mathbf{x}^j$ *then*

$$C\left(\mathbf{x}|X\right) = \frac{1}{n-d}\sum_{j=1}^n \left(a_j C\left(\mathbf{x}|X_j\right) + (1-a_j) C\left(\mathbf{x} - \mathbf{x}^j|X_j\right)\right).$$

(v) *If* $\mathbf{y} = \sum_{j=1}^n a_j \mathbf{x}^j$ *then*

$$D_{\mathbf{y}}\left(C\left(\cdot|X\right)\right) = \sum_{j=1}^n a_j \left(C\left(\cdot|X_j\right) - C\left(\cdot - \mathbf{x}^j|X_j\right)\right).$$

(vi) *If every subset of ℓ elements in X is linearly independent then* $C\left(\cdot|X\right) \in C^{n-\ell-1}(\mathbb{R}^d)$. *Moreover, it is a polynomial of degree* $\leq n - d$ *on any region not intersected by X-planes.*

Proof. The proof of (i) follows directly from the decomposition of the cube $[0,1]^n$ into simplices $\{\sigma_\pi : \pi \in \mathcal{P}_n\}$ and the definition of X_π. For (ii) we define the difference operator

$$\left(\nabla_{\mathbf{y}} f\right)(\mathbf{x}) = f(\mathbf{x}) - f\left(\mathbf{x} - \mathbf{y}\right)$$

for every $\mathbf{y} \in \mathbb{R}^d$ and, generally, for any finite set $X \subset \mathbb{R}^d$, we set

$$\nabla_X = \prod_{\mathbf{x} \in X} \nabla_{\mathbf{x}} \,.$$

We claim that

$$\left(\nabla_X f\right)(\cdot) = \sum_{V \subseteq X} (-1)^{\#V} f\left(\cdot - \sum_{\mathbf{x} \in V} \mathbf{x}\right).$$

This is easily proved by induction on the cardinality of X. Specifically, we consider the set $X \cup \{\mathbf{y}\}$, which contains one more element than X, and suppose that $V \subseteq X \cup \{\mathbf{y}\}$. Then either $V \subseteq X$ or $V = V' \cup \{\mathbf{y}\}$, where $V' \subseteq V$, and so we obtain

$$\sum_{V \subseteq X \cup \{\mathbf{y}\}} (-1)^{\#V} f\left(\cdot - \sum_{\mathbf{x} \in V} \mathbf{x}\right)$$

$$= \sum_{V \subseteq X} (-1)^{\#V} f\left(\cdot - \sum_{\mathbf{x} \in V} \mathbf{x}\right) - \sum_{V \subseteq X} (-1)^{\#V} f\left(\cdot - \mathbf{y} - \sum_{\mathbf{x} \in V} \mathbf{x}\right).$$

Next, we observe that for every $f \in C_0(\mathbb{R}^d)$ we have

$$\int_{\mathbb{R}^d} f(\mathbf{x}) \nabla_{\mathbf{y}} (O(\cdot|X)) (\mathbf{x}) d\mathbf{x} = \int_{\mathbb{R}^d} (f(\mathbf{x}) - f(\mathbf{x} + \mathbf{y})) O(\mathbf{x}|X) \, dx$$

$$= \int_{\mathbb{R}_+} (f(X\mathbf{t}) - f(X\mathbf{t} + \mathbf{y})) \, dt.$$

When $\mathbf{y} = \mathbf{x}^1$ this last integral simplifies to

$$\int_{[0,1] \times \mathbb{R}_+^{n-1}} f(X\mathbf{t}) \, dt.$$

Successively repeating this computation, for each vector in X, we get

$$\int_{\mathbb{R}^d} f(\mathbf{x}) \nabla_X (O(\cdot|X)) (\mathbf{x}) d\mathbf{x} = \int_{[0,1]^n} f(X\mathbf{t}) \, dt.$$

For (iii) we again consider any $f \in C_0(\mathbb{R}^d)$ and note that we can group the terms in the finite sum

$$\sum_{\mathbf{k} \in \mathbb{Z}_+^n} f(X\mathbf{k})$$

as

$$\sum_{\mathbf{j} \in \mathbb{Z}^d} \left(\sum_{\{\mathbf{k} : X\mathbf{k} = \mathbf{j}\}} f(X\mathbf{k}) \right) = \sum_{\mathbf{j} \in \mathbb{Z}^d} t(\mathbf{j}|X) f(\mathbf{j}).$$

Thus we have

$$\int_{\mathbb{R}^d} f(\mathbf{x}) \left(\sum_{\mathbf{j} \in \mathbb{Z}^d} t(\mathbf{j}|X) C(\mathbf{x} - \mathbf{j}|X) \right) d\mathbf{x}$$

$$= \sum_{\mathbf{j} \in \mathbb{Z}^d} t(\mathbf{j}|X) \int_{\mathbb{R}^d} f(\mathbf{x} + \mathbf{j}) C(\mathbf{x}|X) \, d\mathbf{x}$$

$$= \int_{\mathbb{R}^d} \sum_{\mathbf{j} \in \mathbb{Z}_+^n} f(\mathbf{x} + X\mathbf{j}) C(\mathbf{x}|X) \, d\mathbf{x} = \sum_{\mathbf{j} \in \mathbb{Z}_+^n} \int_{[0,1]^n} f(X\mathbf{t} + X\mathbf{j}) \, dt$$

$$= \sum_{\mathbf{j} \in \mathbb{Z}_+^n} \int_{[0,1]^n + \mathbf{j}} f(X\mathbf{t}) \, dt = \int_{\mathbb{R}_+^n} f(X\mathbf{t}) \, dt.$$

For the proof of (v), it suffices to prove that for any $\mathbf{u} \in X$

(4.75) $D_{\mathbf{u}}\left(C\left(\cdot | X\backslash\{\mathbf{u}\}\right)\right) = C\left(\cdot | X\backslash\{\mathbf{u}\}\right) - C\left(\cdot - \mathbf{u} | X\backslash\{\mathbf{u}\}\right).$

To this end, let $f \in C(\mathbb{R}^d)$. Then for $\ell = 1, \ldots, n$,

$$\int_{\mathbb{R}^d} f(\mathbf{x}) D_{\mathbf{x}^\ell}\left(C\left(\cdot | X\right)\right)(\mathbf{x}) d\mathbf{x} = -\int_{\mathbb{R}^d} \left(D_{\mathbf{x}^\ell} f\right)(\mathbf{x}) C\left(\mathbf{x} | X\right) d\mathbf{x}$$

$$= -\int_{[0,1]^n} \frac{d}{dt_\ell}\left(f\left(X\mathbf{t}\right)\right) d\mathbf{t}$$

$$= \int_{[0,1]^{n-1}} \left(f\left(X_\ell \mathbf{t}\right) - f\left(\mathbf{x}_\ell + X_\ell \mathbf{t}\right)\right) d\mathbf{t}$$

$$= \int_{\mathbb{R}^d} f(\mathbf{x}) \left(C\left(\mathbf{x} | X_\ell\right) - C\left(\mathbf{x} - \mathbf{x}_\ell | X_\ell\right)\right) d\mathbf{x}.$$

There remains the proof of (iv). First, let us assume that $\mathbf{0} \notin [X]$. In this case, we can make use of (ii) in Theorem 4.17 and (ii) of Theorem 4.10. In the computation that we are about to perform, it is convenient for us to index the constants $a_1, \ldots, a_n$ in (iv) by an element $\mathbf{y} \in X$; that is, we set $a_i = a_{\mathbf{x}^i}, i = 1, \ldots, n$. Therefore, for any set $V \subseteq X$ we have

$$\mathbf{x} - \mathbf{x}_V = \sum_{\mathbf{y} \notin V} a_{\mathbf{y}} \mathbf{y} + \sum_{\mathbf{y} \in V} \left(a_{\mathbf{y}} - 1\right) \mathbf{y},$$

where we define

$$\mathbf{x}_V = \sum_{\mathbf{y} \in V} \mathbf{y},$$

and also

$$C\left(\mathbf{x} | X\right) = \sum_{V \subseteq X} (-1)^{\#V} O\left(\mathbf{x} - \mathbf{x}_V | X\right)$$

$$= \frac{1}{n-d} \sum_{V \subseteq X} (-1)^{\#V} \left(\sum_{\mathbf{y} \notin V} a_{\mathbf{y}} O\left(\mathbf{x} - \mathbf{x}_V | X\backslash\{\mathbf{y}\}\right) \right.$$

$$\left. + \sum_{\mathbf{y} \in V} \left(a_{\mathbf{y}} - 1\right) O\left(\mathbf{x} - \mathbf{x}_V | X\backslash\{\mathbf{y}\}\right) \right).$$

Now, pick a typical $\mathbf{y} \in X$ and consider the contribution of the terms $a_{\mathbf{y}}$ and $a_{\mathbf{y}} - 1$ in the formula above. These are, respectively,

$$\frac{1}{n-d} \sum_{V \subseteq X\backslash\{\mathbf{y}\}} (-1)^{\#V} O\left(\mathbf{x} - \mathbf{x}_V | X\backslash\{\mathbf{y}\}\right) = \frac{1}{n-d} C\left(\mathbf{x} | X\backslash\{\mathbf{y}\}\right),$$

and

$$-\frac{1}{n-d}\sum_{V\subseteq X\backslash\{\mathbf{y}\}}(-1)^{\#V}O\left(\mathbf{x}-\mathbf{y}-\mathbf{x}_V|X\backslash\{\mathbf{y}\}\right) = -\frac{1}{n-d}C\left(\mathbf{x}-\mathbf{y}|X\backslash\{\mathbf{y}\}\right).$$

Hence, in total, we get

$$C\left(\mathbf{x}|X\right) = \frac{1}{n-d}\sum_{\mathbf{y}\in X}\left(a_{\mathbf{y}}C\left(\mathbf{x}|X\backslash\{\mathbf{y}\}\right) + \left(1-a_{\mathbf{y}}\right)C\left(\mathbf{x}-\mathbf{y}|X\backslash\{\mathbf{y}\}\right)\right).$$

There remains the case that $\mathbf{0}\in[X]$. In this case, choose signs $\epsilon_i\in\{-1,1\}$, $i=1,\ldots,n$, such that the vectors

$$\hat{\mathbf{x}}^i := \epsilon_i\mathbf{x}^i, \qquad i=1,\ldots,n,$$

have the property that

$$\mathbf{0}\notin[\hat{X}].$$

It is easy to see, for instance, from (i) in Theorem 4.17, that

$$C(\mathbf{x}-\mathbf{u}|\hat{X}) = C\left(\mathbf{x}|X\right),$$

where $\mathbf{u} := \Sigma\{\mathbf{x}^i : \epsilon_i = -1,\ i=1,\ldots,n\}$. Hence we may apply formula (iv) of Theorem 4.17 to the set $\hat{X}$. For this purpose, we note that

$$(4.76) \qquad \mathbf{x}-\mathbf{u} = \sum_{i=1}^{n}\hat{a}_i\hat{\mathbf{x}}^i$$

where

$$(4.77) \qquad \hat{a}_i := \begin{cases} a_i, & \epsilon_i = 1, \\ 1-a_i, & \epsilon_i = -1, \end{cases}$$

and, of course, we have by definition that

$$(4.78) \qquad \hat{\mathbf{x}}_i = \begin{cases} \mathbf{x}_i, & \epsilon_i = 1, \\ -\mathbf{x}_i, & \epsilon_i = -1. \end{cases}$$

Thus

$$C(\mathbf{x}|X) = \frac{1}{n-d}\sum_{i=1}^{n}\left(\hat{a}_iC(\mathbf{x}-\mathbf{u}|\hat{X}_i) + \left(1-\hat{a}_i\right)C(\mathbf{x}-\hat{\mathbf{x}}^i-\mathbf{u}|\hat{X}_i)\right).$$

We break up the sum into those terms where $\hat{a}_i = a_i$ and the remainder where $\hat{a}_i = 1-a_i$. Using (4.76)–(4.78) to simplify the resulting expression, we get

$$C(\mathbf{x}|X) = \frac{1}{n-d}\sum_{i=1}^{n}\left(a_iC\left(\mathbf{x}|X_i\right) + \left(1-a_i\right)C\left(\mathbf{x}-\mathbf{x}^i|X_i\right)\right).$$

The property (vi) follows easily from (ii) above and the similar property for the truncated power spline given in (iv) of Theorem 4.10.   □

We remark that de Boor and DeVore [deBD] use (i) of Theorem 4.17 to define the cube spline. Most of the results of Theorem 4.17 are due to de Boor and Höllig [deBH].

THEOREM 4.18. *Let* $X = \{\mathbf{x}^1, \ldots, \mathbf{x}^n\} \subset \mathbb{R}^d\backslash\{\mathbf{0}\}$ *and* rank $X = d$. *Then, for any* $\mathbf{y} \in \mathbb{R}^d$ *and*

$$\mathbf{x} = \sum_{i=1}^{n} a_i \mathbf{x}^i$$

$$D\mathbf{y}(C(\cdot|X))(\mathbf{x})$$
$$= \frac{1}{n-d-1} \sum_{i=1}^{n} (a_i D\mathbf{y}(C(\cdot|X_i))(\mathbf{x}^i) + (1 - a_i) D\mathbf{y}(C(\cdot|X_i))(\mathbf{x} - \mathbf{x}^i)).$$

Proof. Differentiate both sides of formula (iv) of Theorem 4.17 with respect to $a_\ell, 1 \leq \ell \leq n$, to obtain the equation

$$D_{\mathbf{x}^\ell}(C(\cdot|X))(\mathbf{x})$$
$$(4.79) \quad = \frac{1}{n-d} \left(\sum_{j=1}^{n} (a_j D_{\mathbf{x}^\ell}(C(\cdot|X_j))(\mathbf{x}) + (1 - a_j) D_{\mathbf{x}^\ell}(C(\cdot|X_j))(\mathbf{x} - \mathbf{x}^j) \right)$$
$$+ \frac{1}{n-d} (C(\mathbf{x}|X_\ell) - C(\mathbf{x} - \mathbf{x}^\ell|X_\ell)).$$

Using (v) of Theorem 4.17 (specialized to the case $\mathbf{y} = \mathbf{x}^\ell$) in the above equation and simplifying the resulting expression gives us the formula

$$D_{\mathbf{x}^\ell}(C(\cdot|X_i))(\mathbf{x})$$
$$= \frac{1}{n-d-1} \sum_{i=1}^{n} (a_i D_{\mathbf{x}^\ell}(C(\cdot|X_i))(\mathbf{x}) + (1 - a_i) D_{\mathbf{x}^\ell}(C(\cdot|X_i))(\mathbf{x} - \mathbf{x}^i)).$$

Since rank $X = d$, there are constants $u_1, \ldots, u_n$ such that

$$\mathbf{y} = \sum_{j=1}^{n} u_j \mathbf{x}^j.$$

Now, multiply both sides of this equation by u_ℓ and sum over $\ell = 1, \ldots, n$ to obtain the result.   □

The last formula states that the cube spline satisfies a *refinement equation* (see Chapter 2).

THEOREM 4.19. *Let* $X = \{\mathbf{x}^1, \dots, \mathbf{x}^n\} \subset \mathbb{Z}^d \backslash \{\mathbf{0}\}$. *Then*

$$(4.80) \qquad C(\cdot|X) = \sum_{\mathbf{j} \in \mathbb{Z}^d} a_{\mathbf{j}} C(2 \cdot -\mathbf{j}|X),$$

where

$$(4.81) \qquad \sum_{\mathbf{j} \in \mathbb{Z}^d} a_{\mathbf{j}} \mathbf{z}^{\mathbf{j}} = 2^{d-n} \Pi_{j=1}^n \left(1 + \mathbf{z}^{\mathbf{x}^j}\right), \qquad \mathbf{z} \in \mathbb{C}^d.$$

Proof. We prove this result by induction on n. For the case $n = 1$ and $X = \{\mathbf{x}^1\}$ we see that

$$a_{\mathbf{j}} = \begin{cases} 0, & \mathbf{j} \in \mathbb{Z}^d \backslash \{\mathbf{0}, \mathbf{x}^1\}, \\ 2^{d-1}, & \mathbf{j} \in \{\mathbf{0}, \mathbf{x}^1\}. \end{cases}$$

Moreover, in this case, for any $f \in C(\mathbb{R}^d)$ we have

$$\int_{\mathbb{R}^d} f(\mathbf{x}) C(\mathbf{x}|X)\, d\mathbf{x} = \int_0^1 f(t\mathbf{x}^1)\, dt$$

$$= \frac{1}{2} \left(\int_0^1 f\left(\frac{t}{2}\mathbf{x}^1\right) dt + \int_0^1 f\left(\frac{t\mathbf{x}^1 + \mathbf{x}^1}{2}\right) dt \right)$$

$$= 2^{-d} \int_0^1 \sum_{\mathbf{j} \in \mathbb{Z}^d} a_{\mathbf{j}} f\left(\frac{t\mathbf{x}^1 + \mathbf{j}}{2}\right) dt$$

$$= 2^{-d} \sum_{\mathbf{j} \in \mathbb{Z}^d} a_{\mathbf{j}} \int_{\mathbb{R}^d} f\left(\frac{\mathbf{x} + \mathbf{j}}{2}\right) C(\mathbf{x}|X)\, d\mathbf{x}$$

$$= \int_{\mathbb{R}^d} f(\mathbf{x}) \left(\sum_{\mathbf{j} \in \mathbb{Z}^d} a_{\mathbf{j}} C(2\mathbf{x} - \mathbf{j}|X) \right) d\mathbf{x}.$$

Now, suppose that (4.80) is true for all sets X with n elements. Consider the set $X' = X \cup \{\mathbf{x}^{n+1}\}$. Let $\{a_{\mathbf{j}}^{n+1}\}_{\mathbf{j} \in \mathbb{Z}^d}$, $\{a_{\mathbf{j}}^n\}_{\mathbf{j} \in \mathbb{Z}^d}$ be the coefficients corresponding to X' and X, respectively. Then (4.81) implies that

$$(4.82) \qquad a_{\mathbf{j}}^{n+1} = \frac{1}{2}\left(a_{\mathbf{j}}^n + a_{\mathbf{j}-\mathbf{x}^{n+1}}^n\right), \qquad \mathbf{j} \in \mathbb{Z}^d.$$

Also we need to observe that

$$(4.83) \qquad C(\mathbf{x}|X') = \int_0^1 C(\mathbf{x} - t\mathbf{x}^{n+1}|X)\, dt,$$

a fact obtained directly from (i) of Theorem 4.17. Hence for any $f \in C(\mathbb{R}^d)$, we have

$$
\begin{aligned}
\int_{\mathbb{R}^d} f(\mathbf{x}) C(\mathbf{x}|X')\,d\mathbf{x} &= \int_{\mathbb{R}^d} f(\mathbf{x}) \left(\int_0^1 C\left(\mathbf{x} - t\mathbf{x}^{n+1}|X\right)\,dt \right) d\mathbf{x} \\
&= \int_0^1 \left(\int_{\mathbb{R}^d} f\left(\mathbf{x} + t\mathbf{x}^{n+1}\right) \sum_{\mathbf{j} \in \mathbb{Z}^d} a_{\mathbf{j}}^n C\left(2\mathbf{x} - \mathbf{j}|X\right)\,d\mathbf{x} \right) dt \\
&= \int_{\mathbb{R}^d} f(\mathbf{x}) \left(\sum_{\mathbf{j} \in \mathbb{Z}^d} a_{\mathbf{j}}^n \int_0^1 C\left(2\mathbf{x} - 2t\mathbf{x}^{n+1} - \mathbf{j}|X\right)\,dt \right) d\mathbf{x} \\
&= \int_{\mathbb{R}^d} f(\mathbf{x}) \left(\sum_{\mathbf{j} \in \mathbb{Z}^d} a_{\mathbf{j}}^n \frac{1}{2} \left(C\left(2\mathbf{x} - \mathbf{j}|X'\right) \right. \right. \\
&\qquad\qquad \left. \left. + C\left(2\mathbf{x} - \mathbf{j} - \mathbf{x}^{n+1}|X'\right) \right) \right) d\mathbf{x} \\
&= \int_{\mathbb{R}^d} f(\mathbf{x}) \sum_{\mathbf{j} \in \mathbb{Z}^d} a_{\mathbf{j}}^{n+1} C\left(2\mathbf{x} - \mathbf{j}|X'\right)\,d\mathbf{x} \qquad \square
\end{aligned}
$$

The above result was taken from Dahmen and Micchelli [DM2], where it is used to develop a stationary subdivision procedure for computing linear combinations of integer translates of cube splines.

4.7. Correspondence on the historical development of B-splines.

The development of the multivariate B-spline is an exciting chapter in the history of spline functions. Some of the milestones in the study of B-splines were mentioned in [DM5], especially as they appear in the statistical literature. Perhaps more than anyone else, I.J. Schoenberg pushed the theory of spline functions forward. He described some of his unpublished ideas on bivariate B-splines in a letter he sent to P.J. Davis on May 31, 1965, which we referred to in the first section of this chapter. Below we have provided a copy of this letter.

In a seemingly unrelated paper, Motzkin and Schoenberg [MSc] studied a class of multivariate entire functions which, in their terminology, have "affine lineage." In the univariate case these functions were an indispensable tool in Curry and Schoenberg's study of distributions that are limits of B-splines. They presented these ideas in [CS]. Later, in [DM6] we were able to connect the Motzkin–Schoenberg functions of affine lineage to limits of multivariate B-splines. Schoenberg's apparent enthusiasm for this discovery is expressed in an open letter he sent to several colleagues. This letter also appears below.

Sometime later, with Schoenberg's encouragement, we sent a copy of paper [Mic2] on computing multivariates to H.B. Curry. In his response to our preprint, Curry explains the background of his collaboration with Schoenberg during World War II and his contribution to their discovery of the univariate B-spline [CS].

A letter from I.J. Schoenberg to P.J. Davis dated May 31, 1965:

Professor Philip J. Davis
Division of Applied Mathematics
Brown University
Providence, RI

Madison, Wisconsin
May 31, 1965

Dear Phil:

Lately I have been busy with euclidean polyhedra from the point of view of distance geometry, but on and off I return to spline functions. The other day I noticed that the triangle formulae in the complex plane generalize nicely to higher order derivatives. The source is again the Hermite–Gennochi formula

$$(1) \quad f(x_0, x_1, \ldots, x_n) = \int \cdots_{\tau_n} \int F^{(n)}(x_0 t_0 + x_1 t_1 + \cdots + x_n t_n) dt_1 \cdots dt_n$$

where $t_0 = 1 - t_1 - \cdots - t_n$ and τ_n where the integration is to be carried out over the simplex

$$\tau_n : t_1 \geq 0, \ldots, t_n \geq 0, \ \sum_1^k t_\nu \leq 1.$$

A projection onto the real axis allows us to show that the kernel $M(x; x_0, x_1, \ldots, x_n)$ in the fundamental formula

$$(2) \quad f(x_0, x_1, \ldots, x_n) = \frac{1}{n!} \int_{x_0}^{x_n} f^{(n)}(x) M(x; x_0, \ldots, x_n) dx, \ (x_0 < \cdots < x_n),$$

may be interpreted as follows: Interpret the x-axis as one of the coordinate axis of n-dimensional space. Erect at $x_0, x_1, \ldots, x_n$ n hyperplane orthogonal to the x-axis at these points and select at will in each of these hyperplanes a point, with the proviso that the simplex having these points as vertices should have n-dimensional volume unity. If we project the volume of this simplex on the x-axis we obtain the linear density function $M(x : x_0, \ldots, x_n)$ appearing in (2).

I now pass to the complex domain. Let $z_0, z_1, \ldots, z_n$ be distinct points (not all on a line) of the complex plane, and let $f(z)$ be regular in the convex hull Π of these points. Again we erect at $z_0, z_1, \ldots, z_n$ the orthogonal complements of the plane (these are of dimension $n - 2$) and select in each of these a point so that the simplex has n-dim. volume unity. We now project the volume of this simplex onto the plane and denote the surface density function by

$$M(x, y; z_0, \ldots, z_n).$$

Then the following formula holds

$$f(z_0, z_1, \ldots, z_n) = \frac{1}{n!} \int \int_\Pi f^{(n)}(z) M(x, y; z_0, \ldots, z_n) dx dy.$$

For $n = 2$ this gives the old triangle formula

$$f(z_0, z_1, z_2) = \frac{1}{2A} \int \int_T f''(z)\,dx\,dy,$$

where A is the area of the triangle T of vertices z_0, z_1, z_2 (P.J. Davis, *Triangle formulas in the complex plane*, Math. of Comp., 1964, p. 569-577).

Joining all parts of points z_j, z_k $(j < k)$ by segments, the polygon Π is disserted into disjoint polygons in each of which $M(x, y)$ is a polynomial in x and y of joint degree $n - 2$, while $M(x, y) = 0$ outside Π. Moreover the function $M(x, y)$ has continuous partial derivatives of all orders $\leq n-3$.[1] Moreover, these properties always determine $M(x, y)$ *uniquely* up to a constant factor. Another property is this: $M(x, y) > 0$ inside Π and

$$\log M(x, y) \qquad ((x, y) \text{ inside } \Pi)$$

is a *concave* function. In particular $M(x, y)$ has exactly one maximum point. Thus for $n = 3$ the surface $z = M(x, y)$ is a pyramid.

In case all the points $z_0, z_1, \ldots, z_n$ are on the boundary of the polygon Π then $M(x, y)$ can be represented inside Γ by means of the truncated power function x_+^{n-2}. Here is an example: For $n = 4$ and

$$z_0 = 2, \; z_1 = 1 + i, \; z_2 = -1 + i, \; z_3 = -1 - i, \; z_4 = 1 - i$$

$M(x, y)$ is up to a positive factor, which I did not determine, identical *inside* Π to

(2)
$$\begin{aligned}
2 + 4x - 4x^2 - 6y^2 + 6(x - 1)_+^2 + (x + 3y - 2)_+^2 + 3(y - x)_+^2 \\
+ (x - 3y - 2)_+^2 + 3(-y - x)_+^2.
\end{aligned}$$

I am very much interested in these 2-dimensional frequency functions $M(x, y; z_0, \ldots, z_n)$ because I suspect that the limits of much frequency functions (with the z_j depending on n and as $n \to \infty$) in the usual reuse of probability theory, ought to be interesting frequency functions. The similar equation on the line led to the Polya frequency function. What will one get in the plane, if anything?

I hope these remarks did not bore you. With greetings and good wishes.

Yours,

Iso

(I.J. Schoenberg)

I add a sketch of the second degree $M(x, y)$ $(n = 4)$ for the case when $z_0, \ldots, z_4$ are the vertices of a regular pentagon. All vertical sections are, of

---

[1] This assumes that the points $z_0, \ldots, z_n$ are in "general position," i.e., no three are collinear.

course, ordinary spline functions. I sent copies of this letter to H.B. Curry, S. Karlin, and T.S. Motzkin.

A letter from I.J. Schoenberg to friends sent in 1979:

November 3, 1979

Dear Colleague:

Let me bring you the good news that B-splines and Polya frequency functions are now being studied in higher dimensions. This was recently done by Wolfgang Dahmen, of Bonn, and Charles Micchelli, of IBM. Let me remind you that Polya frequency functions—de Bruijn called them "Polyamials"—were characterized in four different but equivalent ways:

    I. As limits of B-splines with appropriate knots as their degree tends to infinity,

    II. As totally positive functions (frequency functions),

    III. By their variation diminishing property on convolution,

    IV. By their Laplace transforms being reciprocals of entire functions of the Laguerre–Polya class.

At the same time a B-spline M_n of degree $n-1$, having n knots, was known to be the orthogonal projection onto the real axis of an n-dimensional simplex S, of volume 1.

Now Dahmen and Micchelli are projecting S onto the lower-dim. space R^s, obtaining the s-dimensional B-spline $M_{n,s}$. Keeping s fixed and letting n tend to infinity, they obtain as limits of $M_{n,s}$ the s-dimensional Polya frequency functions f_s, and also their Laplace transforms as the reciprocals of a class E_s of entire functions of s variables. This class E_s turns out to be identical with the class of "lineal" functions studied by Motzkin and myself in 1952. Lineal functions were an extension of results of Laguerre and Polya, and B-splines $M_{n,s}$ and Polya frequency functions f_s were not even mentioned in 1952. Now we know where this class E_s is coming from.

This work extends to R^s the characterizations I and IV. Very likely also II and III will appear by means of restrictions of f_s to appropriate lines or planes. I am very happy with these developments.

With regards and best wishes,

Yours,

I.J. Schoenberg

We may describe this subject as asymptotic properties of an n-simplex as n tends to infinity.

A letter from Haskell B. Curry dated November 17, 1979:

The Pennsylvania State University
215 McAllister Building
University Park, Pennsylvania 16802

College of Science   Area Code 814-865-7527
Department of Mathematics   November 17, 1979

Dr. Charles Micchelli
Thomas J. Watson Research Center, IBM
P.O. Box 218
Yorktown Heights, N.Y. 10598

Dear Fellow-worker:

The copy you sent me of your report "On a numerically efficient method for computing multivariate B-splines" reached me here about November 1; but I was not able to acknowledge it right away because I am still convalescing from a spell of illness which I had last summer. I still have very little time to tend to scientific matters. However, I am very glad that you sent me this document. To be sure I have not been active in that field for more than thirty years, and my connection with it was somewhat of an accident. During part of World War II Iso Schoenberg and I were both working for the War Department at Aberdeen Proving Ground, attempting to do what we could to help in the war effort. While so engaged a paper by Schoenberg on splines came across my desk. After examining it, I came to the conclusion that the matter could be generalized. Iso had considered only the case where the nodes, or "knots" were equally spaced; I noticed that the approach could be extended to arbitrary nodes (as in the Lagrange interpolation formula) by using algebraic techniques instead of the Fourier transform. Iso suggested that we write a joint paper. I wrote up the algebraic part and submitted it to him. He, however, was evidently interested in other matters; at any rate he did nothing about it for some 15 years. In due time it was published; first in the proceedings of the Army Research Centre at U. Wisconsin, then in Israel. Unfortunately the paper was published with my name first, as if I had been the prime mover; actually, my part of it was very small. However, he would not listen to any other arrangement.

When the war ended, I came back to Penn State, and resumed research in my original field, which is a very abstract and specialized field of mathematical logic. I have not done any work in that field since. But I have been quite interested in seeing what you and others have been doing with the field which has opened up.

Sincerely yours,

Haskell B. Curry

References

[CS]   H. B. CURRY AND I. J. SCHOENBERG, *Polya frequency functions* IV. *The fundamental spline function and their limits*, J. d'Anal. Math., 17(1966), 71–107.

[Dah]   W. DAHMEN, *On multivariate B-splines.* SIAM J. Numer. Anal., 17(1980), 179–190.

[DM1]   W. DAHMEN AND C. A. MICCHELLI, *On the linear independence of multivariate B-splines* I: *Triangulation of simploids.* SIAM J. Numer. Anal., 19(1982), 992–1012.

[DM2]   W. DAHMEN AND C. A. MICCHELLI, *Recent progress in multivariate splines*, in Approximation Theory IV, C. K. Chui, L. L. Schumaker, and J. D. Ward, eds., Academic Press, New York, 1983, 27–121.

[DM3]   W. DAHMEN AND C. A. MICCHELLI, *On the linear independence of multivariate B-splines* II: *Complete configuration.* Math. Comp., 41(1983), 141–164.

[DM4]   W. DAHMEN AND C. A. MICCHELLI, *Subdivision algorithms for the generation of box spline surfaces*, Comput. Aided Geom. Design, 1(1984), 115–129.

[DM5]   W. DAHMEN AND C. A. MICCHELLI, *Statistical encounters with B-splines*, in Function Estimates, J. S. Marron, ed., Contemporary Mathematics 59, American Mathematical Society, Providence, RI, 1985.

[DM6]   W. DAHMEN AND C. A. MICCHELLI, *On limits of multivariate B-splines*, J. d'Anal. Math., 39(1981), 256–278.

[deB]   C. DE BOOR, *Splines as linear combinations of B-splines: A survey*, in Approximation Theory II, G. G. Lorentz, C. K. Chui, and L. L. Schumaker, eds., Academic Press, New York, 1976, 1–47.

[deBD]   C. DE BOOR AND R. DEVORE, *Approximation by smooth multivariate splines*, Trans. Amer. Math. Soc., 276(1983), 775–785.

[dBH]   C. DE BOOR AND K. HÖLLIG, *B-splines from parallelepipeds.* J. d'Anal. Math., 42(1982/83), 99–115.

[H]   K. HÖLLIG, *Multivariate splines*, SIAM J. Numer. Anal., 41(1982), 1013–1031.

[Mic1]   C. A. MICCHELLI, *A constructive approach to Kergin interpolation in R^k: Multivariate B-splines and Lagrange interpolation.* Rocky Mountain J. Math., 10(1980), 485–497.

[Mic2]   C. A. MICCHELLI, *On a numerically efficient method for computing multivariate B-splines.* in Multivariate Approximation Theory, W. Schempp, and K. Zeller, eds., Birkhaüser, Basel, 1979, 211–248.

[MS]   V. D. MILMAN AND G. SCHECHTMAN, *Asymptotic Theory of Finite Dimensional Normed Spaces*, Lecture Notes in Mathematics, Vol. 1200, Springer-Verlag, Berlin, 1980.

[MSc]   T. S. MOTZKIN AND I. J. SCHOENBERG, *On lineal entire functions of n complex variables*, Proc. Amer. Math. Soc., 3(1952), 517–526.

[NW]   G. L. NEMHAUSER AND L. A. WOLSEY, *Integer and Combinatorial Optimization*, John Wiley and Sons, New York, 1988.

[RV]   A.W. ROBERTS AND D.F. VARBERG, *Convex Functions*, Pure and
         Applied Mathematics, Vol 57, Academic Press, New York, 1973.

[Sch]   A. SCHRIJVER, *Theory of Linear and Integer Programming*, John Wiley
         and Sons, New York, 1986.

CHAPTER 5

Recursive Algorithms for Polynomial Evaluation

5.0. Introduction.

Properties of the univariate Bernstein polynomials motivate the material presented this chapter. Specifically, it has been known for a long time that, given any point on the Bernstein–Bézier curve, there is a simple recurrence that terminates at that point. This observation naturally connects to the recurrence formula for the Bernstein polynomials themselves and also to dual bases, the binomial theorem, and polarization of polynomial identities. All of these ideas are amplified in this chapter to include univariate B-spline representation of a polynomial curve over an interval between two consecutive knots. But it is in the multivariate context that these ideas blossom into a diverse mathematical theory for the recursive computation of polynomials by pyramid schemes. Several examples of such schemes are given and one example is singled out by its unique symmetry properties. This scheme, for H.P. Seidel's B-patches, specializes in the univariate case to the B-spline bases on an interval between consecutive knots. At the end of this chapter, using the recurrence formula for the multivariate B-spline developed in Chapter 4, we identify B-patches with B-splines.

5.1. The de Casteljau recursion.

The results presented in this chapter closely follow the material in Cavaretta and Micchelli [CM] and Dahmen, Micchelli, and Seidel [DMS]. Specifically, we study recurrence formulas for the generation of polynomial curves and surfaces and explore their relationship to the multivariate B-spline of Chapter 4.

We begin as in Chapter 1, with the Bernstein–Bézier polynomials. In Chapter 1 our concern was with the *de Casteljau subdivision* and its generalizations. Here, we focus on the *de Casteljau recurrence formula* given in (1.68) as

$$(5.1) \qquad \mathbf{c}_r^\ell = (1-x)\mathbf{c}_r^{\ell-1} + x\mathbf{c}_{r+1}^{\ell-1}, \qquad r = 0, 1, \ldots, n-\ell, \ \ell = 1, \ldots, n.$$

It was pointed out without proof, after our discussion following (1.68), that this recurrence formula terminates at the value of the Bernstein–Bézier curve *at* x, corresponding to the control points $\mathbf{c}_r := \mathbf{c}_r^0$, $r = 0, 1, \ldots, n$, viz.

(5.2) $$\mathbf{c}_0^n = \sum_{j=0}^{n} \mathbf{c}_j b_j^n(x),$$

where

(5.3) $$b_j^n(x) = \binom{n}{j} x^j (1-x)^{n-j}, \qquad j = 0, 1, \ldots, n.$$

(See Fig. 5.1 for the case of cubic planar curves.) Actually, we shall now show that

(5.4) $$\mathbf{c}_r^\ell = \sum_{j=0}^{\ell} \mathbf{c}_{r+j} b_j^\ell(x), \qquad r = 0, 1, \ldots, n - \ell, \ \ell = 0, 1, \ldots, n,$$

so that, in particular, (5.2) holds. The proof of formula (5.4) requires the recurrence formula for the Bernstein–Bézier polynomials, namely

(5.5)
$$b_j^n(x) = (1-x)b_j^{n-1}(x) + xb_{j-1}^{n-1}(x), \qquad j = 0, 1, \ldots, n,$$
$$b_{-1}^{n-1}(x) = b_n^{n-1}(x) = 0, \qquad n = 1, 2, \ldots,$$

initialized with $b_0^0(x) := 1$. This recurrence formula follows directly from (5.3). We prove (5.4) by induction on ℓ. Thus we suppose that $\ell \geq 1$ and that (5.4) has been proved for ℓ replaced by $\ell - 1$. Using the recurrence formulas (5.1)

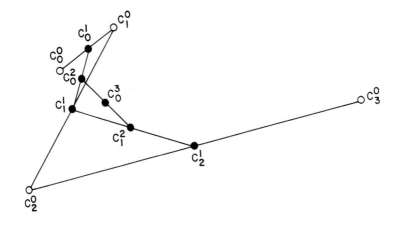

FIG. 5.1.   *de Casteljau evaluation of a cubic curve.*

and (5.5) (with n replaced by ℓ) we get

$$\mathbf{c}_r^\ell = (1-x)\sum_{j=0}^{\ell-1}\mathbf{c}_{r+j}b_j^{\ell-1}(x) + x\sum_{j=1}^{\ell}\mathbf{c}_{r+j}b_{j-1}^{\ell-1}(x)$$

$$= \sum_{j=0}^{\ell}\mathbf{c}_{r+j}\left\{(1-x)b_j^{\ell-1}(x) + xb_{j-1}^{\ell-1}(x)\right\}$$

$$= \sum_{j=0}^{\ell}\mathbf{c}_{r+j}b_j^{\ell}(x).$$

This computation advances the induction step to ℓ and proves (5.4). In particular choosing $\ell = n$ in (5.4) gives us the formula

$$\mathbf{c}_0^n = (C\mathbf{b})(x) = \sum_{j=0}^{n}\mathbf{c}_j b_j^n(x),$$

that is, $\mathbf{c}_0^n$ is the value of the Bernstein–Bézier curve $C\mathbf{b}$ at x.

A convenient way to visualize the recurrence formula (5.1) is in a *recursive triangle*; see Fig. 5.2 for the case of polynomials of degree three.

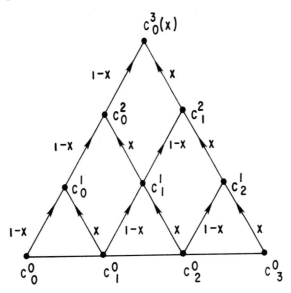

FIG. 5.2. *Up recurrence formula for the Bernstein–Bézier cubic curve.*

Note that at the base of the triangle are the control points of the Bernstein–Bézier curve $C\mathbf{b}$. The upward pointing arrows each have a label, either $1-x$ or x depending on whether it leans to the right or left. One proceeds up the triangle by means of formula (5.1). Thus the vector at any location in the triangle is a sum of the two vectors at the base of the arrows pointing into its location and weighted by the labels associated with these arrows. At the apex of the triangle emerges the curve $C\mathbf{b}$ at x.

Now, let us reverse these steps. Place the value one at the apex of the triangle, reverse the arrows, and proceed down the triangle, and at the base out pop the Bernstein–Bézier polynomials (see Fig. 5.3). This is just the recurrence formula (5.5).

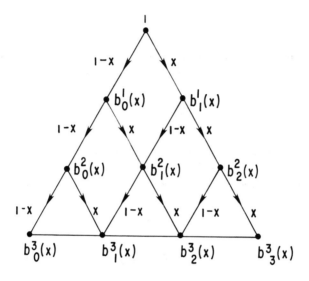

FIG. 5.3.   *Down recurrence formula for the Bernstein–Bézier cubic curve.*

The upward and downward recurrence formulas are tied together by a *conservation law*. By this we mean that the quantity

$$\sum_{r=0}^{n-\ell+1} c_r^{\ell-1} b_r^{n-\ell+1}(x)$$

is *independent* of ℓ, $\ell = 1, \ldots, n+1$. The proof is straightforward. We compute the above expression by using first downward recurrence formula (5.5) and then the upward recurrence formula (5.1), viz.

$$\sum_{r=0}^{n-\ell+1} c_r^{\ell-1} b_r^{n-\ell+1}(x)$$

$$= \sum_{r=0}^{n-\ell+1} c_r^{\ell-1} \left\{ (1-x) b_r^{n-\ell}(x) + x b_{r-1}^{n-\ell}(x) \right\}$$

$$= \sum_{r=0}^{n-\ell+1} (1-x)\mathbf{c}_r^{\ell-1} b_r^{n-\ell}(x) + \sum_{r=-1}^{n-\ell} x\mathbf{c}_{r+1}^{\ell-1} b_r^{n-\ell}(x)$$

$$= \sum_{r=0}^{n-\ell} \left\{ (1-x)\mathbf{c}_r^{\ell-1} + x\mathbf{c}_{r+1}^{\ell-1} \right\} b_r^{n-\ell}(x)$$

$$= \sum_{r=0}^{n-\ell} \mathbf{c}_r^{\ell} b_r^{n-\ell}(x).$$

In summary, we have shown that

$$(5.6) \qquad \sum_{r=0}^{n-\ell+1} \mathbf{c}_r^{\ell-1} b_r^{n-\ell+1}(x) = \sum_{r=0}^{n-\ell} \mathbf{c}_r^{\ell} b_r^{n-\ell}(x), \quad \ell = 1, \ldots, n.$$

5.2.   Blossoming Bernstein–Bézier polynomials: The univariate case.

Much of this chapter is devoted to multivariate extensions of recursive triangles. Before we get into this matter, we continue to highlight, in the context of the de Casteljau recurrence formula, other basic principles for recursive triangles. The two that we have in mind are *blossoming* (or *polarization*) of polynomial identities and *duality*. We begin with blossoming.

A function $P : \mathbb{R}^n \rightarrow \mathbb{R}$ is said to be *multiaffine* provided that for each $j = 1, \ldots, n$, $P(x_1, x_2, \ldots, x_n)$ is an affine function of x_j. Let E_n be the set of all vectors in $\mathbb{R}^n$ whose components are either zero or one. E_n consists of 2^n vectors and for each $\mathbf{a} = (a_1, a_2, \ldots, a_n)^T \in E_n$ the equation

$$L_{\mathbf{a}}(\mathbf{x}) := \prod_{a_i=1} x_i \prod_{a_i=0} (1 - x_i), \qquad \mathbf{x} = (x_1, x_2, \ldots, x_n)^T \in \mathbb{R}^n,$$

defines a multiaffine function. Moreover, each multiaffine function P can be written as a unique linear combination of these functions. In fact, we have the (multilinear) interpolation formula

$$(5.7) \qquad\qquad P = \sum_{\mathbf{a} \in E_n} P(\mathbf{a}) L_{\mathbf{a}}.$$

This formula follows by induction on n and the univariate interpolation formula

$$P(x_1, x_2, \ldots, x_n) = (1 - x_1)P(0, x_2, \ldots, x_n) + x_1 P(1, x_2, \ldots, x_n).$$

Specializing formula (5.7), we see that $P(x, x, \ldots, x)$ is a polynomial of at most degree n. The next lemma provides a converse to this fact.

LEMMA 5.1. *Given any polynomial p of degree at most n there exists a unique multiaffine symmetric function P such that*

$$P(x, x, \ldots, x) = p(x), \quad x \in \mathbb{R}.$$

Proof. When P is symmetric, formula (5.7) becomes

$$P(\mathbf{x}) = \sum_{k=0}^{n} P(\mathbf{v}^k)\ell_k(\mathbf{x}), \qquad \mathbf{x} \in \mathbb{R}^n,$$

where

$$\mathbf{v}^k = (\overbrace{1, 1, \ldots, 1}^{k}, 0, \ldots, 0)^T, \ k = 0, 1, \ldots, n,$$

and

$$\ell_k(\mathbf{x}) := \sum_{\#\{i: a_i = 1\} = k} L_{\mathbf{a}}(\mathbf{x}), \qquad k = 0, 1, \ldots, n,$$

a symmetric multiaffine function. A little thought yields the fact that $\#\{\mathbf{a} : \mathbf{a} \in E_n, \#\{i : a_i = 1\} = k\} = \binom{n}{k}$, from which it follows that

$$(5.8) \qquad\qquad \ell_k(x, x, \ldots, x) = b_k^n(x), \qquad k = 0, 1, \ldots, n.$$

Hence, if P is a symmetric multiaffine function such that $P(x, x, \ldots, x) = 0$, $x \in \mathbb{R}$, then P is identically zero. This proves that given any polynomial p of degree at most n, written in Bernstein–Bézier form

$$p = \sum_{k=0}^{n} c_k b_k^n,$$

the unique symmetric multiaffine function such that $P(x, x, \ldots, x) = p(x)$ is given by

$$P(\mathbf{x}) = \sum_{k=0}^{n} c_k \ell_k(\mathbf{x}), \qquad \mathbf{x} \in \mathbb{R}^n. \qquad \square$$

We say that the symmetric multiaffine function P described in Lemma 5.1 is the *blossom* of the polynomial p.

There is a corollary to Lemma 5.1 that should be emphasized here. Namely, given a polynomial curve $\mathbf{p}$ of degree n in Bernstein–Bézier form, viz. $\mathbf{p} = C\mathbf{b}$, its control points are given by

$$\mathbf{c}_j = \mathbf{P}(\underbrace{1, 1, \ldots, 1}_{j}, 0, 0, \ldots, 0), \ j = 0, 1, \ldots, n,$$

because it was proved above that

$$(5.9) \qquad \mathbf{p}(x) = \mathbf{P}(x, x, \ldots, x) = \sum_{j=0}^{n} \mathbf{P}(\underbrace{1, 1, \ldots, 1}_{j}, 0, 0, \ldots, 0) b_j^n(x).$$

Next we describe a recursive triangular scheme to blossom a polynomial curve given in Bernstein–Bézier form. To this end, we go back to the recursive triangle

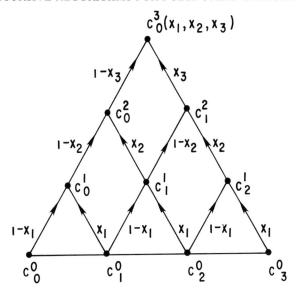

FIG. 5.4. *Blossoming de Casteljau's triangle for a cubic curve.*

in Fig. 5.2 and relabel all the labels at each level of the recurrence as indicated in the recursive triangle above.

The claim is that the apex of the triangle in Fig. 5.4 is the blossom of the Bernstein–Bézier cubic obtained as the vertex of Fig. 5.2. Generally, corresponding to (5.1) we consider the companion recurrence formula

$$(5.10) \quad d_r^\ell = (1 - x_\ell)d_r^{\ell-1} + x_\ell d_{r+1}^{\ell-1}, \ r = 0, 1, \ldots, n - \ell, \ \ell = 1, \ldots, n.$$

We shall show that d_0^n is, indeed, the blossom of c_0^n. For this purpose, we need only verify that d_0^n is a multiaffine symmetric function of $x_1, \ldots, x_n$, since it obviously agrees with c_0^n when $x_1 = \cdots = x_n = x$. Obviously, each d_r^k is an affine function of x_k and so the core of the matter is to show that they are symmetric functions. For $\ell = 1$ there is nothing to prove. Now, assume the result is true for all $k < \ell$ with $\ell \geq 2$ and consider the function $d_r^\ell(x_1, \ldots, x_\ell)$.

From the recurrence formula (5.10) and the induction hypothesis it is clear that d_r^ℓ is a symmetric function of $x_1, \ldots, x_{\ell-1}$. Hence it remains to show that, for each $j = 1, \ldots, \ell - 1$, it is a symmetric function of x_j and x_ℓ. As a function of these two variables it is a bilinear function and therefore its symmetry is equivalent to showing that its value at $(x_j, x_\ell) = (0, 1)$ and at $(x_j, x_\ell) = (1, 0)$ are the same. That is, the proof hinges on showing that

$$(5.11) \quad \begin{aligned} &d_r^\ell(x_1, \ldots, x_{j-1}, 0, x_{j+1}, \ldots, x_{\ell-1}, 1) \\ &= d_r^\ell(x_1, \ldots, x_{j-1}, 1, x_{j+1}, \ldots, x_{\ell-1}, 0). \end{aligned}$$

We evaluate the left-hand side of (5.11) by specializing (5.10) to $x_\ell = 1$ and obtain

(5.12)
$$\begin{aligned}
\mathbf{d}_r^\ell(x_1, &\ldots, x_{j-1}, 0, x_{j+1}, \ldots, x_{\ell-1}, 1) \\
&= \mathbf{d}_{r+1}^{\ell-1}(x_1, \ldots, x_{j-1}, 0, x_{j+1}, \ldots, x_{\ell-1}) \\
&= \mathbf{d}_{r+1}^{\ell-1}(x_1, \ldots, x_{j-1}, x_{j+1}, \ldots, x_{\ell-1}, 0) \\
&= \mathbf{d}_{r+1}^{\ell-2}(x_1, \ldots, x_{j-1}, x_{j+1}, \ldots, x_{\ell-1}).
\end{aligned}$$

In the last two equations, we first used the symmetry guaranteed by the induction hypothesis and then the recurrence formula (5.10) at level $\ell - 1$. Computing the right-hand side of (5.11) in the same manner gives

(5.13)
$$\begin{aligned}
\mathbf{d}_r^\ell(x_1, &\ldots, x_{j-1}, 1, x_{j+1}, \ldots, x_{\ell-1}, 0) \\
&= \mathbf{d}_r^{\ell-1}(x_1, \ldots, x_{j-1}, 1, x_{j+1}, \ldots, x_{\ell-1}) \\
&= \mathbf{d}_r^{\ell-1}(x_1, \ldots, x_{j-1}, x_{j+1}, \ldots x_{\ell-1}, 1) \\
&= \mathbf{d}_{r+1}^{\ell-2}(x_1, \ldots, x_{j-1}, x_{j+1}, \ldots, x_{\ell-1}).
\end{aligned}$$

Comparing (5.12) and (5.13) proves (5.11) and therefore our claim that $\mathbf{d}_0^n(x_1, \ldots, x_n)$ blossoms $\mathbf{c}_0^n(x)$.

Next we explore the role of duality in the present setup. Associated with every sequence of $n + 1$ *linearly independent* polynomials $p_j, j = 0, 1, \ldots, n$, of degree at most n are unique polynomials $q_j, j = 0, 1, \ldots, n$, also linearly independent, and at most of degree n such that

(5.14)
$$(x - t)^n = \sum_{j=0}^{n} q_j(t) p_j(x), \qquad x, t \in \mathbb{R}.$$

We will call $\mathcal{Q} = \{q_j : 0 \le j \le n\}$ the *dual basis* associated with $\mathcal{P} = \{p_j : 0 \le j \le n\}$ and $(\mathcal{P}, \mathcal{Q})$ a duality pair. The philosophy behind dual pairs is that sometimes it is easier to study properties of a dual basis $\mathcal{Q}$ through the binomial identity (5.14) and then obtain properties of $\mathcal{P}$. For the Bernstein–Bézier polynomials, the dual basis is given by

$$q_j^b(t) = (-1)^{n-j} \binom{n}{j}^{-1} b_{n-j}^n(t), \qquad j = 0, 1, \ldots, n.$$

This follows from the binomial theorem and the formula $x - t = (1 - x)(-t) + x(1 - t)$.

A nice application of dual pairs is an alternate proof of (5.9). The blossom of the polynomial $p_t(x) := (x - t)^n$ is $P_t(x_1, \ldots, x_n) = (x_1 - t) \cdots (x_n - t)$ and, consequently, $P_t(1, \ldots, 1, 0, \ldots, 0) = q_j^b(t)$, where one occurs j times. Therefore, we may rewrite (5.14) in the equivalent form

(5.15)
$$p_t(x) = \sum_{j=0}^{n} P_t(\underbrace{1, \ldots, 1}_{j}, 0, \ldots, 0) b_j(x).$$

For any $n+1$ distinct points $t_0, \ldots, t_n$ the polynomials $p_{t_0}, \ldots, p_{t_n}$ are linearly independent. One way to see this is to notice that their dual bases are the linearly independent Lagrange polynomials

$$\ell_j(x) = \frac{w(t)}{w'(t_j)(t - t_j)}, \qquad j = 0, 1, \ldots, n,$$

$$w(t) = \prod_{j=0}^{n} (t - t_j),$$

since

(5.16) $$(x - t)^n = \sum_{j=0}^{n} (x - t_j)^n \ell_j(t).$$

Now represent any polynomial p as

$$p = \sum_{j=0}^{n} a_j p_{t_j},$$

for some constants $a_0, a_1, \ldots, a_n$, and obtain its blossom as

$$P = \sum_{j=0}^{n} a_j P_{t_j}.$$

Hence we get from (5.15) that

(5.17) $$p = \sum_{j=0}^{n} P(\underbrace{1, \ldots, 1}_{j}, 0, \ldots 0) b_j,$$

which is equation (5.9).

5.3.   Blossoming B-spline series: The univariate case.

Let us apply the above ideas to the B-spline basis. Throughout our discussion we concentrate on the *dual basis* for B-splines and only at the end will the identification with B-splines be made. This is the approach that we shall follow later in the multivariate case.

We begin with any set of $2n$ scalar values $t_{-n+1}, \ldots, t_n$ such that

(5.18) $$\{t_{-n+j}, \ldots, t_0\} \cap \{t_j\} = \emptyset, \qquad j = 1, \ldots, n.$$

These points are neither arranged in any particular order nor do they need to be distinct.

Encouraged by the identity (3.77) we consider the polynomials of degree n

(5.19) $\quad N_j(t) = N_j^n(t) := (t_{-n+j+1} - t) \cdots (t_j - t), \qquad j = 0, 1, \ldots, n.$

First let us verify that these polynomials are linearly independent. To this end, suppose there are constants $a_0, a_1, \ldots, a_n$ such that

$$\sum_{j=0}^{n} a_j N_j(t) = 0, \qquad t \in \mathbb{R}.$$

Since $N_0(t_1) \neq 0$, $N_1(t_1) = \cdots = N_n(t_1) = 0$, we conclude that $a_0 = 0$.

We delete t_{-n+1} and t_1 from our set $\{t_{-n+1}, \ldots, t_n\}$ introduce the subset of points defined by

$$\hat{t}_j := \begin{cases} t_j, & j = -n+2, \ldots, 0, \\ t_{j+1}, & j = 1, \ldots, n-1, \end{cases}$$

and the functions $\hat{N}_j(t) = (\hat{t}_{-n+j+2} - t) \cdots (\hat{t}_j - t)$, $j = 0, 1, \ldots, n-1$. Then one can easily check that $N_j(t) = (t_1 - t)\hat{N}_{j-1}(t)$, $j = 1, \ldots, n$, and so

$$\sum_{j=0}^{n-1} a_{j+1} \hat{N}_j(t) = 0, \qquad t \in \mathbb{R}.$$

Since the new points $\hat{t}_{-n+2}, \ldots, \hat{t}_{n-1}$ satisfy (5.18) with n replaced by $n-1$ we can now proceed, by induction on n, to conclude that $a_0 = \cdots = a_n = 0$ and consequently, $N_0, N_1, \ldots, N_n$ are linearly independent when (5.18) is satisfied.

With this result in hand we consider the dual basis for $N_0, \ldots, N_n$, which we denote by $M_0, \ldots, M_n$. Our goal is to develop upward and downward recurrence formulas for the dual basis, develop blossoming relations, and finally identify the dual basis with B-splines.

We begin by observing that the common zeros of the polynomials N_i^{n+1} and N_{i+1}^{n+1} are exactly the zeros of N_i^n, $i = 0, 1, \ldots, n$. For this reason, we get the following formula:

(5.20)
$$(x - t)N_i^n(t) = \frac{x - t_{i+1}}{t_{-n+i} - t_{i+1}} N_i^{n+1}(t)$$
$$+ \frac{x - t_{-n+i}}{t_{i+1} - t_{-n+i}} N_{i+1}^{n+1}(t), \qquad i = 0, 1, \ldots, n.$$

Let us use this formula to find the recurrence formula for the dual basis. We compute as follows:

$$\sum_{i=0}^{n+1} M_i^{n+1}(x) N_i^{n+1}(t) = (x - t)(x - t)^n$$
$$= \sum_{i=0}^{n} M_i^n(x)(x - t) N_i^n(t)$$

$$= \sum_{i=0}^{n} M_i^n(x) \left\{ \frac{x - t_{i+1}}{t_{-n+i} - t_{i+1}} N_i^{n+1}(t) + \frac{x - t_{-n+i}}{t_{i+1} - t_{-n+i}} N_{i+1}^{n+1}(t) \right\}$$

$$= \sum_{i=0}^{n+1} \left\{ \frac{x - t_{i+1}}{t_{-n+i} - t_{i+1}} M_i^n(x) + \frac{x - t_{-n+i-1}}{t_i - t_{-n+i-1}} M_{i-1}^n(x) \right\} N_i^{n+1}(t),$$

where in the last equation we have set $M_{-1}^n(x) = M_{n+1}^n(x) = 0$. Comparing both sides of the first and last sum above we obtain the recurrence formula

$$\begin{aligned}
(5.21) \qquad M_i^{n+1}(x) \;=\; & \frac{x - t_{i+1}}{t_{-n+i} - t_{i+1}} M_i^n(x) \\[2mm]
& + \frac{x - t_{-n+i-1}}{t_i - t_{-n+i-1}} M_{i-1}^n(x), \qquad i = 0, 1, \ldots, n+1,
\end{aligned}$$

initialized with $M_0^0(x) = 0$.

For the upward recurrence formula we conserve sums, as with the Bernstein–Bézier polynomials (see (5.6)) and compute as follows:

$$\sum_{i=0}^{n-\ell+1} \mathbf{c}_i^{\ell-1} M_i^{n-\ell+1}(x)$$

$$= \sum_{i=0}^{n-\ell+1} \frac{x - t_{i+1}}{t_{-n+\ell+i} - t_{i+1}} \mathbf{c}_i^{\ell-1} M_i^{n-\ell}(x) + \sum_{i=-1}^{n-\ell} \frac{x - t_{-n+\ell+i}}{t_{i+1} - t_{-n+\ell+i}} \mathbf{c}_{i+1}^{\ell-1} M_i^{n-\ell}(x)$$

$$= \sum_{i=0}^{n-\ell} \left\{ \frac{x - t_{i+1}}{t_{-n+\ell+i} - t_{i+1}} \mathbf{c}_i^{\ell-1} + \frac{x - t_{-n+\ell+i}}{t_{i+1} - t_{-n+\ell+i}} \mathbf{c}_{i+1}^{\ell-1} \right\} M_i^{n-\ell}(x).$$

Thus the upward recurrence formula is

$$(5.22) \qquad \mathbf{c}_i^\ell = \frac{x - t_{i+1}}{t_{-n+\ell+i} - t_{i+1}} \mathbf{c}_i^{\ell-1} + \frac{x - t_{-n+\ell+i}}{t_{i+1} - t_{-n+\ell+i}} \mathbf{c}_{i+1}^{\ell-1},$$

where $i = 0, 1, \ldots, n - \ell$, $\ell = 1, \ldots, n$. Moreover, since

$$\sum_{i=0}^{n-\ell+1} \mathbf{c}_i^{\ell-1} M_i^{n-\ell+1}(x) = \sum_{i=0}^{n-\ell} \mathbf{c}_i^\ell M_i^{n-\ell}(x), \quad \ell = 1, 2, \ldots, n,$$

the apex of (5.22) is

$$(5.23) \qquad \mathbf{c}_0^n = \sum_{j=0}^{n} \mathbf{c}_j^0 M_j^n(x).$$

To express the control points of the polynomial on the right-hand side of (5.23) in terms of its blossom we note, using the same notation as employed in formula (5.15), that

$$P_t(t_{-n+i+1}, \ldots, t_i) = N_i(t), \qquad i = 0, 1, \ldots, n.$$

Hence, as in the proof of (5.9) for the Bernstein–Bézier representation, it follows that for any polynomial p with blossom P we have

$$(5.24) \qquad\qquad p = \sum_{i=0}^{n} P(t_{-n+i+1}, \ldots, t_i) M_i^n.$$

Therefore, if $\mathbf{d}_0^n(\mathbf{x})$ is the blossom of $c_0^n(x)$ in (5.23), we have

$$(5.25) \qquad\qquad \mathbf{c}_i^0 = \mathbf{d}_0^n(t_{-n+i+1}, \ldots, t_i), \qquad i = 0, 1, \ldots, n.$$

Moreover, the fact that $\mathbf{d}_0^n(\mathbf{x})$ can be computed by the recurrence formula

$$\mathbf{d}_r^{\ell} = \frac{x_{\ell} - t_{r+1}}{t_{-n+\ell+r} - t_{r+1}} \mathbf{d}_r^{\ell-1} + \frac{x_{\ell} - t_{-n+\ell+r}}{t_{r+1} - t_{-n+\ell+r}} \mathbf{d}_{r+1}^{\ell-1},$$

where $r = 0, 1, \ldots, n - \ell$, $\ell = 1, \ldots, n$ can be proved in the same way that we proved (5.10) blossoms (5.1).

Our final remark about the duality pair $(\mathcal{N}, \mathcal{M})$ identifies $M_0, \ldots, M_n$ as a set of B-splines. From now on we assume the points in the set $\{t_{-n+1}, \ldots, t_n\}$ are *ordered*,

$$t_{-n+1} \leq \cdots \leq t_0 < t_1 \leq \cdots \leq t_n.$$

and we *add two more points* to it, one to the left of t_{-n+1} and one to the right of t_n, viz.

$$t_{-n} \leq \cdots \leq t_0 < t_1 \leq \cdots \leq t_{n+1}.$$

Corresponding to this larger set there are $n + 1$ B-splines of degree n. In the notation of Chapters 3 and 4 they are $S(x|t_{-n+i}, \ldots, t_{i+1})$, $i = 0, 1, \ldots, n$, and each is nonzero on the interval $[t_0, t_1]$.

Next we express each $x \in \mathbb{R}$ in the form

$$x = \frac{x - t_{i+1}}{t_{-n+i} - t_{i+1}} t_{-n+i} + \frac{t_{-n+i} - x}{t_{-n+i} - t_{i+1}} t_{i+1},$$

and specialize Theorem 4.4 to the B-spline $S(x|t_{-n+i}, \ldots, t_{i+1})$ to obtain the recurrence formula

$$nS(x|t_{-n+i}, \ldots, t_{i+1}) = \frac{x - t_{i+1}}{t_{-n+i} - t_{i+1}} S(x|t_{-n+i+1}, \ldots, t_{i+1})$$

$$+ \frac{t_{-n+i} - x}{t_{-n+i} - t_{i+1}} S(x|t_{-n+i}, \ldots, t_i).$$

Let $R_i^n(x) := n!(t_{i+1} - t_{-n+i})S(x|t_{-n+i}, \ldots, t_{i+1})$, $i = 0, 1, \ldots, n$, and rewrite the above identity for $n \geq 1$ in the form

$$R_i^n(x) = \frac{x - t_{i+1}}{t_{-n+i+1} - t_{i+1}} R_i^{n-1}(x) + \frac{x - t_{-n+i}}{t_i - t_{-n+i}} R_{i-1}^{n-1}(x), \qquad i = 0, 1, \ldots, n.$$

Note also that $R_0^0(x) = 1$, and $R_{-1}^0(x) = R_1^0(x) = 0$ for $x \in (t_0, t_1)$. Thus, it follows by induction on n that

$$M_i^n(x) = n!(t_{i+1} - t_{-n+i})S(x|t_{-n+i}, \ldots, t_{i+1}), \qquad i = 0, 1, \ldots, n,$$

for x in the interval (t_0, t_1).

Consequently, we have established that $(\mathcal{N}, \mathcal{R})$ form a duality pair on (t_0, t_1). This observation is nothing more than the identity of Marsden, (3.77), restricted to the interval (t_0, t_1). So what have we accomplished? At the least we have again confirmed the point of view, also central to Chapter 3, that one way to understand B-splines is through duality. Chapter 4, on the other hand, approaches B-splines through geometric principles. This gave us a technique to develop a multivariate theory of B-splines. Can these approaches be unified for multivariate functions? To answer this question we need to identify the appropriate multivariate recurrence formulas and dual polynomials. The principles of polarization will be our guide and Theorem 4.4 will allow us to make the connection with multivariate B-splines. This is our agenda for the remainder of the chapter. Let us now use the univariate material presented so far as a foundation and turn our attention to recurrence formulas for multivariate polynomials.

5.4. Multivariate blossoming.

It is convenient to work with homogeneous polynomials. All results have inhomogeneous versions, most of which follow in a straightforward manner from their homogeneous counterparts. The space of real homogeneous polynomials of degree n in s variables will be denoted by $H_n(\mathbb{R}^s)$. We make use of the set $\Gamma = \Gamma_{s,n} \subset \mathbb{Z}_+^s$ given by

$$\Gamma_{s,n} = \left\{ \mathbf{k} \in \mathbb{Z}_+^s : |\mathbf{k}|_1 = n \right\},$$

where $\mathbf{k} = (k_1, \ldots, k_s)^T$ and $|\mathbf{k}|_1 = k_1 + \cdots + k_s$, to index the monomial basis of $H_n(\mathbb{R}^s)$. Thus every $p \in H_n(\mathbb{R}^s)$ can be written in the form

$$(5.26) \qquad p(\mathbf{x}) = \sum_{\mathbf{k} \in \Gamma_{s,n}} p_{\mathbf{k}} \mathbf{x}^{\mathbf{k}}$$

for some scalars $p_{\mathbf{k}}, \mathbf{k} \in \Gamma_{s,n}$, and

$$(5.27) \qquad \dim H_n(\mathbb{R}^s) = \#\Gamma_{s,n} = \binom{n+s-1}{n}.$$

The notion of *lineal polynomials* is basic to our study of multivariate polynomial recurrence formulas. We define this concept next. For every set of n vectors

$$(5.28) \qquad X = \{ \mathbf{x}_\ell : \ell = 1, \ldots, n \} \subset \mathbb{R}^s,$$

we define the lineal polynomial determined by X as

$$(5.29) \qquad a(\mathbf{x}) := a(\mathbf{x}|X) = \prod_{\ell=1}^{n} (\mathbf{x}, \mathbf{x}_\ell), \qquad \mathbf{x} \in \mathbb{R}^s,$$

where $(\mathbf{x}, \mathbf{y})$ denotes the standard inner product of $\mathbf{x}$ and $\mathbf{y}$ in $\mathbb{R}^s$. Clearly $a(\cdot|X) \in H_n(\mathbb{R}^s)$ and a fundamental question is to decide when a set of lineal polynomials spans $H_n(\mathbb{R}^s)$. To answer this question, we need to first extend our previous discussion of blossoming of univariate polynomials to the multivariate case.

Let $\mathbf{e}_1, \ldots, \mathbf{e}_s$ be the standard basis for $\mathbb{R}^s$, that is, $(\mathbf{e}_i)_j = \delta_{ij}$, $i, j = 1, \ldots, s$. Then every linear function Q on $\mathbb{R}^s$ can be written as

$$(5.30) \qquad Q(x_1, \ldots, x_s) = \sum_{i=1}^{s} Q(\mathbf{e}_i) x_i, \quad (x_1, \ldots, x_s)^T \in \mathbb{R}^s.$$

We need some additional notation to extend the univariate interpolation formula (5.7) to multilinear functions in n vector variables $\mathbf{y}_1, \ldots, \mathbf{y}_n$ in $\mathbb{R}^s$. We use I for the set $\{\mathbf{i} = (i_1, \ldots, i_n)^T : 1 \leq i_\ell \leq s, \ell = 1, \ldots, n\}$ and also set $\mathbf{y} = (\mathbf{y}_1, \ldots, \mathbf{y}_n)^T$ for the column vector in $\mathbb{R}^{ns}$ whose components are successively built from the components of $\mathbf{y}_1, \ldots, \mathbf{y}_n$ by setting $(\mathbf{y})_{(k-1)s+i} = (\mathbf{y}_k)_i$, $k = 1, \ldots, n, i = 1, \ldots, s$. A basic multilinear function is defined by the formula

$$(5.31) \qquad F_{\mathbf{i}}(\mathbf{y}) = \prod_{r=1}^{n} (\mathbf{y}_r)_{i_r}$$

where $\mathbf{y}_r = ((\mathbf{y}_r)_1, \ldots, (\mathbf{y}_r)_s)^T$, $r = 1, \ldots, n$, and $\mathbf{i} = (i_1, \ldots, i_n)^T$ is a typical vector in I. Also, it is nice to have available the vector

$$\mathbf{e}_{\mathbf{i}} = (\mathbf{e}_{i_1}, \ldots, \mathbf{e}_{i_s})^T$$

in $\mathbb{R}^{ns}$, associated with each such $\mathbf{i}$. Then, every multilinear function has the form

$$(5.32) \qquad P = \sum_{\mathbf{i} \in I} P(\mathbf{e}_{\mathbf{i}}) F_{\mathbf{i}}.$$

This can be proved, inductively on n, using (5.30) as a starting point.

A simplification occurs in (5.32) when P is symmetric. To this end, we let $\mathcal{P}_{s,n}$ denote the space of symmetric multilinear functions in n vector variables in $\mathbb{R}^s$. For every $\mathbf{k} \in \Gamma_{s,n}$ we define a vector in $\mathbb{R}^{ns}$ by setting

$$(5.33) \qquad \mathbf{v}_{\mathbf{k}} = (\underbrace{\mathbf{e}_1, \ldots, \mathbf{e}_1}_{k_1}, \ldots, \underbrace{\mathbf{e}_s, \ldots, \mathbf{e}_s}_{k_s})^T,$$

where $\mathbf{k} = (k_1, \ldots, k_s)^T$. For every $\mathbf{i} \in I$, we let $d_j(\mathbf{i})$ be the number of times that an integer $j, 1 \leq j \leq s$, appears as a component of $\mathbf{i}$, that is,

$$(5.34) \qquad d_j(\mathbf{i}) = \# \{\ell : 1 \leq \ell \leq n \text{ such that } i_\ell = j\}.$$

Then $\mathbf{d}(\mathbf{i}) = (d_1(\mathbf{i}), \ldots, d_s(\mathbf{i}))^T \in \Gamma_{s,n}$ and for every $P \in \mathcal{P}_{s,n}$

$$P(\mathbf{e}_{\mathbf{i}}) = P(\mathbf{v}_{\mathbf{d}(\mathbf{i})}).$$

We also remark that the mapping $\Gamma_{s,n} : \mathbf{k} \to \mathbf{r}(\mathbf{k})$, given by

$$\mathbf{r}(\mathbf{k}) = (\underbrace{1, \ldots, 1}_{k_1}, \ldots, \underbrace{s, \ldots, s}_{k_s})^T \in \Gamma_{s,n}$$

is a right inverse of $\mathbf{d}$, that is, $\mathbf{d}(\mathbf{r}(\mathbf{k})) = \mathbf{k}$ while $\mathbf{r}(\mathbf{d}(\mathbf{i}))$ is the permutation that puts the components of $\mathbf{i}$ in nondecreasing order. Hence, for $P \in \mathcal{P}_{s,n}$ (5.32) becomes

$$(5.35) \qquad\qquad P = \sum_{\mathbf{k} \in \Gamma_{s,n}} P(\mathbf{v_k}) L_{\mathbf{k}},$$

where $L_{\mathbf{k}}$ is the multilinear function defined by the equation

$$(5.36) \qquad\qquad L_{\mathbf{k}} = \sum_{\mathbf{d}(\mathbf{i}) = \mathbf{k}} F_{\mathbf{i}}.$$

Moreover, since $F_{\mathbf{i}}(\mathbf{e_j}) = \delta_{\mathbf{ij}}$ for all $\mathbf{i}, \mathbf{j} \in I$, it likewise follows that $L_{\mathbf{k}}(\mathbf{v_j}) = \delta_{\mathbf{kj}}$ for all $\mathbf{k}, \mathbf{j} \in \Gamma_{s,n}$. Since, for every $\mathbf{i} \in I$,

$$F_{\mathbf{i}}(\mathbf{x}, \ldots, \mathbf{x}) = \mathbf{x}^{\mathbf{d}(\mathbf{i})},$$

appropriately specializing formula (5.36) gives us the fact that

$$L_{\mathbf{k}}(\mathbf{x}, \ldots, \mathbf{x}) = \#\{\mathbf{i} : \mathbf{i} \in I, \mathbf{d}(\mathbf{i}) = \mathbf{k}\} \mathbf{x}^{\mathbf{k}}.$$

Moreover, a little thought shows that

$$\#\{\mathbf{i} : \mathbf{i} \in I, \ \mathbf{d}(\mathbf{i}) = \mathbf{k}\} = \binom{n}{\mathbf{k}},$$

and so

$$(5.37) \qquad\qquad L_{\mathbf{k}}(\mathbf{x}, \ldots, \mathbf{x}) = \binom{n}{\mathbf{k}} \mathbf{x}^{\mathbf{k}}.$$

(These monomials on homogeneous space play the role of Bernstein–Bézier polynomials on affine space.) Hence we obtain from (5.35) the formula

$$P(\mathbf{x}, \ldots, \mathbf{x}) = \sum_{\mathbf{k} \in \Gamma_{s,n}} P(\mathbf{v_k}) \binom{n}{\mathbf{k}} \mathbf{x}^{\mathbf{k}},$$

from which we conclude that whenever $P \in \mathcal{P}_{s,n}$ and $P(\mathbf{x}, \mathbf{x}, \ldots, \mathbf{x}) = 0$ then $P = 0$. This leads us to the following extension of Lemma 5.1.

LEMMA 5.2. *Given any $p \in H_n(\mathbb{R}^s)$, there exists a unique $P \in \mathcal{P}_{s,n}$ such that*

$$(5.38) \qquad\qquad P(\mathbf{x}, \mathbf{x}, \ldots, \mathbf{x}) = p(\mathbf{x}), \qquad \mathbf{x} \in \mathbb{R}^s,$$

and, moreover, $\dim \mathcal{P}_{n,s} = \dim H_n(\mathbb{R}^s)$.

Proof. There is not much more to say except that given a typical element p of $H_n(\mathbb{R}^s)$ written in the form

$$p(\mathbf{x}) = \sum_{\mathbf{k} \in \Gamma_{s,n}} \binom{n}{\mathbf{k}} c_\mathbf{k} \mathbf{x}^\mathbf{k},$$

the function

$$P := \sum_{\mathbf{k} \in \Gamma_{s,n}} c_\mathbf{k} L_\mathbf{k}$$

is the unique $P \in \mathcal{P}_{s,n}$, which satisfies (5.38).   □

As an application of these ideas, we present a proof of the formula in Lemma 1.5 of Chapter 1. We restate it here, since we will use it often in the remainder of this chapter.

LEMMA 5.3. *Let C be an $n \times s$ matrix. Then*

$$(5.39) \qquad (C\mathbf{x})_1 \cdots (C\mathbf{x})_n = \sum_{\mathbf{k} \in \Gamma_{s,n}} \frac{\mathbf{x}^\mathbf{k}}{\mathbf{k}!} \operatorname{per} C(\mathbf{k}), \qquad \mathbf{x} \in \mathbb{R}^s.$$

Proof. We let $\mathbf{c}_1, \ldots, \mathbf{c}_n$ be the rows of the matrix C. These are vectors in $\mathbb{R}^s$. Call the function on the left-hand side of (5.39) $R(\mathbf{c}_1, \ldots, \mathbf{c}_n)$. Of course, it depends on $\mathbf{x}$, but as a function of $\mathbf{c}_1, \ldots, \mathbf{c}_n$ it is in $\mathcal{P}_{s,n}$. Referring back to definition (5.33), we see that for every $\mathbf{k} \in \Gamma_{s,n}$

$$R(\mathbf{v_k}) = \mathbf{x}^\mathbf{k},$$

and hence comparing (5.39) to (5.35) shows that (5.39) is equivalent to

$$(5.40) \qquad \operatorname{per} C(\mathbf{k}) = \mathbf{k}! L_\mathbf{k}(\mathbf{c}_1, \ldots, \mathbf{c}_n).$$

Note that both sides of (5.40) are multilinear symmetric functions on $\mathbb{R}^s$. Therefore, by Lemma 5.2, to verify this formula it suffices to check it for $n \times s$ matrices C, all of whose rows are equal to some vector $\mathbf{c}$. Suppose C has this special form. Then, by the definition of a permanent it follows that $\operatorname{per} C(\mathbf{k}) = n! \mathbf{c}^\mathbf{k}$. Thus, using equation (5.37), we see that (5.40) indeed holds when $\mathbf{c}_1 = \ldots = \mathbf{c}_n = \mathbf{c}$.   □

5.5.   Linear independence of lineal polynomials.

We are now ready to answer the question we raised earlier about when a family of lineal polynomials span $H_n(\mathbb{R}^s)$.

PROPOSITION 5.1. *For every $\mathbf{k} \in \Gamma_{s,n}$ let $V_\mathbf{k} = \{\mathbf{v}_{\mathbf{k},\ell}, \ \ell = 1, \ldots, n\}$ be a set of vectors in $\mathbb{R}^s$ and define*

$$\mathbf{u}_\mathbf{k} = \left(\mathbf{v}_{\mathbf{k},1}, \ldots, \mathbf{v}_{\mathbf{k},n} \right)^T.$$

Then the set of lineal polynomials

$$\{a(\cdot|V_{\mathbf{k}}) : \mathbf{k} \in \Gamma_{s,n}\}$$

spans $H_n(\mathbb{R}^s)$ if and only if whenever $P \in \mathcal{P}_{s,n}$ and $P(\mathbf{u_k}) = 0$ for all $\mathbf{k} \in \Gamma_{s,n}$ then $P = 0$.

Proof. Let $V^{\mathbf{k}}$ be the $n \times s$ *matrix whose rows are* $\mathbf{v}_{\mathbf{k},1}, \ldots, \mathbf{v}_{\mathbf{k},n}$. Then from (5.39) we have

$$a(\mathbf{x}|V_{\mathbf{k}}) = \sum_{\mathbf{j} \in \Gamma_{s,n}} \frac{\mathbf{x^j}}{\mathbf{j}!} \operatorname{per} V^{\mathbf{k}}(\mathbf{j}).$$

Therefore, the linear polynomials $a(\cdot|V_{\mathbf{k}}), k \in \Gamma_{s,n}$ are linearly dependent if and only if

$$\det \left(\operatorname{per} V^{\mathbf{k}}(\mathbf{j})\right)_{\mathbf{k}, \mathbf{j} \in \Gamma_{s,n}} = 0.$$

Using formula (5.40), we see that this is the case if and only if there are constants $f_{\mathbf{k}}, \mathbf{k} \in \Gamma_{s,n}$, not all zero, such that the multilinear symmetric function

(5.41) $$Q := \sum_{\mathbf{k} \in \Gamma_{s,n}} f_{\mathbf{k}} L_{\mathbf{k}}$$

vanishes at $\mathbf{u_k}$. Since we have already established that span $\{L_{\mathbf{k}} : \mathbf{k} \in \Gamma_{s,n}\} = H_n(\mathbb{R}^s)$ we see that there is a nontrivial $P \in \mathcal{P}_{s,n}$ with $P(\mathbf{u_k}) = 0$, all $\mathbf{k} \in \Gamma_{s,n}$ if and only if there are constants $f_{\mathbf{j}}, \mathbf{j} \in \Gamma_{s,n}$, not all zero, for which Q in (5.41) vanishes on all $\mathbf{u_k}, \mathbf{k} \in \Gamma_{s,n}$. ☐

A simple example of the above result is the case that each set $V_{\mathbf{k}}$ consists of one vector $\mathbf{v_k} \in \mathbb{R}^s$ repeated n times, that is,

$$V_{\mathbf{k}} = \{\mathbf{v_k}, \ldots, \mathbf{v_k}\}.$$

In this case

$$a(\mathbf{x}|V_{\mathbf{k}}) = (\mathbf{x}, \mathbf{v_k})^n, \qquad \mathbf{k} \in \Gamma_{s,n},$$

and these lineal polynomials span $H_n(\mathbb{R}^s)$ if and only if there is no $p \in H_n(\mathbb{R}^s)\backslash\{0\}$ that vanishes on all the points $\mathbf{v_k}, \mathbf{k} \in \Gamma_{s,n}$.

To obtain a more substantial example we begin with a rectangular array $\mathcal{X}$ of vectors in $\mathbb{R}^s$,

$$\mathcal{X} = \{\mathbf{x}^{i,j}; \ 1 \le i \le s, \ 1 \le j \le n\},$$

(see Fig. 5.5).

For each $\mathbf{k} \in \Gamma_{s,n}$ we choose a subset $X_{\mathbf{k}}$ of $\mathcal{X}$ consisting of n elements by setting

(5.42) $$X_{\mathbf{k}} = \{\mathbf{x}^{i,j} : \ 1 \le j \le k_i, \ i = 1, \ldots, s\}.$$

It is to be understood that if $k_i = 0$ for some $i = 1, \ldots, s$ no vector of the form $\mathbf{x}^{i,j}, \ j = 1, \ldots, n$ appears in $X_{\mathbf{k}}$. We can visualize $X_{\mathbf{k}}$ as a *histogram* by associating every element in $\mathcal{X}$ with an integer vector (i, j) in the rectangle $[1, s] \times [1, n]$ (see Fig. 5.6).

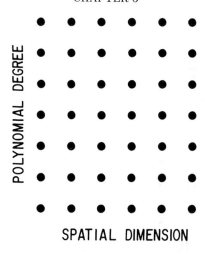

FIG. 5.5.   *Rectangular array of vectors.*

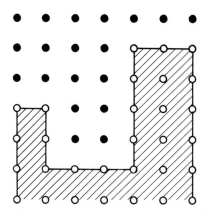

FIG. 5.6.   *Histogram of vectors.*

Thus, every $\mathbf{k} \in \Gamma_{s,n}$ corresponds to the histogram in $[1, s] \times [1, n]$ given by $\{(i, j) : 1 \leq j \leq k_i, \; i = 1, \ldots, s\}$ and $X_{\mathbf{k}}$ selects all vectors in this histogram.

THEOREM 5.1. *Let* $\mathcal{X} = \{\mathbf{x}^{i,j} : 1 \leq i \leq s, \; 1 \leq j \leq n\}$ *be an array of vectors in* $\mathbb{R}^s$ *with* $s \geq 2$ *and for every* $\mathbf{k} = (k_1, \ldots, k_s)^T \in \Gamma_{s,n}$ *set*

$$X_{\mathbf{k}} = \{\mathbf{x}^{i,j} : \; 1 \leq j \leq k_i, \; i = 1, \ldots, s\}.$$

Suppose that for every $\mathbf{j} = (j_1, \ldots, j_s)^T$ *with* $|\mathbf{j}|_1 \leq n - 1$ *the vectors* $\mathbf{x}^{1,j_1+1}, \ldots, \mathbf{x}^{s,j_s+1}$ *are linearly independent. Then the lineal polynomials* $a_{\mathbf{k}} := a(\cdot | X_{\mathbf{k}})$, $\mathbf{k} \in \Gamma_{s,n}$ *are linearly independent.*

Proof. The proof is by simultaneous induction on s and n. The easiest case to consider is $n = 1$ and any $s \geq 2$, since it is obvious then that the linear functions $(\mathbf{x}, \mathbf{x}^{1,1}), \ldots, (\mathbf{x}, \mathbf{x}^{s,1})$ are linearly independent on $\mathbb{R}^s$ if and only if $\mathbf{x}^{1,1}, \ldots, \mathbf{x}^{s,1}$ are linearly independent. The next case to consider is $s = 2$ and any $n \geq 2$. Here the lineal polynomials are given as

$$a_{\mathbf{k}}(\mathbf{x}) = \prod_{\ell=1}^{k_1} (\mathbf{x}, \mathbf{x}^{1,\ell}) \prod_{m=1}^{k_2} (\mathbf{x}, \mathbf{x}^{2,m}), \qquad \mathbf{k} = (k_1, k_2)^T \in \Gamma_{2,n}.$$

Suppose that there are scalars $c_{\mathbf{k}}, \mathbf{k} \in \Gamma_{2,n}$ such that

$$(5.43) \qquad \sum_{\mathbf{k} \in \Gamma_{2,n}} c_{\mathbf{k}} a_{\mathbf{k}}(\mathbf{x}) = 0, \qquad \mathbf{x} \in \mathbb{R}^2.$$

There is a $\mathbf{y} \in \mathbb{R}^2$ such that $(\mathbf{y}, \mathbf{x}^{1,1}) = 0$ and $(\mathbf{y}, \mathbf{x}^{2,n}) \neq 0$, since by hypothesis $\mathbf{x}^{1,1}$ and $\mathbf{x}^{2,n}$ are linearly independent. Therefore, $a_{\mathbf{k}}(\mathbf{y}) = 0$ for all $\mathbf{k} \in \Gamma_{2,n}$ except for $\mathbf{k}_0 := (0, n)^T$. Moreover, for that choice of $\mathbf{k}$, $a_{\mathbf{k}_0}(\mathbf{y}) \neq 0$ because $\mathbf{x}^{1,1}$ and $\mathbf{x}^{2,j}$ are linearly independent for $j = 1, 2, \ldots, n$. Thus, evaluating (5.43) for $\mathbf{x} = \mathbf{y}$ gives $c_{\mathbf{k}_0} = 0$.

We now remove the two vectors $\mathbf{x}^{1,1}, \mathbf{x}^{2,n}$ from our collection and introduce the new vectors

$$\mathbf{y}^{i,j} := \begin{cases} \mathbf{x}^{1,j+1}, & i = 1, \quad 1 \leq j \leq n-1, \\ \mathbf{x}^{2,j}, & i = 2, \quad 1 \leq j \leq n-1. \end{cases}$$

Every lineal polynomial $a_{\mathbf{k}}$ with $\mathbf{k} \neq \mathbf{k}_0$ contains $(\mathbf{x}, \mathbf{x}^{1,1})$ as a factor so that

$$a_{\mathbf{k}}(\mathbf{x}) = (\mathbf{x}, \mathbf{x}^{1,1}) a(\mathbf{x} | Y_{\mathbf{k}-\mathbf{e}_1}), \qquad \mathbf{k} \in \Gamma_{2,n} \setminus \{\mathbf{k}_0\},$$

where $\mathbf{e}_1 = (1, 0)^T$. Substituting this formula into (5.43) and cancelling the factor $(\mathbf{x}, \mathbf{x}^{1,1})$ gives the formula

$$\sum_{\mathbf{j} \in \Gamma_{2,n-1}} c_{\mathbf{j}+\mathbf{e}_1} a(\mathbf{x} | Y_{\mathbf{j}}) = 0, \qquad \mathbf{x} \in \mathbb{R}^2.$$

Thus, by the induction hypothesis applied to the array $\mathcal{Y} = \{\mathbf{y}^{i,j} : i = 1, 2, j = 1, \ldots, n-1\}$, we obtain $c_{\mathbf{j}+\mathbf{e}_1} = 0$ for all $|\mathbf{j}|_1 \leq n-2$ and so $c_{\mathbf{k}} = 0$ for all $\mathbf{k} \in \Gamma_{2,n-1}$. This proves the linear independence in the case that $s = 2$ and any $n \geq 2$.

We now deal with the final case when $s \geq 3$ and $n \geq 2$. Again, we suppose for some scalars $c_{\mathbf{k}}, \mathbf{k} \in \Gamma_{s,n}$,

$$(5.44) \qquad \sum_{\mathbf{k} \in \Gamma_{s,n}} c_{\mathbf{k}} a_{\mathbf{k}}(\mathbf{x}) = 0, \qquad \mathbf{x} \in \mathbb{R}^s.$$

We let $\Gamma^0_{s,n} = \{\mathbf{k} \in \Gamma_{s,n} :, \; k_1 = 0\}$, $\Gamma^+_{s,n} = \Gamma_{s,n} \backslash \Gamma^0_{s,n}$ and break up the sum in (5.44) as

$$(5.45) \qquad \sum_{\mathbf{k} \in \Gamma^0_{s,n}} c_{\mathbf{k}} a_{\mathbf{k}}(\mathbf{x}) + \sum_{\mathbf{k} \in \Gamma^+_{s,n}} c_{\mathbf{k}} a_{\mathbf{k}}(\mathbf{x}) = 0, \qquad \mathbf{x} \in \mathbb{R}^s.$$

Let T be any $s \times s$ nonsingular matrix. Our hypothesis on the array $\mathcal{X}$ remains valid for the array

$$\{T\mathbf{x}^{i,j} : 1 \le i \le s, \; 1 \le j \le n\},$$

where T is any nonsingular $s \times s$ matrix. Therefore, we may rotate $\mathbf{x}^{1,1}$ into $\mathbf{e}_1 = (1, 0, \dots, 0)^T$ and assume, without loss of generality, that in fact $\mathbf{x}^{1,1} = \mathbf{e}_1$. We let $P : \mathbb{R}^{s-1} \to \mathbb{R}^s$ be the projection defined by setting

$$P\mathbf{y} = (0, y_1, \dots, y_{s-1})^T, \qquad \mathbf{y} = (y_1, \dots, y_{s-1})^T \in \mathbb{R}^{s-1}.$$

Then $(P\mathbf{y}, \mathbf{x}^{1,1}) = 0$ and so $a_{\mathbf{k}}(P\mathbf{y}) = 0$ whenever $\mathbf{y} \in \mathbb{R}^{s-1}$ and $\mathbf{k} \in \Gamma^+_{s,n}$. This means that (5.45) implies that

$$(5.46) \qquad \sum_{\mathbf{k} \in \Gamma^0_{s,n}} c_{\mathbf{k}} a_{\mathbf{k}}(P\mathbf{y}) = 0, \qquad \mathbf{y} \in \mathbb{R}^{s-1}.$$

Next we let $Q : \mathbb{R}^s \to \mathbb{R}^{s-1}$ be defined as

$$Q\mathbf{x} = (x_2, \dots, x_s)^T, \qquad \mathbf{x} = (x_1, \dots, x_s)^T \in \mathbb{R}^s,$$

and introduce the vectors in $\mathbb{R}^{s-1}$

$$\mathbf{u}^{i,j} = Q\mathbf{x}^{i+1,j}, \qquad 1 \le i \le s - 1, \qquad 1 \le j \le n.$$

From our hypothesis on $\mathcal{X}$ it follows that for every $\mathbf{k} = (k_1, \dots, k_{s-1})^T \in \mathbb{Z}^{s-1}_+$ with $|\mathbf{k}|_1 \le n-1$ the vectors $\mathbf{u}^{1,k_1+1}, \dots, \mathbf{u}^{1,k_{s-1}+1}$ are linearly independent, and consequently, by the induction hypothesis the polynomials $\{a(\cdot|U_{\mathbf{k}}) : \mathbf{k} \in \Gamma_{s-1,n}\}$ are linearly independent on $\mathbb{R}^{s-1}$. Moreover, since

$$a(P\mathbf{y}|X_{\mathbf{k}}) = a(\mathbf{y}|U_{\mathbf{k}}), \quad \mathbf{k} \in \Gamma^0_{s,n},$$

(5.46) implies $c_{\mathbf{k}} = 0$, $\mathbf{k} \in \Gamma^0_{s,n}$.

Now let us look at the other scalars $c_{\mathbf{k}}, \mathbf{k} \in \Gamma^+_{s,n}$, appearing in equation (5.45). For $\mathbf{k} \in \Gamma^+_{s,n}$, the lineal polynomial $a(\cdot|X_{\mathbf{k}})$ contains $(\mathbf{x}, \mathbf{x}^{1,1})$ as a factor. Hence, as in the case $s = 2$, we define the vectors

$$\mathbf{y}^{i,j} = \begin{cases} \mathbf{x}^{1,j+1}, & i = 1, j = 1, \dots, n-1, \\ \mathbf{x}^{i,j}, & 2 \le i \le s, j = 1, \dots, n-1, \end{cases}$$

and observe that for $\mathbf{k} \in \Gamma^+_{s,n}$

$$a(\mathbf{x}|X_{\mathbf{k}}) = (\mathbf{x}, \mathbf{x}^{1,1}) a(\mathbf{x}|Y_{\mathbf{k}-\mathbf{e}_1}),$$

where $\mathbf{e}_1 = (1, 0, \ldots, 0)^T$. Therefore (5.45) reduces to the equation

$$(5.47) \qquad \sum_{\mathbf{j} \in \Gamma_{s,n-1}} c_{\mathbf{j}+\mathbf{e}_1} \, a(\mathbf{x}|Y_{\mathbf{j}}) = 0, \qquad \mathbf{x} \in \mathbb{R}^s,$$

and, since our hypothesis implies that the vectors $y^{1,k_1+1}, \ldots, y^{s,k_s+1}$ are linearly independent whenever $\mathbf{k} = (k_1, \ldots, k_s)^T$ with $|\mathbf{k}|_1 \le n-2$, our induction hypothesis gives us the linear independence of the lineal polynomials $a(\cdot|Y_{\mathbf{j}})$, $\mathbf{j} \in \Gamma_{s,n-1}$. Consequently, we also conclude that $c_{\mathbf{j}+\mathbf{e}_1} = 0$ for $\mathbf{j} \in \Gamma_{s,n-1}$. This means all the scalars appearing in equation (5.45) are zero since $\Gamma_{s,n}^+ = \Gamma_{s,n-1} + \mathbf{e}_1$.   □

Let us pause a moment in our multivariate development and look back to the univariate case and connect the theorem above to what was said there concerning B-splines. Let $\{t_{-n+1}, \ldots, t_n\}$ be a set of scalars. We create from the elements of this set the vectors in $\mathbb{R}^2$, given by

$$(5.48) \qquad \begin{aligned} \mathbf{x}^{1,j} &= (1, -t_j)^T, & j &= 1, \ldots, n, \\ \mathbf{x}^{2,j} &= (1, -t_{-j+1})^T, & j &= 1, \ldots, n. \end{aligned}$$

Clearly, for any $\mathbf{k} = (k_1, k_2)^T \in \mathbb{Z}_+^2$ with $k_1 + k_2 \le n-1$ the vectors $\mathbf{x}^{1,k_1+1}$ and $\mathbf{x}^{2,k_2+1}$ are linearly independent exactly when $t_{k_1+1} \ne t_{-k_2}$, which is precisely condition (5.18). The corresponding lineal polynomials are given by

$$a_{(k,n-k)}((-t, -1)) = (t_1 - t) \cdots (t_k - t)(t_0 - t) \cdots (t_{-n+k+1} - t),$$

which are precisely the polynomial $N_k(t)$ defined in (5.19). We proved in this case that the dual polynomials are B-splines. Our goal is to extend this result to $\mathbb{R}^s, s > 2$.

5.6. B-patches.

We begin by introducing, under the hypothesis of Theorem 5.1, the dual polynomials $\mathcal{B} = \{b_{\mathbf{k}} : \mathbf{k} \in \Gamma_{s,n}\}$ for the set $\mathcal{A} = \{a(\cdot|X_{\mathbf{k}}) : \mathbf{k} \in \Gamma_{s,n}\}$ of lineal polynomials. In analogy with the univariate case, they are defined by the formula

$$(5.49) \qquad (\mathbf{x}, \mathbf{y})^n = \sum_{\mathbf{k} \in \Gamma_{s,n}} a_{\mathbf{k}}(\mathbf{x}) b_{\mathbf{k}}(\mathbf{y}), \qquad \mathbf{x}, \mathbf{y} \in \mathbb{R}^s,$$

and therefore are guaranteed to be linearly independent.

We call each $b_{\mathbf{k}}, \mathbf{k} \in \Gamma_{s,n}$ a B-patch and the collection $\mathcal{B}$ a B-basis for $H_n(\mathbb{R}^s)$. To obtain a recurrence formula for B-patches we use the linear independence of the vectors $\mathbf{x}^{1,k_1+1}, \ldots, \mathbf{x}^{s,k_s+1}$ for $|\mathbf{k}|_1 \le n-1$ to define linear functions $u_{i,\mathbf{k}}$ on $\mathbb{R}^s$ by requiring that

$$(5.50) \qquad \mathbf{x} = \sum_{i=1}^{s} u_{i,\mathbf{k}}(\mathbf{x}) \mathbf{x}^{i,k_i+1},$$

for all $\mathbf{x} \in \mathbb{R}^s$. The explicit form of the lineal polynomials $a_{\mathbf{k}}(\mathbf{x})$ leads us immediately to the recurrence formula

$$(5.51) \qquad (\mathbf{x}, \mathbf{y}) a_{\mathbf{k}}^n(\mathbf{x}) = \sum_{i=1}^{s} u_{i,\mathbf{k}}(\mathbf{y}) a_{\mathbf{k}+\mathbf{e}_i}^{n+1}(\mathbf{x}).$$

Thus, from the duality relation (5.49) and equations (5.50) and (5.51), we have

$$\sum_{\mathbf{k} \in \Gamma_{s,n+1}} a_{\mathbf{k}}^{n+1}(\mathbf{x}) b_{\mathbf{k}}^{n+1}(\mathbf{y}) = (\mathbf{x}, \mathbf{y})(\mathbf{x}, \mathbf{y})^n$$

$$= \sum_{\mathbf{k} \in \Gamma_{s,n}} (\mathbf{x}, \mathbf{y}) a_{\mathbf{k}}^n(\mathbf{x}) b_{\mathbf{k}}^n(\mathbf{y})$$

$$= \sum_{\mathbf{k} \in \Gamma_{s,n}} \sum_{i=1}^{s} u_{i,\mathbf{k}}(\mathbf{y}) a_{\mathbf{k}+\mathbf{e}_i}^{n+1}(\mathbf{x}) b_{\mathbf{k}}^n(\mathbf{y})$$

$$= \sum_{\mathbf{k} \in \Gamma_{s,n+1}} \left(\sum_{i=1}^{s} u_{i,\mathbf{k}-\mathbf{e}_i}(\mathbf{y}) b_{\mathbf{k}-\mathbf{e}_i}^n(\mathbf{y}) \right) a_{\mathbf{k}}^{n+1}(\mathbf{x}).$$

To ensure the validity of our change of indices in the sum above we impose the convention that $b_{\mathbf{k}} = 0$ for $\mathbf{k} \in \mathbb{Z}^s \backslash \Gamma_{s,n}$. Now, comparing coefficients of $a_{\mathbf{k}}^{n+1}(\mathbf{x})$, $\mathbf{k} \in \Gamma_{s,n+1}$ above and assuming their linear independence, we get the formula

$$b_{\mathbf{k}}^{n+1}(\mathbf{y}) = \sum_{i=1}^{s} u_{i,\mathbf{k}-\mathbf{e}_i}(\mathbf{y}) b_{\mathbf{k}-\mathbf{e}_i}^n(\mathbf{y}), \qquad \mathbf{k} \in \Gamma_{s,n+1}.$$

We state this computation formally in the next proposition.

PROPOSITION 5.2. *Let* $\mathcal{X}_n = \{\mathbf{x}^{i,j} : 1 \le i \le s, \ 1 \le j \le n\}$ *be an array of vectors in* $\mathbb{R}^s$ *such that for each* $\mathbf{k} \in \Gamma_{s,n-1}$ *the vectors* $\mathbf{x}^{1,k_1+1}, \dots, \mathbf{x}^{s,k_s+1}$ *are linearly independent. Define homogeneous polynomials* $b_{\mathbf{k}}^\ell$, $\mathbf{k} \in \Gamma_{s,\ell}$, $\ell = 0, 1, \dots, n$ *by the formula*

$$(\mathbf{x}, \mathbf{y})^\ell = \sum_{\mathbf{k} \in \Gamma_{s,\ell}} b_{\mathbf{k}}^\ell(\mathbf{x}) a_{\mathbf{k}}^\ell(\mathbf{x}),$$

where $\{a_{\mathbf{k}}^\ell : \mathbf{k} \in \Gamma_{s,\ell}\}$ *are the lineal polynomials corresponding to the array* $\mathcal{X}_\ell = \{\mathbf{x}^{i,j} : 1 \le i \le s, \ 1 \le j \le \ell\}$ *with the convention that* $b_{\mathbf{k}} = 0$, $\mathbf{k} \in \mathbb{Z}^s \backslash \Gamma_{s,n}$ *and* $a_{\mathbf{0}}^0 = b_{\mathbf{0}}^0 = 1$. *Then we obtain the recurrence formula*

$$(5.52) \qquad b_{\mathbf{k}}^\ell(\mathbf{x}) = \sum_{i=1}^{s} u_{i,\mathbf{k}-\mathbf{e}_i}(\mathbf{x}) b_{\mathbf{k}-\mathbf{e}_i}^{\ell-1}(\mathbf{x}), \qquad \mathbf{k} \in \Gamma_{s,\ell}.$$

In the sense used in the univariate case, this is the down recurrence formula for B-patches. The up recurrence formula is given next.

PROPOSITION 5.3. *Let $\mathcal{X}_n$ be an array of vectors in $\mathbb{R}^s$ which satisfies the hypothesis of Proposition 5.1. For any scalars $c_\mathbf{k}$, $\mathbf{k} \in \Gamma_{s,n}$, we define homogeneous polynomials $c_\mathbf{k}^\ell(\mathbf{x})$, $\mathbf{k} \in \Gamma_{s,n-\ell}$, $\ell = 0, 1, \ldots, n$, by the recurrence formula*

$$(5.53) \qquad c_\mathbf{k}^{\ell+1}(\mathbf{x}) = \sum_{i=1}^{s} u_{i,\mathbf{k}}(\mathbf{x}) c_{\mathbf{k}+\mathbf{e}_i}^\ell(\mathbf{x}), \qquad \ell = 0, 1, \ldots, n-1,$$

where

$$(5.54) \qquad c_\mathbf{k}^0(\mathbf{x}) = c_\mathbf{k}, \qquad \mathbf{k} \in \Gamma_{s,n}.$$

Then the sum

$$\sum_{\mathbf{k} \in \Gamma_{s,n-\ell}} c_\mathbf{k}^\ell(\mathbf{x}) b_\mathbf{k}^{n-\ell}(\mathbf{x})$$

is conserved for $\ell = 0, 1, \ldots, n$. In particular, we have

$$(5.55) \qquad c_0^n(\mathbf{x}) = \sum_{\mathbf{k} \in \Gamma_{s,n}} c_\mathbf{k} b_\mathbf{k}^n(\mathbf{x}).$$

Proof. We compute for $\ell = 0, 1, \ldots, n-1$ the sum below using (5.53), viz.

$$\sum_{\mathbf{k} \in \Gamma_{s,n-\ell-1}} c_\mathbf{k}^{\ell+1}(\mathbf{x}) b_\mathbf{k}^{n-\ell-1}(\mathbf{x})$$

$$= \sum_{\mathbf{k} \in \Gamma_{s,n-\ell-1}} \sum_{i=1}^{s} u_{i,\mathbf{k}}(\mathbf{x}) c_{\mathbf{k}+\mathbf{e}_i}^\ell(\mathbf{x}) b_\mathbf{k}^{n-\ell-1}(\mathbf{x})$$

$$= \sum_{\mathbf{k} \in \Gamma_{s,n-\ell}} \left(\sum_{i=1}^{s} u_{i,\mathbf{k}-\mathbf{e}_i}(\mathbf{x}) b_\mathbf{k}^{n-\ell-1}(\mathbf{x}) \right) c_\mathbf{k}^\ell(\mathbf{x}).$$

Proposition 5.2 states that the sum in the parentheses is $b_\mathbf{k}^{n-\ell}(\mathbf{x})$ and so we get, in summary, the equation

$$\sum_{\mathbf{k} \in \Gamma_{s,n-\ell-1}} c_\mathbf{k}^{\ell+1}(\mathbf{x}) b_\mathbf{k}^{n-\ell-1}(\mathbf{x}) = \sum_{\mathbf{k} \in \Gamma_{s,n-\ell}} c_\mathbf{k}^\ell(\mathbf{x}) b_\mathbf{k}^{n-\ell}(\mathbf{x}). \qquad \square$$

The recurrence formulas (5.53) and (5.54) gives a *pyramid scheme* for computing any homogeneous polynomial of degree n expressed in terms of B-patches. It generalizes the corresponding univariate recurrence formulas for Bernstein–Bezier polynomials and B-splines presented earlier. Also, when the array $\mathcal{X}_n = \{\mathbf{x}^{i,j} : 1 \le i \le s, \ 1 \le j \le n\}$ has the special property that $\mathbf{x}^{i,j} = \mathbf{x}_i$ for all i, j we get $a_\mathbf{k}(\mathbf{x}) = (\mathbf{x}, \mathbf{x}_1)^{k_1} \cdots (\mathbf{x}, \mathbf{x}_s)^{k_s}$. So, by the multinomial theorem,

$$(5.56) \qquad b_\mathbf{k}(\mathbf{x}) = \binom{n}{\mathbf{k}} u_1^{k_1}(\mathbf{x}) \cdots u_s^{k_s}(\mathbf{x}), \qquad \mathbf{k} \in \Gamma_{s,n}.$$

These are the (homogeneous) multivariate Bernstein–Bezier polynomials and so (5.52) and (5.53)–(5.54) are the familiar up and down recurrence formulas for Bernstein–Bezier polynomials (see Figs. 5.7 and 5.8, respectively).

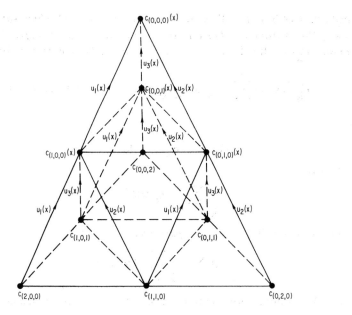

FIG. 5.7.   *Up recurrence formula for quadratic bivariate Bernstein–Bézier polynomials.*

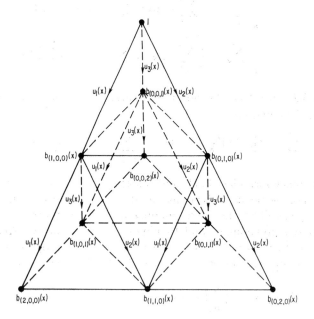

FIG. 5.8.   *Down recurrence formula for quadratic bivariate Bernstein–Bézier polynomials.*

Our next result explains how to express the B-patch coefficients of a polynomial in $H_n(\mathbb{R}^s)$ in terms of its polar form.

PROPOSITION 5.4. *Let* $\mathcal{X} = \{\mathbf{x}^{i,j} : 1 \leq i \leq s, \ 1 \leq j \leq n\}$ *be an array of vectors in* $\mathbb{R}^s$ *that satisfies the hypothesis of Proposition 5.1. Suppose* p *is an arbitrary homogeneous polynomial given in B-patch form*

$$(5.57) \qquad p(\mathbf{x}) = \sum_{\mathbf{k} \in \Gamma_{s,n}} c_{\mathbf{k}} b_{\mathbf{k}}(\mathbf{x}), \qquad \mathbf{x} \in \mathbb{R}^s,$$

for some scalars $c_{\mathbf{k}}, \mathbf{k} \in \Gamma_{s,n}$. *Define the vector* $\mathbf{q_k} \in \mathbb{R}^{sn}$ *by the equation*

$$\mathbf{q_k} = (\mathbf{x}^{1,1}, \ldots, \mathbf{x}^{1,k_1}, \ldots, \mathbf{x}^{s,1}, \ldots, \mathbf{x}^{s,k_s})^T.$$

Then

$$(5.58) \qquad c_{\mathbf{k}} = P(\mathbf{q_k}), \qquad \mathbf{k} \in \Gamma_{s,n}.$$

Proof. For a fixed $\mathbf{x} \in \mathbb{R}^s$, the polynomial $p_{\mathbf{x}}(\mathbf{y}) = (\mathbf{x}, \mathbf{y})^n$ has the polar form

$$P_{\mathbf{x}}(\mathbf{y}_1, \ldots, \mathbf{y}_n) = (\mathbf{x}, \mathbf{y}_1) \cdots (\mathbf{x}, \mathbf{y}_n)$$

and therefore $a(\mathbf{x}|X_{\mathbf{k}}) = P_{\mathbf{x}}(\mathbf{q_k})$. Substituting these equations in (5.49) gives us the formula

$$p_{\mathbf{x}}(\mathbf{y}) = \sum_{\mathbf{k} \in \Gamma_{k,n}} P_{\mathbf{x}}(\mathbf{q_k}) b_{\mathbf{k}}(\mathbf{y}).$$

Since span $\{p_{\mathbf{x}} : \mathbf{x} \in \mathbb{R}^s\} = H_n(\mathbb{R}^s)$, we get (5.58).   □

The polar form of a polynomial in B-patch form can be computed by an up recurrence formula; this is presented next. (See Fig. 5.9 for the special case of Bernstein–Bézier polynomials.)

THEOREM 5.2. *Let* $\mathcal{X} = \{\mathbf{x}^{i,j} : 1 \leq i \leq s, \ 1 \leq j \leq n\}$ *be an array of vectors that satisfies the hypothesis of Proposition 5.1. For any scalars* $c_{\mathbf{k}}, \mathbf{k} \in \Gamma_{s,n}$ *and vectors* $\mathbf{x}_1, \ldots, \mathbf{x}_n \in \mathbb{R}^s$, *we define* $C_{\mathbf{k}}^{\ell}(\mathbf{x}_1, \ldots, \mathbf{x}_\ell)$, *for* $\ell = 1, \ldots, n$, *by the recurrence formula*

$$(5.59) \qquad \begin{aligned} & C_{\mathbf{k}}^{\ell+1}(\mathbf{x}_1, \ldots, \mathbf{x}_{\ell+1}) \\ & = \sum_{i=1}^{s} u_{i,\mathbf{k}}(\mathbf{x}_{\ell+1}) C_{\mathbf{k}+\mathbf{e}_i}^{\ell}(\mathbf{x}_1, \ldots, \mathbf{x}_\ell), \qquad \mathbf{k} \in \Gamma_{s,n-\ell-1}, \end{aligned}$$

where

$$(5.60) \qquad C_{\mathbf{k}}^0 = c_{\mathbf{k}}, \qquad \mathbf{k} \in \Gamma_{s,n}.$$

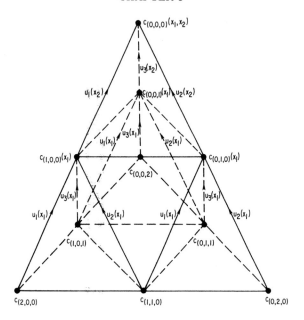

FIG. 5.9. *Blossoming the up recurrence formula for quadratic Bernstein–Bézier polynomials.*

Then $C_{\mathbf{0}}^{n} \in \mathcal{P}_{s,n}$ and

$$C_{\mathbf{0}}^{n}(\mathbf{x}, \ldots, \mathbf{x}) = \sum_{\mathbf{k} \in \Gamma_{s,n}} c_{\mathbf{k}} b_{\mathbf{k}}(\mathbf{x}), \qquad \mathbf{x} \in \mathbb{R}^{s}.$$

Proof. The only thing that is not immediately obvious is the fact that $C_{\mathbf{k}}^{\ell}(\mathbf{x}_1, \ldots, \mathbf{x}_{\ell})$ is a *symmetric* function of its arguments, $\ell = 1, \ldots, n$. Using (5.59) we will prove this by induction on ℓ.

By (5.59) it suffices to prove that $C_{\mathbf{k}}^{\ell+1}(\mathbf{x}_1, \ldots, \mathbf{x}_{\ell+1})$ is a symmetric function of $\mathbf{x}_j$ and $\mathbf{x}_{\ell}$, for $j = 1, \ldots, \ell$, that is,

(5.61)
$$\begin{aligned}
C_{\mathbf{k}}^{\ell+1}(\mathbf{x}_1, \ldots, \mathbf{x}_{j-1}, \mathbf{x}_{\ell+1}, \mathbf{x}_{j+1}, \ldots, \mathbf{x}_{\ell}, \mathbf{x}_j) \\
= C_{\mathbf{k}}^{\ell+1}(\mathbf{x}_1, \ldots, \mathbf{x}_{\ell+1}), \qquad \mathbf{k} \in \Gamma_{s,n-\ell-1}.
\end{aligned}$$

We first evaluate the left-hand side of (5.61) by using our recurrence formula (5.59), viz.

$$\begin{aligned}
& C_{\mathbf{k}}^{\ell+1}(\mathbf{x}_1, \ldots, \mathbf{x}_j, \mathbf{x}_{\ell+1}, \mathbf{x}_{j+1}, \ldots, \mathbf{x}_{\ell}, \mathbf{x}_j) \\
& = \sum_{i=1}^{s} u_{i,\mathbf{k}}(\mathbf{x}_j) C_{\mathbf{k}+\mathbf{e}_i}^{\ell}(\mathbf{x}_1, \ldots, \mathbf{x}_{j-1}, \mathbf{x}_{\ell+1}, \mathbf{x}_{j+1}, \ldots, \mathbf{x}_{\ell})
\end{aligned}$$

$$= \sum_{i=1}^{s} u_{i,\mathbf{k}}(\mathbf{x}_j) C_{\mathbf{k}+\mathbf{e}_i}^{\ell}(\mathbf{x}_1, \ldots, \mathbf{x}_{j-1}, \mathbf{x}_{j+1}, \ldots, \mathbf{x}_\ell, \mathbf{x}_{\ell+1})$$

$$= \sum_{i=1}^{s} \sum_{m=1}^{s} u_{i,\mathbf{k}}(\mathbf{x}_j) u_{m,\mathbf{k}+\mathbf{e}_i}(\mathbf{x}_{\ell+1}) C_{\mathbf{k}+\mathbf{e}_i+\mathbf{e}_m}^{\ell-1}(\mathbf{x}_1, \ldots, \mathbf{x}_{j-1}, \mathbf{x}_{j+1}, \ldots, \mathbf{x}_\ell).$$

The right side of this equation has the same form with the variables $\mathbf{x}_j$ and $\mathbf{x}_{\ell+1}$ interchanged since the term $C_{\mathbf{k}+\mathbf{e}_i+\mathbf{e}_m}^{\ell-1}$ is symmetric in i and m. Thus, it suffices to show that the bilinear expression

$$u_{i,\mathbf{k}}(\mathbf{x}) u_{m,\mathbf{k}+\mathbf{e}_i}(\mathbf{y}) + u_{m,\mathbf{k}}(\mathbf{x}) u_{i,\mathbf{k}+\mathbf{e}_m}(\mathbf{y})$$

is symmetric in $\mathbf{x}$ and $\mathbf{y}$. Equivalently, if we let

$$U_{i,m}(\mathbf{x}, \mathbf{y}) := u_{i,\mathbf{k}}(\mathbf{x}) u_{m,\mathbf{k}+\mathbf{e}_i}(\mathbf{y}) - u_{i,\mathbf{k}}(\mathbf{y}) u_{m,\mathbf{k}+\mathbf{e}_i}(\mathbf{x})$$

then we must show that

$$(5.62) \qquad\qquad U_{i,m}(\mathbf{x}, \mathbf{y}) = -U_{m,i}(\mathbf{x}, \mathbf{y}).$$

To this end, we need to relate the linear function $u_{i,\mathbf{k}+\mathbf{e}_m}, i = 1, \ldots, s$ to $u_{i,\mathbf{k}}, i = 1, \ldots, s$. Since the sets of vectors that determine them (see (5.50)) have all the vectors $\mathbf{x}^{i,k_i+1}$, $i = 1, \ldots, s, i \neq m$ in common, it follows that

$$(5.63) \qquad u_{i,\mathbf{k}+\mathbf{e}_m}(\mathbf{x}) = (1 - \delta_{im}) u_{i,\mathbf{k}}(\mathbf{x}) + A_{im} u_{m,\mathbf{k}}(\mathbf{x}), \qquad \mathbf{x} \in \mathbb{R}^s,$$

where

$$A_{i,m} := u_{i,\mathbf{k}+\mathbf{e}_m}(\mathbf{x}^{m,k_m+1}).$$

We prove equation (5.63) by showing that both sides of this equation agree on the linearly independent vectors $\mathbf{x}^{j,k_j+1}$, $j = 1, \ldots, s$. The details are straightforward and are based on the fact that

$$(5.64) \qquad\qquad u_{i,\mathbf{k}}(\mathbf{x}^{j,k_j+1}) = \delta_{ij}, \qquad i, j = 1, \ldots, s$$

and

$$(5.65) \qquad\qquad u_{i,\mathbf{k}+\mathbf{e}_m}(\mathbf{x}^{j,k_j+1}) = \delta_{ij}, \qquad j \neq m,$$

which are both obvious from our definition (5.50) of the linear functions $u_{i,\mathbf{k}}$ and $u_{i,\mathbf{k}+\mathbf{e}_m}$.

Now, on the left side of the equation in (5.63) for $\mathbf{x} = \mathbf{x}^{j,k_j+1}$, $j \neq m$ we get, by (5.65), δ_{ij}, and on the right side, using (5.64), we get $(1 - \delta_{im})\delta_{ij} = \delta_{ij} - \delta_{im}\delta_{ij} = \delta_{ij}$, which are the same. When $j = m$ on the left side we get A_{im}, while on the right side we get $(1 - \delta_{im})\delta_{im} + A_{im} = \delta_{im} - \delta_{im} + A_{im} = A_{im}$. This establishes (5.63) and we can use it to compute $U_{i,m}(\mathbf{x}, \mathbf{y})$. A substitution of (5.63) into our definition of $U_{i,m}$ and a simplification gives us the formula

$$U_{i,m}(\mathbf{x}, \mathbf{y}) = (1 - \delta_{im})(u_{i,\mathbf{k}}(\mathbf{x}) u_{m,\mathbf{k}}(\mathbf{y}) - u_{i,\mathbf{k}}(\mathbf{y}) u_{m,\mathbf{k}}(\mathbf{x})),$$

from which (5.62) is obvious.   □

5.7. Pyramid schemes.

Among a large class of polynomial recurrence formulas Theorem 5.2 characterizes B-patches. To explain this, we begin with a general collection of vectors

$$(5.66) \qquad \mathcal{X} = \{x_{i,\mathbf{k}}^\ell, \ i = 1, \dots, s, \ \ell = 0, 1, \dots, n, \ \mathbf{k} \in \Gamma_{s,n-\ell}\},$$

from which we will define *a pyramid scheme*. We require that for each $\ell = 0, 1, \dots, n, \mathbf{k} \in \Gamma_{s,n-\ell}$, the vectors $x_{1,\mathbf{k}}^\ell, \dots, x_{s,\mathbf{k}}^\ell$ are linearly independent. Then, we can define linear functionals $u_{i,\mathbf{k}}^\ell(x), \ i = 1, \dots, s$ by requiring that for all $\mathbf{x} \in \mathbb{R}^s$

$$(5.67) \qquad\qquad \mathbf{x} = \sum_{i=1}^{s} u_{i,\mathbf{k}}^\ell(\mathbf{x}) x_{i,\mathbf{k}}^\ell.$$

Associated with $\mathcal{X}$ is a pyramid scheme, or up recurrence formula, given for $\ell = 1, \dots, n$, by the formula

$$c_{\mathbf{k}}^\ell(\mathbf{x}) = \sum_{i=0}^{s} u_{i,\mathbf{k}}^\ell(\mathbf{x}) c_{\mathbf{k}+\mathbf{e}_i}^{\ell-1}(\mathbf{x}), \qquad \mathbf{k} \in \Gamma_{s,n-\ell},$$

$$(5.68) \qquad\qquad c_{\mathbf{k}}^0 = c_{\mathbf{k}}, \qquad \mathbf{k} \in \Gamma_{s,n}.$$

The apex $c_{\mathbf{0}}^n(\mathbf{x})$ of this pyramid is clearly a homogeneous polynomial. The corresponding down recurrence formula is given by the formula

$$p_{\mathbf{0}}^0 = 1$$

$$(5.69) \quad p_{\mathbf{k}}^\ell(\mathbf{x}) = \sum_{i=1}^{s} u_{i,\mathbf{k}-\mathbf{e}_i}^\ell(\mathbf{x}) p_{\mathbf{k}-\mathbf{e}_i}^{\ell-1}(\mathbf{x}), \qquad \mathbf{k} \in \Gamma_{s,\ell}, \ \ell = 1, \dots, n,$$

where, by our usual convention, $p_{\mathbf{k}}^\ell = 0$ if $\mathbf{k} \in \mathbb{Z}^s \backslash \Gamma_{s,\ell}$.

The up and down pyramid recurrence formulas conserve sums, that is, as before the function

$$\sum_{\mathbf{k} \in \Gamma_{s,n-\ell}} c_{\mathbf{k}}^\ell(\mathbf{x}) p_{\mathbf{k}}^{n-\ell}(\mathbf{x})$$

does not depend on $\ell = 0, 1, \dots, n$. Therefore, in particular, the apex of the pyramid is

$$c_{\mathbf{0}}^n(\mathbf{x}) = \sum_{\mathbf{k} \in \Gamma_{s,n}} c_{\mathbf{k}} p_{\mathbf{k}}^n(\mathbf{x}), \qquad \mathbf{x} \in \mathbb{R}^s.$$

A basic unsolved problem is to describe the totality of homogeneous polynomials, which can be an apex polynomial of a given pyramid scheme in terms of the array $\mathcal{X}$ that defines the scheme.

Every pyramid scheme has an associated pyramid scheme for generating multilinear functions, namely for $\ell = 1, \ldots, n$,

$$
(5.70) \quad C_{\mathbf{k}}^{\ell}(\mathbf{x}_1, \ldots, \mathbf{x}_\ell) = \sum_{i=0}^{s} u_{i,\mathbf{k}}^{\ell}(\mathbf{x}_\ell) C_{\mathbf{k}+\mathbf{e}_i}^{\ell-1}(\mathbf{x}_1, \ldots, \mathbf{x}_{\ell-1}), \qquad \mathbf{k} \in \Gamma_{s,n-\ell}
$$

$$
C_{\mathbf{k}}^0 = c_{\mathbf{k}}, \qquad \mathbf{k} \in \Gamma_{s,n}.
$$

As it turns out, the B-patch scheme is the only one that is symmetric. This result follows next.

THEOREM 5.3. *Let* $\mathcal{X} = \{\mathbf{x}_{i,j}^{\ell} : i = 1, \ldots, s, \ell = 1, \ldots, n, \mathbf{k} \in \Gamma_{s,n-\ell}\}$ *be a collection of vectors in some hyperplane* $\{\mathbf{x} : \mathbf{x} \in \mathbb{R}^s, (\mathbf{a}, \mathbf{x}) = 1\}$ *such that* $\mathbf{x}_{1,\mathbf{k}}^{\ell}, \ldots, \mathbf{x}_{s,\mathbf{k}}^{\ell}$ *are linearly independent for* $\ell = 1, \ldots, n$, *and* $\mathbf{k} \in \Gamma_{s,n-\ell}$. *Then a necessary and sufficient condition for the multiaffine forms* $C_{\mathbf{k}}^{\ell}, \mathbf{k} \in \Gamma_{s,n-\ell}, \ell = 0, 1, \ldots, n$ *to be symmetric for all scalars* $c_{\mathbf{k}}, \mathbf{k} \in \Gamma_{s,n}$ *is that the vector* $\mathbf{x}_{i,\mathbf{k}}^{\ell}$ *depends only on* k_i. *That is, there exists an array* $Y = \{\mathbf{y}^{i,j} : 1 \leq i \leq s, 1 \leq j \leq n\}$ *such that*

$$
\mathbf{x}_{i,\mathbf{k}}^{\ell} = \mathbf{y}^{i,k_i+1}, \qquad i = 1, \ldots, s, \qquad \ell = 1, \ldots, n, \qquad \mathbf{k} \in \Gamma_{s,n-\ell}.
$$

Proof. The sufficiency of this result is covered by Theorem 5.2. For the necessity, we first show that for each $\ell = 2, \ldots, n$, $\mathbf{k} \in \Gamma_{s,n-\ell}$, $i, j = 1, \ldots, s$ and $\mathbf{x}, \mathbf{y} \in \mathbb{R}^s$ we have the formula

$$
(5.71) \quad \begin{aligned} & u_{i,\mathbf{k}}^{\ell}(\mathbf{x})u_{j,\mathbf{k}+\mathbf{e}_i}^{\ell-1}(\mathbf{y}) + u_{j,\mathbf{k}}^{\ell}(\mathbf{x})u_{i,\mathbf{k}+\mathbf{e}_j}^{\ell-1}(\mathbf{y}) \\ &= u_{i,\mathbf{k}}^{\ell}(\mathbf{y})u_{j,\mathbf{k}+\mathbf{e}_i}^{\ell-1}(\mathbf{x}) + u_{j,\mathbf{k}}^{\ell}(\mathbf{y})u_{i,\mathbf{k}+\mathbf{e}_j}^{\ell-1}(\mathbf{x}). \end{aligned}
$$

To this end, call the left-hand side of equation (5.71) $V_{i,j,\mathbf{k}}^{\ell}(\mathbf{x}, \mathbf{y})$. Thus equation (5.71) says that this function is symmetric in $\mathbf{x}$ and $\mathbf{y}$. Using the up recurrence formula twice, as in the proof of Theorem 5.2, we get the equation

$$
(5.72) \quad \begin{aligned} & C_{\mathbf{k}}^{\ell}(\mathbf{x}_1, \ldots, \mathbf{x}_{\ell-2}, \mathbf{y}, \mathbf{x}) \\ &= \sum_{i=1}^{s} u_{i,\mathbf{k}}^{\ell}(\mathbf{x}) C_{\mathbf{k}+\mathbf{e}_i}^{\ell-1}(\mathbf{x}_1, \ldots, \mathbf{x}_{\ell-2}, \mathbf{y}) \\ &= \sum_{j=1}^{s}\sum_{i=1}^{s} u_{i,\mathbf{k}}^{\ell}(\mathbf{x}) u_{j,\mathbf{k}+\mathbf{e}_i}^{\ell-1} C_{\mathbf{k}+\mathbf{e}_i+\mathbf{e}_j}^{\ell-2}(\mathbf{x}_1, \ldots, \mathbf{x}_{\ell-2}) \\ &= \frac{1}{2}\sum_{j=1}^{s}\sum_{i=1}^{s} V_{i,j,\mathbf{k}}^{\ell}(\mathbf{x}, \mathbf{y}) C_{\mathbf{k}+\mathbf{e}_i+\mathbf{e}_j}^{\ell-2}(\mathbf{x}_1, \ldots, \mathbf{x}_{\ell-2}). \end{aligned}
$$

The left-hand side of (5.72) is a symmetric function of $\mathbf{y}$ and $\mathbf{x}$. Thus, to prove (5.71), we must show, given any $s \times s$ symmetric matrix T, that for some choice of $\mathbf{x}_1, \ldots, \mathbf{x}_{\ell-2}$ and scalars $c_{\mathbf{k}}, \mathbf{k} \in \Gamma_{s,n}$,

$$
C_{\mathbf{k}+\mathbf{e}_i+\mathbf{e}_j}^{\ell-2}(\mathbf{x}_1, \ldots, \mathbf{x}_{\ell-2}) = T_{ij}, \qquad i, j = 1, \ldots, s.
$$

By definition, we see that for each level $\ell = 1, \ldots, s$

$$(5.73) \qquad u^\ell_{i,\mathbf{k}}(\mathbf{x}^\ell_{j,\mathbf{k}}) = \delta_{ij}, \quad i, j = 1, \ldots, s, \qquad \mathbf{k} \in \Gamma_{s,n-\ell}.$$

Hence, it follows that for any $\mathbf{k} \in \Gamma_{s,n-\ell}$

$$C^\ell_\mathbf{k}(\mathbf{x}_1, \ldots, \mathbf{x}_{\ell-1}, \mathbf{x}^\ell_{i,\mathbf{k}}) = C^{\ell-1}_{\mathbf{k}+\mathbf{e}_i}(\mathbf{x}_1, \ldots, \mathbf{x}_{\ell-1})$$

for any $\mathbf{x}_1, \ldots, \mathbf{x}_{\ell-1} \in \mathbb{R}^s$. Repeating this equation $\ell - 1$ more times, we get for any $i_1, \ldots, i_\ell \in \{1, \ldots, s\}$ the formula

$$C^\ell_\mathbf{k}\left(\mathbf{x}^1_{i_\ell, \mathbf{k}+\mathbf{e}_{i_1}+\cdots+\mathbf{e}_{i_\ell}}, \mathbf{x}^2_{i_{\ell-1}, \mathbf{k}+\mathbf{e}_{i_1}+\cdots+\mathbf{e}_{i_{\ell-2}}}, \ldots, \mathbf{x}^{\ell-1}_{i_2, \mathbf{k}+\mathbf{e}_{i_1}}, \mathbf{x}^\ell_{i_1,\mathbf{k}}\right)$$
$$= c_{\mathbf{k}+\mathbf{e}_{i_1}+\cdots+\mathbf{e}_{i_\ell}}.$$

Thus we can fix any $i_1, \ldots, i_{\ell-2}$ and choose $i_{\ell-1} = i, i_\ell = j$ and set $\mathbf{r} := \mathbf{k}+\mathbf{e}_{i_1} + \cdots + \mathbf{e}_{i_{\ell-2}}$. Then we choose scalars so that $(c_{\mathbf{r}+\mathbf{e}_i+\mathbf{e}_j})_{i,j=1,\ldots,s}$ is the prescribed $s \times s$ symmetric matrix. This establishes equation (5.71).

Our next assertion is that

$$(5.74) \qquad u^\ell_{i,\mathbf{k}}(\mathbf{x}^{\ell-1}_{m,\mathbf{k}+\mathbf{e}_j}) = 0, \quad \text{if } m \in \{1, \ldots, s\}\backslash\{i,j\}.$$

For the proof of (5.74), we distinguish between two cases.

Case 1. $i = j$. In this case, (5.71) simplifies to the equation

$$u^\ell_{i,\mathbf{k}}(\mathbf{x})u^{\ell-1}_{i,\mathbf{k}+\mathbf{e}_i}(\mathbf{y}) = u^\ell_{i,\mathbf{k}}(\mathbf{y})u^{\ell-1}_{i,\mathbf{k}+\mathbf{e}_i}(\mathbf{x}).$$

Now choose $\mathbf{x} = \mathbf{x}^{\ell-1}_{m,\mathbf{k}+\mathbf{e}_i}$ and $\mathbf{y} = \mathbf{x}^{\ell-1}_{i,\mathbf{k}+\mathbf{e}_i}$ in the above equation. By the definition of these linear functions (see (5.67)) we have $u^{\ell-1}_{i,\mathbf{k}+\mathbf{e}_i}(\mathbf{y}) = 1$ and $u^{\ell-1}_{i,\mathbf{k}+\mathbf{e}_i}(\mathbf{x}) = 0$, since $m \neq i$. Hence, we get

$$(5.75) \qquad u^\ell_{i,\mathbf{k}}(\mathbf{x}^{\ell-1}_{m,\mathbf{k}+\mathbf{e}_i}) = 0, \qquad m \in \{1, \ldots, s\}\backslash\{i\}.$$

Case 2. $i \neq j$. Here we choose $\mathbf{x} = \mathbf{x}^{\ell-1}_{m,\mathbf{k}+\mathbf{e}_j}$ and $\mathbf{y} = \mathbf{x}^{\ell-1}_{j,\mathbf{k}+\mathbf{e}_i}$ so that

$$u^{\ell-1}_{i,\mathbf{k}+\mathbf{e}_j}(\mathbf{x}) = 0, \qquad u^{\ell-1}_{j,\mathbf{k}+\mathbf{e}_i}(\mathbf{y}) = 1.$$

Also, from (5.75) of Case 1, we have

$$u^\ell_{i,\mathbf{k}}(\mathbf{y}) = u^\ell_{j,\mathbf{k}}(\mathbf{x}) = 0.$$

Substituting these equations into (5.71) and simplifying gives us the equation

$$u^\ell_{i,\mathbf{k}}(\mathbf{x}) = 0,$$

which agrees with (5.74). Thus, in all cases, equation (5.74) holds.

Now fix an $m \neq j$. Then definition (5.67) (with $x = \mathbf{x}^{\ell-1}_{m,\mathbf{k}+\mathbf{e}_j}$) and (5.74) yields the formula

$$\mathbf{x}^{\ell-1}_{m,\mathbf{k}+\mathbf{e}_j} = u^{\ell}_{m,\mathbf{k}}(\mathbf{x}^{\ell-1}_{m,\mathbf{k}+\mathbf{e}_j})\mathbf{x}^{\ell}_{m,\mathbf{k}}.$$

Taking the inner product of both sides of this equation with $\mathbf{a}$, we conclude that $u^{\ell}_{m,\mathbf{k}}(\mathbf{x}^{\ell-1}_{m,\mathbf{k}+\mathbf{e}_j}) = 1$ and therefore we obtain the relation

$$\mathbf{x}^{\ell-1}_{m,\mathbf{k}+\mathbf{e}_j} = \mathbf{x}^{\ell}_{m,\mathbf{k}}.$$

We use this equation repeatedly in the following way. We choose $\mathbf{k} = k_m\mathbf{e}_m$ above and, for every $m \notin \{j_1, \ldots, j_{s-1}\}$, we have

$$\mathbf{x}^{n-k_m-k_{j_1}-\cdots-k_{j_{s-1}}}_{m,k_m\mathbf{e}_m+k_{j_1}\mathbf{e}_{j_1}+\cdots+k_{j_{s-1}}\mathbf{e}_{j_{s-1}}} = \cdots = \mathbf{x}^{n-k_m}_{m,k_m\mathbf{e}_m}.$$

Choosing $\{j_1, \ldots, j_{s-1}\} = \{1, \ldots, s\}\setminus\{m\}$, it follows that for $\mathbf{k} \in \Gamma_{s,n-\ell}$

$$\mathbf{x}^{\ell}_{m,\mathbf{k}} = \mathbf{x}^{n-k_m}_{m,k_m\mathbf{e}_m}.$$

Now we define

$$\mathbf{y}^{i,j} = \mathbf{x}^{n-j+1}_{i,(j-1)\mathbf{e}_i}, \qquad i = 1, \ldots, s, \qquad j = 1, \ldots, n$$

and obtain

$$\mathbf{x}^{\ell}_{i,\mathbf{k}} = \mathbf{y}^{i,k_i+1}. \qquad \square$$

5.8. M-patches.

Before we return to B-patches we wish to describe another example of a pyramid scheme. The scheme is selected by assuming the labels $u^{\ell}_{i,\mathbf{k}}$, $i = 1, \ldots, s$ are all the same at any given level ℓ, that is, are not dependent on a location $\mathbf{k}$ within the level.

Specifically, we start with the array

(5.76) $$\mathcal{X} = \{\mathbf{x}^{i,j} : i = 1, \ldots, s, j = 1, \ldots, n\},$$

as with B-patches, but now we choose the set of vectors

$$\{\mathbf{x}^{1,\ell}, \ldots, \mathbf{x}^{s,\ell}\},$$

assumed to be linearly independent, and define our linear forms by the requirement that

(5.77) $$\mathbf{x} = \sum_{i=1}^{s} u^{\ell}_i(\mathbf{x})\mathbf{x}^{i,\ell}, \qquad \mathbf{x} \in \mathbb{R}^s.$$

This leads us to the down recurrence formula

(5.78)

$$M_{\mathbf{0}}^{0}(\mathbf{x}) = 0$$

$$M_{\mathbf{k}}^{\ell}(\mathbf{x}) = \sum_{i=1}^{s} u_i^{\ell}(\mathbf{x}) M_{\mathbf{k}-\mathbf{e}_i}^{\ell-1}(\mathbf{x}), \qquad \mathbf{k} \in \Gamma_{s,n-\ell}.$$

(See Fig. 5.10 for the case of bivariate quadratic M-patches.) We call the homogeneous polynomials $M_{\mathbf{k}}^{n}(\mathbf{x})$, $\mathbf{k} \in \Gamma_{s,n}$, M-patches and, as before, we set $M_{\mathbf{k}}^{\ell} = 0$ for all $\mathbf{k} \in \mathbb{Z}^{s} \backslash \Gamma_{s,n-\ell}$.

To obtain the generating function for M-patches, it is convenient to introduce for each ℓ, linearly independent vectors $\mathbf{y}^{1,\ell}, \ldots, \mathbf{y}^{s,\ell}$ such that

$$u_i^{\ell}(\mathbf{x}) = (\mathbf{y}^{i,\ell}, \mathbf{x}), \qquad \mathbf{x} \in \mathbb{R}^{s}.$$

Let R_{ℓ} be the matrix

$$R_{\ell} = \begin{pmatrix} (\mathbf{y}^{1,\ell})_1 & \cdots & (\mathbf{y}^{s,\ell})_1 \\ \vdots & & \vdots \\ (\mathbf{y}^{1,\ell})_s & \cdots & (\mathbf{y}^{s,\ell})_s, \end{pmatrix},$$

and observe that $u_i^{\ell}(\mathbf{x}) = (R_{\ell}^{T}\mathbf{x})_i$, $i = 1, \ldots, s$. Then, by equation (5.77), R_{ℓ}^{-1}

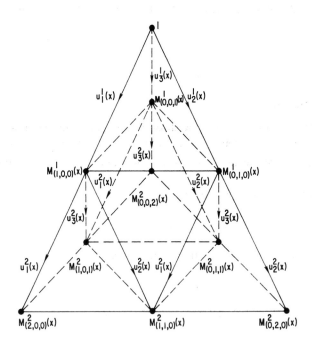

FIG. 5.10. *Down recurrence formula for quadratic M-patches.*

is given by

$$R_\ell^{-1} = \begin{pmatrix} (\mathbf{x}^{1,\ell})_1 & \cdots & (\mathbf{x}^{1,\ell})_s \\ \vdots & & \vdots \\ (\mathbf{x}^{s,\ell})_1 & \cdots & (\mathbf{x}^{s,\ell})_s \end{pmatrix}.$$

This setup leads us to the next proposition.

PROPOSITION 5.5. *Let* $\mathcal{X} = \{\mathbf{x}^{i,j} : 1 \leq i \leq s, \ 1 \leq j \leq n\}$ *be an array of vectors in* $\mathbb{R}^s$ *such that the vectors* $\mathbf{x}^{1,\ell}, \ldots, \mathbf{x}^{s,\ell}$ *are linearly independent for each* $\ell = 1, \ldots, n$. *Then*

$$(5.79) \qquad \prod_{j=1}^n (\mathbf{x}, R_j \mathbf{y}) = \sum_{\mathbf{k} \in \Gamma_{s,n}} \mathbf{y}^{\mathbf{k}} M_{\mathbf{k}}(\mathbf{x}), \qquad \mathbf{x}, \mathbf{y} \in \mathbb{R}^s.$$

Moreover, the following are equivalent:

(i) p *is an apex polynomial for the* M-*patches, that is, for some scalars* $c_{\mathbf{k}}$, $\mathbf{k} \in \Gamma_{s,n}$

$$p = \sum_{\mathbf{k} \in \Gamma_{s,n}} c_{\mathbf{k}} M_{\mathbf{k}}.$$

(ii) $p \in \text{span} \ \{\prod_{j=1}^n (\mathbf{x}, R_j \cdot) : \mathbf{x} \in \mathbb{R}^s\}$,
(iii) $p \in \{P(R_1 \cdot, R_2 \cdot, \ldots, R_n \cdot) : P \in \mathcal{P}_{s,n}\}$

Proof. First we prove equation (5.79). To this end, we multiply both sides of (5.78) by $\mathbf{y}^{\mathbf{k}}$ and sum the resulting expressions over $\mathbf{k} \in \Gamma_{s,n-\ell}$. This gives us the formula

$$\sum_{\mathbf{k} \in \Gamma_{s,n-\ell}} \mathbf{y}^{\mathbf{k}} M_{\mathbf{k}}^\ell(\mathbf{x}) = \left(\sum_{i=1}^s u_i^\ell(\mathbf{x}) y_i \right) \sum_{\mathbf{k} \in \Gamma_{s,n-\ell+1}} \mathbf{y}^{\mathbf{k}} M_{\mathbf{k}}^{\ell-1}(\mathbf{x})$$

$$= (\mathbf{x}, R_\ell \mathbf{y}) \sum_{\mathbf{k} \in \Gamma_{s,n-\ell+1}} \mathbf{y}^{\mathbf{k}} M_{\mathbf{k}}^{\ell-1}(\mathbf{x}).$$

With this formula, it follows easily, for $\ell = 0, 1, \ldots, n$, that

$$\prod_{j=1}^\ell (\mathbf{x}, R_j \mathbf{y}) = \sum_{\mathbf{k} \in \Gamma_{s,n-\ell}} \mathbf{y}^{\mathbf{k}} M_{\mathbf{k}}^\ell(\mathbf{x}).$$

and, in particular, (5.79) follows.

Claim (ii) follows directly from (5.79), while for (iii) we recall by Lemma 5.3 and formula (5.40) that

$$\prod_{j=1}^n (\mathbf{x}, R_j \mathbf{y}) = \sum_{\mathbf{k} \in \Gamma_{s,n}} \mathbf{x}^{\mathbf{k}} L_{\mathbf{k}}(R_1 \mathbf{y}, \ldots, R_n \mathbf{y}).$$

We already proved that span $\{L_{\mathbf{k}} : \mathbf{k} \in \Gamma_{s,n}\} = \mathcal{P}_{s,n}$ and so the above formula shows that

$$\text{span} \left\{ \prod_{j=1}^{n} (\mathbf{x}, R_j \cdot) : \mathbf{x} \in \mathbb{R}^s \right\}$$
$$= \text{span} \{P(R_1 \cdot, \ldots, R_n \cdot) : P \in \mathcal{P}_{s,n}\},$$

which establishes (iii).   □

From the fact that dim $\mathcal{P}_{s,n} = \dim H_n(\mathbb{R}^s)$ we obtain the next corollary.

COROLLARY 5.1.   Let $\mathcal{X} = \{\mathbf{x}^{i,j} : 1 \leq i \leq s, \ 1 \leq j \leq n\}$ be an array of vectors that satisfies the hypothesis of Proposition 5.4. Then the M-patches $\{M_{\mathbf{k}} : \mathbf{k} \in \Gamma_{s,n}\}$ are linearly independent if and only if whenever $P \in \mathcal{P}_{s,n}$ and $P(R_1\mathbf{x}, \ldots, R_n\mathbf{x}) = 0$ for all $\mathbf{x} \in \mathbb{R}^s$, it follows that $P = 0$.

We apply this corollary to prove two facts about M-patches. First, we go back to formula (5.79) and observe that the M-patches corresponding to matrices $R_1, \ldots, R_n$ are unchanged if we replace these matrices by the symmetric matrices $\frac{1}{2}(R_1 + R_1^T), \ldots, \frac{1}{2}(R_n + R_n^T)$, respectively. Also, observe that if for some i, $1 \leq i \leq n$, R_i is a singular matrix, so that $R_i\mathbf{y} = 0$ for some $\mathbf{y} \in \mathbb{R}^s \backslash \{\mathbf{0}\}$, then by (5.79)

$$\sum_{\mathbf{k} \in \Gamma_{s,n}} \mathbf{y}^{\mathbf{k}} M_{\mathbf{k}}(\mathbf{x}) = 0.$$

Thus the nonsingularity of the matrices $R_1, \ldots, R_n$ is a necessary condition for the linear independence of the corresponding M-patches.

The next proposition provides necessary and sufficient conditions for the linear independence of *quadratic* M-patches on $\mathbb{R}^s$.

PROPOSITION 5.6.   Let R_1, R_2 be two $s \times s$ matrices and suppose $\lambda_1, \ldots, \lambda_s$ are the eigenvalues of the matrix $Q = R_1 R_2^{-1}$. Then the corresponding quadratic M-patches are linearly independent if and only if

(5.80) $\lambda_i + \lambda_j \neq 0, \qquad i, j = 1, \ldots, s.$

Proof. The set of symmetric bilinear functions on $\mathbb{R}^s$ correspond to $s \times s$ symmetric matrices. Thus, Corollary 5.1 says that the quadratic M-patches for R_1 and R_2 are linearly independent if and only if whenever B is an $s \times s$ symmetric matrix such that $(R_1\mathbf{x}, BR_2\mathbf{x}) = 0$ for all $\mathbf{x} \in \mathbb{R}^s$ then $B = 0$. That is to say, whenever $BQ + Q^T B = 0$ it follows that $B = 0$. By a theorem of Schur, cf. Stoer and Bulirsch [SB, p. 328], there is a nonsingular $s \times s$ matrix A such that $Q = ADA^{-1}$ where

$$D = \begin{pmatrix} \lambda_1 & \times & \cdots & \times \\ 0 & \lambda_2 & \ddots & \vdots \\ \vdots & \ddots & & \times \\ 0 & & \cdots & \lambda_s \end{pmatrix}$$

is an upper triangular matrix. Now suppose that B is an $s \times s$ symmetric matrix such that $BQ + Q^T B = 0$. Multiply both sides of this equation, from the left by A^T and from the right by A to get the equation

$$GD + DG = 0$$

where $G = A^T BA$. Since D is upper triangular it follows that $G = 0$ (or equivalently $B = 0$) if and only if (5.80) is valid.   □

The next theorem extends this result to M-patches of degree n.

THEOREM 5.4. *Let $R_1, \ldots, R_n$ be nonsingular $s \times s$ matrices such that there are nonsingular $s \times s$ matrices L and K such that the matrices $Q_i = LR_i K$ are lower triangular for $i = 1, \ldots, n$. Let W be the $s \times n$ matrix, with columns $\mathbf{w}_1, \ldots, \mathbf{w}_n$ given by*

$$W = \left(\begin{array}{ccc} (Q_1)_{11} \cdots (Q_n)_{11} \\ (Q_1)_{ss} \cdots (Q_n)_{ss} \end{array} \right).$$

Then the M-patches $\{M_{\mathbf{k}} : k \in \Gamma_{s,n}\}$ are linearly independent if and only if

(5.81) $\mathrm{per}\, W(\mathbf{k}) \neq 0, \quad \mathbf{k} \in \Gamma_{s,n}.$

Proof. First, observe that if $P \in \mathcal{P}_{s,n}$ and $P(R_1 \mathbf{x}, \ldots, R_n \mathbf{x}) = 0$ for all $\mathbf{x} \in \mathbb{R}^s$ then $G(\mathbf{x}_1, \ldots, \mathbf{x}_n) := P(L^{-1}\mathbf{x}_1, \ldots, L^{-1}\mathbf{x}_n)$ has the property that $G(Q_1 \mathbf{x}, \ldots, Q_n \mathbf{x}) = 0$ for all $\mathbf{x} \in \mathbb{R}^s$. Moreover, $G \in \mathcal{P}_{s,n}$ and is zero if and only if P is zero. Thus we may assume without loss of generality that $R_i = Q_i$, $i = 1, \ldots, n$. Now, let any $P \in \mathcal{P}_{s,n}$ be written in the form

$$P(\mathbf{x}_1, \ldots, \mathbf{x}_n) = \sum_{\mathbf{k} \in \Gamma_{s,n}} P(\mathbf{v}_{\mathbf{k}}) L_{\mathbf{k}}(\mathbf{x}_1, \ldots, \mathbf{x}_n),$$

where

$$L_{\mathbf{k}}(\mathbf{x}_1, \ldots, \mathbf{x}_n) = \sum_{\mathbf{d}(\mathbf{i})=\mathbf{k}} F_{\mathbf{i}}(\mathbf{x}_1, \ldots, \mathbf{x}_n),$$

and

$$F_{\mathbf{i}}(\mathbf{x}_1, \ldots, \mathbf{x}_n) = (\mathbf{x}_1)_{i_1} \cdots (\mathbf{x}_n)_{i_n}$$

(see equations (5.31) and (5.36)). Therefore, we conclude that

$$P(Q_1 \mathbf{x}, \ldots, Q_n \mathbf{x}) = \sum_{\mathbf{k} \in \Gamma_{s,n}} P(\mathbf{v}_{\mathbf{k}}) L_{\mathbf{k}}(Q_1 \mathbf{x}, \ldots, Q_n \mathbf{x}),$$

and, since $Q_i, i = 1, \ldots, n$ are lower triangular,

$$L_{\mathbf{k}}(Q_1 \mathbf{x}, \ldots, Q_n \mathbf{x}) = L_{\mathbf{k}}(\mathbf{w}_1, \ldots, \mathbf{w}_n) \mathbf{x}^{\mathbf{k}} + J_{\mathbf{k}}(\mathbf{x}),$$

where $J_{\mathbf{k}}(\mathbf{x})$ is a homogeneous polynomial containing only monomials of lower order than $\mathbf{x}^{\mathbf{k}}$ relative to the lexicographic ordering on $\Gamma_{s,n}$. Thus we see that the polynomials $L_{\mathbf{k}}(Q_1\cdot,\ldots,Q_n\cdot)$, $\mathbf{k} \in \Gamma_{s,n}$, are linearly independent if and only if

$$L_{\mathbf{k}}(\mathbf{w}_1,\ldots,\mathbf{w}_n) \neq 0, \quad \mathbf{k} \in \Gamma_{s,n}.$$

But in view of equation (5.40), this is precisely the condition (5.81). □

5.9. Subdivision by B-patches.

We now return to B-patches and present a result concerning subdivision using these functions as a basis for polynomials. The observation we want to make is best presented for affine (inhomogeneous) B-patches. For almost all of the remainder of the chapter we will work in the space $\Pi_n(\mathbb{R}^{s-1})$ of all polynomials of total degree at most n in $s-1$ variables with $s \geq 2$. We will not clutter our exposition with new notation special to the inhomogeneous case.

Recall that any $p \in \Pi_n(\mathbb{R}^{s-1})$ gives a homogeneous polynomial q, by the familiar device

$$q((t,\mathbf{x})^T) = t^n p(t^{-1}\mathbf{x}), \qquad \mathbf{x} \in \mathbb{R}^{s-1}, \qquad t \in \mathbb{R}.$$

Conversely, every $q \in H_n(\mathbb{R}^s)$ when restricted to $(1,\mathbf{x})^T$, $\mathbf{x} \in \mathbb{R}^{s-1}$ yields a polynomial in $\Pi_n(\mathbb{R}^{s-1})$. For the inhomogeneous B-patches, we start with an array in $\mathbb{R}^{s-1}$

$$\mathcal{X} = \{\mathbf{x}^{i,j} : 1 \leq i \leq s, \ 1 \leq j \leq n\},$$

apply our previous homogeneous theory in $\mathbb{R}^s$ to the array

$$\{(1,\mathbf{x}^{i,j})^T : 1 \leq i \leq s, \ 1 \leq j \leq n\},$$

and then restrict the homogeneous B-patches to the vector $(1,\mathbf{x})^T, \mathbf{x} \in \mathbb{R}^{s-1}$ to form the inhomogeneous B-patches, $b_{\mathbf{k}}(\mathbf{x}), \mathbf{x} \in \mathbb{R}^{s-1}$. Also, the corresponding lineal polynomials $a_{\mathbf{k}}(\mathbf{x}), \mathbf{k} \in \Gamma_{s,n}$ are similarly obtained. To ensure that they are linearly independent we will always demand that for every $\mathbf{k} \in \Gamma_{s,n}$, the vectors $\mathbf{x}^{1,k_1+1},\ldots,\mathbf{x}^{s,k_s+1}$ are *affinely independent*, that is, the simplex they generate has positive volume. Thus the lineal functions associated with the array in $\mathbb{R}^s$ when restricted to the vectors $(1,\mathbf{x})^T, \mathbf{x} \in \mathbb{R}^{s-1}$, become the barycentric components of $\mathbf{x}$ relative to the simplex generated by $\mathbf{x}^{1,k_1+1},\ldots,\mathbf{x}^{s,k_s+1}$. This means that equation (5.49) becomes

$$(1 + (\mathbf{x},\mathbf{y}))^n = \sum_{\mathbf{k}\in\Gamma_{s,n}} a_{\mathbf{k}}(\mathbf{x})b_{\mathbf{k}}(\mathbf{y}), \qquad \mathbf{x},\mathbf{y} \in \mathbb{R}^{s-1}.$$

Also, since $a_{\mathbf{k}}(\mathbf{0}) = 1$, $\mathbf{k} \in \Gamma_{s,n}$, we get, in particular, that

$$(5.82) \qquad\qquad 1 = \sum_{\mathbf{k}\in\Gamma_{s,n}} b_{\mathbf{k}}(\mathbf{x}), \qquad \mathbf{x} \in \mathbb{R}^{s-1}.$$

The result on subdivision for the B-patch basis that we are about to describe deserves some preliminary explanation. To motivate our presentation, let us suppose for the moment the array has the special property that $\mathbf{x}^{i,j} = \mathbf{x}^i$, $i = 1,\ldots,s, j = 1,\ldots,n$. Therefore, it follows that for $\mathbf{x} \in \mathbb{R}^{s-1}$

$$\mathbf{b}_{\mathbf{k}}(\mathbf{x}) = \binom{n}{\mathbf{k}} u_1^{k_1}(\mathbf{x}) \cdots u_s^{k_s}(\mathbf{x}), \qquad \mathbf{k} \in \Gamma_{s,n},$$

where $u_1(\mathbf{x}),\ldots,u_s(\mathbf{x})$ are the barycentric components of $\mathbf{x}$ relative to the simplex generated by $\mathbf{x}^1,\ldots,\mathbf{x}^s$ (see equation (5.56)). These functions are, of course, the Bernstein–Bézier polynomials. We wish to compute a given polynomial $p \in \Pi_n(\mathbb{R}^{s-1})$ expressed in Bernstein–Bézier form

$$p = \sum_{\mathbf{k} \in \Gamma_{s,n}} c_{\mathbf{k}} b_{\mathbf{k}}$$

at some point $\mathbf{y} \in \mathbb{R}^{s-1}$. For this purpose, we pick some affine bijection T on $\mathbb{R}^{s-1}$ that is contractive and has $\mathbf{y}$ as its unique fixed point. For each $\ell = 1, 2, \ldots$, we represent p in its Bernstein–Bézier form relative to the simplex generated by the points $T^\ell \mathbf{x}^1,\ldots,T^\ell \mathbf{x}^s$. We claim that the new control points will all converge to $p(\mathbf{y})$ as $\ell \to \infty$ (see Fig. 5.11). This fact is proved next.

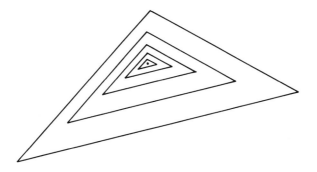

FIG. 5.11. *Subdivision by B-patches.*

THEOREM 5.5. *Let T be a contractive affine bijection on $\mathbb{R}^{s-1}$ with fixed point $\mathbf{y}$. Suppose that $\mathcal{X} = \{\mathbf{x}^{i,j} : 1 \leq i \leq s, \ 1 \leq j \leq n\}$ is an array of points in $\mathbb{R}^{s-1}$ such that for each $\mathbf{k} \in \Gamma_{s,n}$, $\mathbf{x}^{1,k_1+1},\ldots,\mathbf{x}^{s,k_s+1}$ are affinely independent. Given $p \in \Pi_n(\mathbb{R}^{s-1})$ in the B-patch form*

$$p = \sum_{\mathbf{k} \in \Gamma_{s,n}} c_{\mathbf{k}} \mathbf{b}_{\mathbf{k}}.$$

For each $\ell = 0, 1, 2, \ldots$, we write p as

$$p = \sum_{\mathbf{k} \in \Gamma_{s,n}} c_{\mathbf{k},\ell} b_{\mathbf{k},\ell},$$

where $b_{\mathbf{k},\ell}$, $\mathbf{k} \in \Gamma_{s,n}$, *are* B-*patches for the array* $\mathcal{X}_\ell := T^\ell \mathcal{X}$. *Then for all* $\mathbf{k} \in \Gamma_{s,n}$

$$\lim_{\ell \to \infty} c_{\mathbf{k},\ell} = P(\mathbf{y}).$$

The proof of this result is best given in the homogeneous setting. The basic idea of the proof is to identify the form of the matrix that takes the initial control points $\mathbf{c} = (c_{\mathbf{k}} : \mathbf{k} \in \Gamma_{s,n})$ into $\mathbf{c}_\ell = (c_{\mathbf{k},\ell} : \mathbf{k} \in \Gamma_{s,n})$. Therefore, all our matrices will be $N \times N$ where $N := \#\Gamma_{s,n}$. It is convenient to organize all the (homogeneous) B-patches and lineal polynomials into vectors in $\mathbb{R}^N$. For this reason, we define

$$(5.83) \qquad \mathbf{b}(\mathbf{x}|\mathcal{X}) = (b_{\mathbf{k}}(\mathbf{x}) : \mathbf{k} \in \Gamma_{s,n}), \qquad \mathbf{x} \in \mathbb{R}^s,$$

and

$$(5.84) \qquad \mathbf{a}(\mathbf{x}|\mathcal{X}) = (a_{\mathbf{k}}(\mathbf{x}) : \mathbf{k} \in \Gamma_{s,n}), \qquad \mathbf{x} \in \mathbb{R}^s.$$

Here $\mathcal{X}$ is any array in $\mathbb{R}^s$ that satisfies the hypothesis of Theorem 5.1.

We want to see how these vectors transform when $\mathcal{X}$ is replaced by the new array $T\mathcal{X}$. It is the monomial basis in $H_n(\mathbb{R}^s)$ that is the most convenient for the study of this question. Therefore, we define

$$(5.85) \qquad \mathbf{m}(\mathbf{x}) = (m_{\mathbf{k}}(\mathbf{x}) : \mathbf{k} \in \Gamma_{s,n}), \qquad \mathbf{x} \in \mathbb{R}^s,$$

where

$$m_{\mathbf{k}}(\mathbf{x}) = \frac{\mathbf{x}^{\mathbf{k}}}{\mathbf{k}!}, \qquad \mathbf{k} \in \Gamma_{s,n}, \qquad \mathbf{x} \in \mathbb{R}^s.$$

Now let us express $\mathbf{a}(\mathbf{x})$ and $\mathbf{b}(\mathbf{x})$ in monomial basis by introducing $N \times N$ matrices A and B, by setting $\mathbf{a}(\mathbf{x}) = A\mathbf{m}(\mathbf{x})$ and $\mathbf{b}(\mathbf{x}) = B\mathbf{m}(\mathbf{x})$. Obviously, $A = A(\mathcal{X})$ and $B = A(\mathcal{X})$ depend on the array $\mathcal{X}$. We also need for every $s \times s$ matrix R the matrix $\mathcal{E}(R)$ defined by

$$(5.86) \qquad \mathbf{m}(R\mathbf{x}) = \mathcal{E}(R)\mathbf{m}(\mathbf{x}), \qquad \mathbf{x} \in \mathbb{R}^s.$$

Glancing back at the formula preceding Theorem 1.3 we see that

$$(5.87) \qquad (\mathcal{E}(R))_{\mathbf{k}\mathbf{j}} = \frac{1}{\mathbf{k}!} \operatorname{per} R(\mathbf{k},\mathbf{j}), \qquad \mathbf{k}, \mathbf{j} \in \Gamma_{s,n}.$$

In the next lemma, we describe the transformation rules that pertain to sending the array $\mathcal{X}$ to the new array $T\mathcal{X}$.

LEMMA 5.4. *Let* R *be an* $s \times s$ *matrix and* $\mathcal{X}$ *an array in* $\mathbb{R}^s$ *satisfying the hypothesis of Theorem 5.1. Then*

$$(5.88) \qquad A(R\mathcal{X}) = A(\mathcal{X})\mathcal{E}(R^T),$$

and

$$(5.89) \qquad B(R\mathcal{X}) = B(\mathcal{X})\mathcal{E}(R)^{-1}.$$

Proof. From the definition of lineal polynomials we have the equation

$$\mathbf{a}(R^T\mathbf{x}|\mathcal{X}) = \mathbf{a}(\mathbf{x}|R\mathcal{X}) = A(R\mathcal{X})\mathbf{m}(\mathbf{x}),$$

while by the definition of $\mathcal{E}(R^T)$, it follows that

$$\mathbf{a}(R^T\mathbf{x}|\mathcal{X}) = A(\mathcal{X})\mathbf{m}(R^T\mathbf{x})$$
$$= A(\mathcal{X})\mathcal{E}(R^T)\mathbf{m}(\mathbf{x}).$$

This computation proves the first equation (5.88).

For the proof of (5.89), it is convenient to first express the duality relation (5.49) in matrix notation, viz

$$(5.90) \qquad (a(\mathbf{x}|\mathcal{X}), b(\mathbf{y}|\mathcal{X})) = (\mathbf{x}, \mathbf{y})^n, \qquad \mathbf{x}, \mathbf{y} \in \mathbb{R}^s.$$

(Keep in mind that the inner product on the left lives in $\mathbb{R}^N$ and on the right it lives in $\mathbb{R}^{s-1}$.) Expand the right-hand side of equation (5.90) by the multinomial theorem to obtain the equation

$$A(\mathcal{X})^T B(\mathcal{X}) = n!D,$$

where .

$$D = (\mathbf{k}!\delta_{\mathbf{kj}})_{\mathbf{k},\mathbf{j}\in\Gamma_{s,n}}.$$

Thus, replacing $\mathcal{X}$ by $R\mathcal{X}$ in this identity and likewise using (5.88), we see that

$$\begin{aligned}
B(R\mathcal{X}) &= n!A^{-T}(R\mathcal{X})D \\
&= n!A^{-T}(\mathcal{X})\mathcal{E}^{-T}(R^T)D \\
&= B(\mathcal{X})D^{-1}\mathcal{E}^{-T}(R^T)D.
\end{aligned}$$

However, from (5.87) we obtain

$$(5.91) \qquad \mathcal{E}(R^T) = D^{-1}\mathcal{E}(R)^T D,$$

and so equation (5.89) follows.   □

Next we identify a useful form for the control points $\mathbf{c}_1$ obtained after one step of the subdivision algorithm. Specifically, we have, by Lemma 5.4, the equation

$$\begin{aligned}
(\mathbf{c}, B(\chi)\mathbf{m}(\mathbf{x})) &= (\mathbf{c}, b(\mathbf{x}|\mathcal{X})) \\
&= (\mathbf{c}_1, b(\mathbf{x}|R\mathcal{X})) \\
&= (\mathbf{c}_1, B(\mathcal{X})\mathcal{E}(R)^{-1}\mathbf{m}(\mathbf{x})).
\end{aligned}$$

Hence, we conclude that

$$\mathbf{c}_1 = S\mathbf{c},$$

where

$$S^T := B(\mathcal{X})\mathcal{E}(R)B^{-1}(\mathcal{X}).$$

For the ℓth stage of subdivision, it follows in a similar manner that

(5.92) $$\mathbf{c}_\ell = S^\ell \mathbf{c}, \qquad \ell = 1, 2, \ldots,$$

where

(5.93) $$(\mathbf{S}^\ell)^T = B(\chi)\mathcal{E}(R)^\ell B^{-1}(\mathcal{X}), \qquad \ell = 1, 2, \ldots,$$

because the arrays $\mathcal{X}_\ell$ are generated by the recurrence formula, $\mathcal{X}_\ell = T\mathcal{X}_{\ell-1}, \ell = 1, 2, \ldots.$ Thus, the proof of Theorem 5.5 demands an analysis of $\lim_{\ell\to\infty} \mathcal{E}(R)^\ell$. This is done next. But first we need an auxiliary fact about the $N \times N$ matrix $\mathcal{E}(R)$.

LEMMA 5.5. *Let R be an $s \times s$ matrix with eigenvalues $r_1, \ldots, r_s$. Then $\mathcal{E}(R)$ has eigenvalues $k!m_{\mathbf{k}}(\mathbf{r})$, $\mathbf{k} \in \Gamma_{s,n}$, where $\mathbf{r} = (r_1, \ldots, r_s)^T$.*

Proof. From definition (5.86), specialized to the choice $R = I$, it follows that $\mathcal{E}(I) = I$ and also that $\mathcal{E}(R_1 R_2) = \mathcal{E}(R_1)\mathcal{E}(R_2)$. Hence, we have the equation

$$\mathcal{E}(RJR^{-1}) = \mathcal{E}(R)\mathcal{E}(J)\ \mathcal{E}(R)^{-1}.$$

This means we can assume that R has been reduced to its Jordan normal form, viz.

$$J = \begin{pmatrix} J_1 & \cdots & 0 \\ \vdots & \ddots & \vdots \\ 0 & \cdots & J_m \end{pmatrix},$$

where each J_i is a Jordan block. With a little thought, we see that for any $\mathbf{k} \in \Gamma_{s,n}$ and $\mathbf{x} \in \mathbb{R}^s$

$$m_{\mathbf{k}}(J\mathbf{x}) = k!m_{\mathbf{k}}(\mathbf{r})m_{\mathbf{k}}(\mathbf{x}) + f_{\mathbf{k}}(\mathbf{x}),$$

where $f_{\mathbf{k}}$ is a homogeneous polynomial of lower order than $m_{\mathbf{k}}$, relative to lexicographic ordering in $\Gamma_{s,n}$. $\square$

We remark that this lemma leads to the formula

$$\det \mathcal{E}(R) = (\det R)^{n(n+1)/2}.$$

The next lemma is central in our proof of Theorem 5.5.

LEMMA 5.6. *Let R be an $s \times s$ matrix whose largest eigenvalue is one and is simple. Suppose that $\mathbf{x}, \mathbf{y}$ are the corresponding left and right eigenvectors of R, that is, $R\mathbf{x} = \mathbf{x}$, $R^T\mathbf{y} = \mathbf{y}$, normalized so that $(\mathbf{x}, \mathbf{y}) = 1$. Then*

$$\lim_{j\to\infty} \mathcal{E}(R)^j_{\mathbf{k}\mathbf{j}} = j!m_{\mathbf{k}}(\mathbf{x})m_{\mathbf{j}}(\mathbf{y}), \qquad \mathbf{k}, \mathbf{j} \in \Gamma_{s,n}.$$

Proof. From the definition of $\mathcal{E}(R)$, equation (5.91), and our hypothesis it follows that

$$\mathcal{E}(R)\mathbf{m}(\mathbf{x}) = \mathbf{m}(\mathbf{x}), \qquad \mathcal{E}(R)^T(D\mathbf{m}(\mathbf{y})) = D\mathbf{m}(\mathbf{y}).$$

Moreover, from Lemma 5.5 and our hypothesis, we conclude that the largest eigenvalue of $\mathcal{E}(R)$ is also one and simple. We also have by the multinomial theorem the formula

$$(D\mathbf{m}(\mathbf{y}), \mathbf{m}(\mathbf{x})) = (\mathbf{x}, \mathbf{y})^n = 1.$$

Consequently, the lemma follows from the following well-known fact, cf. Stoer and Bulirsch [SB]. □

LEMMA 5.7. *Let G be an $N \times N$ matrix whose highest eigenvalue is one and simple. Suppose that $G\mathbf{u} = \mathbf{u}$ and $G^T\mathbf{v} = \mathbf{v}$, where $\mathbf{u}, \mathbf{v}$ are vectors in $\mathbb{R}^N$, normalized so that $(\mathbf{u}, \mathbf{v}) = 1$. Then*

$$\lim_{\ell \to \infty} G_{ij}^\ell = u_i v_j, \qquad i, j = 1, \ldots, N.$$

We have assembled all the pieces needed to complete the proof of Theorem 5.5. It follows next.

Proof of Theorem 5.5. We introduce the $s \times s$ matrix R by setting

$$R((t, \mathbf{x})^T) = (t, tT(t^{-1}\mathbf{x}))^T, \qquad t \in \mathbb{R}, \mathbf{x} \in \mathbb{R}^{s-1}.$$

By our hypothesis on T, $R((1, \mathbf{y})^T) = (1, \mathbf{y})^T$ and one is a simple eigenvalue of R. Thus it follows from Lemma 5.6 and formula (5.93) that for all $\mathbf{k}, \mathbf{r} \in \Gamma_{s,n}$

$$(5.94) \qquad \lim_{\ell \to \infty} (S^\ell)_{\mathbf{r}\mathbf{k}} = b_{\mathbf{k}}(\mathbf{y}) h_{\mathbf{r}}.$$

The scalars $h_{\mathbf{r}}$ depend on the right eigenvector of R, and also, of course, the array χ. We can identify them by using the following computation. According to (5.82), we have the equation

$$1 = \sum_{\mathbf{k} \in \Gamma_{s,n}} b_{\mathbf{k}}(\mathbf{x}), \qquad \mathbf{x} \in \mathbb{R}^{s-1}.$$

In other words, it is the case that $S\mathbf{e} = \mathbf{e}$ where $\mathbf{e} := (1, 1, \ldots, 1)^T \in \mathbb{R}^N$. Thus, summing both sides of equation (5.94) over $\mathbf{k} \in \Gamma_{s,n}$, we get $h_{\mathbf{r}} = 1$, for $\mathbf{r} \in \Gamma_{s,n}$. This means that

$$(5.95) \qquad \lim_{\ell \to \infty} (S^\ell)_{\mathbf{r},\mathbf{k}} = b_{\mathbf{k}}(\mathbf{y}),$$

and so combining this equation with (5.92) proves the result. □

5.10. B-patches are B-splines.

In our next result, we identify (inhomogeneous) B-patches with multivariate B-splines of Chapter 4. To this end, we recall the notation used in that chapter which states that for any set $V = \{\mathbf{v}^0, \ldots, \mathbf{v}^n\} \subset \mathbb{R}^n$

$$\det(\mathbf{v}^0, \ldots, \mathbf{v}^n) = \begin{vmatrix} 1 & (\mathbf{v}^0)_1 & \cdots & (\mathbf{v}^0)_n \\ 1 & (\mathbf{v}^1)_1 & \cdots & (\mathbf{v}^1)_n \\ \vdots & \vdots & & \vdots \\ 1 & (\mathbf{v}^n)_1 & \cdots & (\mathbf{v}^n)_n \end{vmatrix}.$$

Here we also use, for $j = 0, 1, \ldots, n$, the affine functions of $\mathbf{x} \in \mathbb{R}^n$ given by

(5.96) $\det_j(\mathbf{x}|V) = \det \left(\{\mathbf{v}^0, \ldots, \mathbf{v}^{j-1}, \mathbf{x}, \mathbf{v}^{j+1}, \ldots, \mathbf{v}^n\} \right).$

In particular, it follows that $\det(V) = \det_j(\mathbf{v}^j|V)$, for $j = 0, 1, \ldots, n$. Also note that the barycentric components of $\mathbf{x} \in \mathbb{R}^n$ relative to the simplex $V = [\mathbf{v}^0, \ldots, \mathbf{v}^n]$ are given by the formula

(5.97) $u_j(\mathbf{x}|V) = \dfrac{\det_j(\mathbf{x}|V)}{\det(V)}, \qquad j = 0, 1, \ldots, n.$

Thus, starting with a set X in $\mathbb{R}^{s-1}$ with m points , $m > s$, the recurrence formula for the multivariate B-spline presented in Theorem 4.4 may be rewritten in the following way. For any $V = \{\mathbf{x}^{i_1}, \ldots, \mathbf{x}^{i_s}\} \subseteq X$

(5.98) $S(\mathbf{x}|V) = \chi_{[V]}(\mathbf{x})/|\det(V)|, \qquad \mathbf{x} \in \mathbb{R}^{s-1},$

and for $m > s$

(5.99) $(m - s)S(\mathbf{x}|X) = \displaystyle\sum_{j=1}^{s} \dfrac{\det_j(\mathbf{x}|V)}{\det(V)} S(\mathbf{x}|X \backslash \{\mathbf{x}^{i_j}\}), \qquad \mathbf{x} \in \mathbb{R}^{s-1}.$

Recall that in the univariate case, when we identified B-splines with the dual polynomials N_j, $j = 0, 1, \ldots, n$, corresponding to some set of scalars $\{t_{-n+1}, \ldots, t_n\}$, we constructed the B-splines from the larger set $\{t_{-n}, \ldots, t_{n+1}\}$. We do the same in the multivariate case. Starting with an array

$$\mathcal{X} = \{\mathbf{x}^{i,j} : 1 \le i \le s, \ 1 \le j \le n\}$$

with corresponding B-patches, $b_{\mathbf{k}}(\mathbf{x})$, $\mathbf{k} \in \Gamma_{s,n}$, $\mathbf{x} \in \mathbb{R}^{s-1}$, where $\mathbf{x}^{1,k_1+1}, \ldots, \mathbf{x}^{s,k_s+1}$ are assumed to be affinely independent, we enlarge $\mathcal{X}$ to the array

$$\hat{\mathcal{X}} := \{\mathbf{x}^{i,j} : 1 \le i \le s, \ 1 \le j \le n+1\}.$$

From this larger set, we select subsets

$$Y_{\mathbf{k}} = \{\mathbf{x}^{i,j} : 1 \le i \le s, \ 1 \le j \le k_i + 1, \ i = 1, \ldots, s\},$$

much like we selected subsets from $\mathcal{X}$ to form B-patches. In this case for any $\mathbf{k} \in \Gamma_{s,n}$ it follows that $\#Y_{\mathbf{k}} = n + s$. Therefore, the corresponding B-spline $S(\mathbf{x}|Y_{\mathbf{k}})$ is a piecewise polynomial of degree n (see Theorem 4.1).

We will show that on a certain region of $\mathbb{R}^{s-1}$ $S(\cdot|Y_{\mathbf{k}})$ is proportional to $b_{\mathbf{k}}$. To this end, we let $V_{\mathbf{k}} = [\mathbf{x}^{1,k_1+1}, \ldots, \mathbf{x}^{s,k_s+1}]$ be the simplex generated by the set of vectors $\mathbf{x}^{1,k_1+1}, \ldots, \mathbf{x}^{s,k_s+1}$ and Ω_n the intersection of *all* such simplicies obtained from points in the array $\mathcal{X}$, viz.

$$\Omega_n = \bigcap \{V_{\mathbf{k}} : \ 0 \le k_i \le n, \ i = 1, \ldots, s\}.$$

(See Fig. 5.12.) Note that the set $V_{\mathbf{k}}$ represents the vectors *across the top* of $Y_{\mathbf{k}}$.

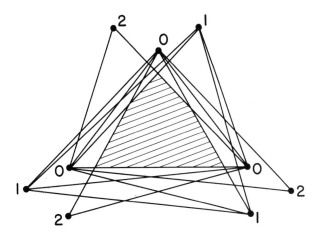

FIG. 5.12.   *The set* Ω_2.

THEOREM 5.6. *Suppose that* $\Omega_n^0 \neq \emptyset$. *Then for every* $\mathbf{k} \in \Gamma_{s,n}$

(5.100) $$b_{\mathbf{k}}(\mathbf{x}) = n!|\det(V_{\mathbf{k}})|S(\mathbf{x}|Y_{\mathbf{k}}), \qquad \mathbf{x} \in \Omega_n.$$

We begin the proof of this theorem with the following lemma.

LEMMA 5.8. *Suppose* $\Omega_n^0 \neq \emptyset$. *Then there is* $\epsilon \in \{-1, 1\}$ *such that*

$$\operatorname{sgn} \det(V_{\mathbf{k}}) = \epsilon$$

for all $\mathbf{k} = (k_1, \ldots, k_s)^T$, $0 \leq k_i \leq n-1$, $i = 1, \ldots, s$.

When $\mathbf{k} \in \Gamma_{s,n}$ and $k_i > 0$ then $Y_{\mathbf{k}-\mathbf{e}_i} = Y_{\mathbf{k}} \setminus \{\mathbf{x}^{i,k_i+1}\}$, otherwise we use the convention that $Y_{\mathbf{k}-\mathbf{e}_i}$ contains no vectors of the form $\mathbf{x}^{i,1}, \ldots, \mathbf{x}^{i,n}$.

Proof. Suppose $\mathbf{k} \in \Gamma_{s,n}$ and for some $i, 1 \leq i \leq s$, $k_i > 0$. By hypothesis the intersection of the two simplices $[V_{\mathbf{k}}]$ and $[V_{\mathbf{k}-\mathbf{e}_i}]$ has a nonempty interior. The hyperplane

$$\det_i(\mathbf{x}|V_{\mathbf{k}}) = \det_i(\mathbf{x}|V_{\mathbf{k}-\mathbf{e}_i}) = 0, \qquad \mathbf{x} \in \mathbb{R}^{s-1}$$

is a common face of both simplices. Since $[V_{\mathbf{k}}]$ and $[V_{\mathbf{k}-\mathbf{e}_i}]$ have a nonempty interior, the vectors $\mathbf{x}^{i,k_i+1} \in [V_{\mathbf{k}}]$ and $\mathbf{x}^{i,k_i} \in [V_{\mathbf{k}-\mathbf{e}_i}]$ must lie on the same side of this hyperplane. Thus we conclude that

$$\operatorname{sgn} \det(V_{\mathbf{k}}) = \operatorname{sgn} \det(V_{\mathbf{k}-\mathbf{e}_i}).$$

From this equation, applied repeatedly to the positive components of $\mathbf{k}$, the result follows.   □

LEMMA 5.9. *Suppose that the hypothesis of Lemma 5.8 holds. Then for every* $\mathbf{k} \in \Gamma_{s,n}$ *such that for some* $i, 1 \leq i \leq s$, $k_i = 0$ *we have*

$$\Omega_n^0 \bigcap [Y_{\mathbf{k}-\mathbf{e}_i}] = \emptyset.$$

The proof of this lemma can be found in [DMS].

Proof of Theorem 5.6. The proof is by induction on n. When $n = 0$, $b_0(\mathbf{x}) = 1$ by convention, and $Y_0 = [\mathbf{x}^{1,1}, \ldots, \mathbf{x}^{s,1}] = V_0$. Thus by equation (5.98) $S(\mathbf{x}|Y_0) = 1/|\det(V_0)|$ on $\Omega_0 = V_0$. Therefore, (5.100) is true in this case.

Now suppose $n > 0$ and call the left-hand side of equation (5.100), $N_{\mathbf{k}}(\mathbf{x})$, that is,

$$N_{\mathbf{k}}(\mathbf{x}) = n!|\det(V_{\mathbf{k}})|S(\mathbf{x}|Y_{\mathbf{k}}).$$

From the recurrence formula (5.99), specialized to the sets $X = Y_{\mathbf{k}}$ and $V = V_{\mathbf{k}}$, we get the equation

$$nS(\mathbf{x}|Y_{\mathbf{k}}) = \sum_{j=1}^{s} \frac{\det_j(\mathbf{x}|V_{\mathbf{k}})}{\det(V_{\mathbf{k}})} S(\mathbf{x}|Y_{\mathbf{k}-\mathbf{e}_j}), \qquad \mathbf{x} \in \mathbb{R}^{s-1},$$

and therefore it follows that

$$N_{\mathbf{k}}(\mathbf{x}) = \sum_{j=1}^{s} \frac{\det_j(\mathbf{x}|V_{\mathbf{k}})|\det(V_{\mathbf{k}})|}{\det(V_{\mathbf{k}})|\det(V_{\mathbf{k}-\mathbf{e}_j})|} N_{\mathbf{k}-\mathbf{e}_j}(\mathbf{x}), \qquad \mathbf{x} \in \mathbb{R}^{s-1}.$$

Using Lemma 5.8, this formula simplifies to the equation

$$N_{\mathbf{k}}(\mathbf{x}) = \sum_{j=1}^{s} \frac{\det_j(\mathbf{x}|V_{\mathbf{k}})}{\det(V_{\mathbf{k}})} N_{\mathbf{k}-\mathbf{e}_j}(\mathbf{x}).$$

On the other hand, from the down recurrence formula for the B-patches (see Proposition 5.1) we have the formula

$$b_{\mathbf{k}}(\mathbf{x}) = \sum_{j=1}^{s} \frac{\det_j(\mathbf{x}|V_{\mathbf{k}-\mathbf{e}_j})}{\det(V_{\mathbf{k}-\mathbf{e}_j})} b_{\mathbf{k}-\mathbf{e}_j}(\mathbf{x}).$$

In this equation, we used the formula for barycentric components given in (5.97). Keeping definition (5.96) in mind, we have that $\det_j(\mathbf{x}|V_{\mathbf{k}-\mathbf{e}_j}) = \det_j(\mathbf{x}|V_{\mathbf{k}})$ whenever $k_j > 0$.

Now suppose the result holds for $n - 1$ where $n \geq 1$. From Lemma 5.9, the sets $[Y_{\mathbf{k}-\mathbf{e}_j}]$ and Ω_n have a disjoint interior, when $k_i = 0$ and so, in this case $N_{\mathbf{k}-\mathbf{e}_j}(\mathbf{x}) = 0$, for $\mathbf{x} \in \Omega_n$. Moreover, $b_{\mathbf{k}-\mathbf{e}_j}(\mathbf{x}) = 0$, when $k_i = 0$. On the other hand, when $k_i > 0$, the induction hypothesis implies that $b_{\mathbf{k}-\mathbf{e}_i}(\mathbf{x}) = N_{\mathbf{k}-\mathbf{e}_i}(\mathbf{x})$ for $\mathbf{x} \in \Omega_n$. Thus in all cases $b_{\mathbf{k}-\mathbf{e}_i}(\mathbf{x}) = N_{\mathbf{k}-\mathbf{e}_i}(\mathbf{x})$, $i = 1, \ldots, s$, $\mathbf{x} \in \Omega_n$, and therefore, by the above two recurrence formulas, $b_{\mathbf{k}}(\mathbf{x}) = N_{\mathbf{k}}(\mathbf{x})$. $\square$

COROLLARY 5.2. *Suppose that Ω_n has a nonempty interior. Then*

$$\sum_{\mathbf{k}\in\Gamma_{s,n}} N_{\mathbf{k}}(\mathbf{x}) = 1, \qquad \mathbf{x} \in \Omega_n,$$

and $\{N_{\mathbf{k}} : \mathbf{k} \in \Gamma_{s,n}\}$ are linearly independent on any region contained in Ω_n.

Proof. Both results follow immediately from the identification of $b_{\mathbf{k}}$ and $N_{\mathbf{k}}$ on Ω_n made in Theorem 5.6.   □

Theorem 5.6 provides a method for construction of linear spaces spanned by B-splines. The interested reader can find a discussion of this important issue in Dahmen, Micchelli, and Seidel [DMS] as well as in the recent papers by Gormaz [G], Gormaz and Laurent [GL], Sauer [S], and Seidel [Se].

References

[CM]   A.S. CAVARETTA AND C.A. MICCHELLI, *Pyramid patches provide potential polynomial paradigms*, in Mathematical Methods in Computer Aided Geometric Design II, T. Lyche and L.L. Schumaker, eds., Academic Press, San Diego, 1992, 69–100.

[DMS]   W. DAHMEN, C.A. MICCHELLI, AND H.P. SEIDEL, *Blossoming begets B-spline bases built better by B-patches*, Math. Comp., 59(1992), 97–115.

[G]   R. GORMAZ, *Floraisons polynomials: Applications a l'etude des B-splines a plusieurs variables*, These, Docteur de L'Universite Joseph Fourier-Grenoble 1, June 1993.

[GL]   R. GORMAZ AND P.J. LAURENT, *Some results on blossoming and multivariate B-splines*, in Multivariate Approximation and Wavelets, K. Jetter and F. Utreras, eds., World Scientific, Singapore, 1993, 147–165.

[S]   T. SAUER, *Multivariate B-splines with (almost) arbitrary knots*, preprint, August 1993.

[Se]   H.P. SEIDEL, *Representing piecewise polynomials as linear combinations of multivariate B-splines*, in Mathematical Methods in Computer Aided Geometric Design II, T. Lyche and L.L. Schumaker, eds., Academic Press, San Diego, 1992, 559–566.

[SB]   J. STOER AND R. BULIRSCH, *Introduction to Numerical Analysis*, Springer-Verlag, New York, 1980.

Index